M11

Steel Pipe—A Guide for Design and Installation

Fifth Edition

Errata April 2018 Incorporated

American Water Works Association

Manual of Water Supply Practices—M11, Fifth Edition

Steel Pipe—A Guide for Design and Installation

Disclaimer

The authors, contributors, editors, and publisher do not assume responsibility for the validity of the content or any consequences of its use. In no event will AWWA be liable for direct, indirect, special, incidental, or consequential damages arising out of the use of information presented in this book. In particular, AWWA will not be responsible for any costs, including, but not limited to, those incurred as a result of lost revenue. In no event shall AWWA's liability exceed the amount paid for the purchase of this book.

Project Manager: Melissa Valentine
Cover Art: Melanie Yamamoto
Production: Stonehill Graphics.
Manuals Specialist: Sue Bach

Library of Congress Cataloging-in-Publication Data

Names: Dechant, Dennis, author. | Bambei, John H., Jr., author. | American Water Works Association.
Title: M11--steel water pipe : a guide for design and installation / by Dennis Dechant and John Bambei.
Other titles: Steel water pipe | Guide for design and installation | Steel pipe--design and installation.
Description: Fifth edition. | Denver, CO : American Water Works Association, [2017] | Originally published as: Steel pipe--design and installation. 1964. | Includes bibliographical references.
Identifiers: LCCN 2017002001 | ISBN 9781625762092
Subjects: LCSH: Water-pipes--Design and construction--Handbooks, manuals, etc. | Pipe, Steel--Design and construction--Handbooks, manuals, etc.
Classification: LCC TC174 .D365 2017 | DDC 628.1/5--dc23 LC record available at https://lccn.loc.gov/2017002001

ISBN 978-162576-209-2
eISBN 978-1-61300-408-1

Printed in the United States of America
American Water Works Association
6666 West Quincy Avenue
Denver, CO 80235-3098
awwa.org

Contents

Figures

This page intentionally blank.

Tables

This page intentionally blank.

Preface

This manual provides a review of experience and theory regarding design of steel pipe used for conveying water, with appropriate references cited. The manual provides general and technical information to be used as an aid in the design and installation of steel pipe. It is a discussion of recommended practice, not an AWWA standard calling for compliance with certain specifications. Application of the principles and procedures discussed in this manual must be based on responsible judgment.

This manual was first authorized in 1943. In 1949, Committee 8310D on Steel Pipe, appointed one of its members, Russell E. Barnard, to act as editor in chief in charge of collecting and compiling the available data on steel pipe. The first draft of the report was completed by January 1957; the draft was reviewed by the committee and other authorities on steel pipe. The first edition of this manual was issued in 1964 with the title *Steel Pipe—Design and Installation*.

The second edition of this manual was approved in June 1984 and published in 1985 with the title *Steel Pipe—A Guide for Design and Installation*. The third edition of the manual was approved in June 1988 and published in 1989. The fourth edition of the manual was approved March 2003 and published in January 2004. This fifth edition was approved August 2016.

Major revisions to this fifth edition are (1) reorganization of the chapters to combine similar content in the same chapters; (2) elimination of some tables which were replaced with formulas and examples; (3) changes in aboveground design and examples to more clearly reflect conditions encountered on a water pipeline; (4) addition of a chapter on thrust design; (5) addition to the fittings chapter to include design of true wyes and crosses, design of crotch plates with higher strength steel, expanded elbow stress design in restrained areas, tangential outlet design was clarified, double outlet design was clarified, strength reduction factors for varying steel strengths of outlets was added, PDV values were clarified to 9000 for test and transient pressures, anchor ring design was added, design of ellipsoidal heads was added, and modified joint harness requirements; (6) added suggested bracing for shipping of pipe; (6) updated the flange bolt torque values and table; (7) buckling of buried pipe was clarified (8) weld details for outlets and crotch plates were added; (9) cement enhanced soil was defined and added; (10) design of welded lap joints was expanded; and (11) Appendixes were added for nomenclature, comparison of increase of E' versus increase of wall thickness, full example of harness ring design, design of harness rod placement for differential settlement, seismic considerations, and useful equations and conversions.

This page intentionally blank.

Acknowledgments

The revision of Manual M11 was reviewed and approved by the Steel Water Pipe Manufacturers Technical Advisory Committee (SWPMTAC). The Steel Water Pipe Manufacturers Technical Advisory Committee Task Group that updated Manual M11 had the following personnel at the time of revision:

Dennis Dechant, *Task Group Chairman*

S.A. Arnaout, Hanson Pressure Pipe, Grand Prairie, Texas
H.H. Bardakjian, Consultant, Glendale, Calif.
M. Bauer, Tnemec Company Inc., North Kansas City, Mo.
J. Buratto, Lifelast Inc., Pflugerville, Texas
R.J. Card, Consultant, Sugar Hill, Ga.
R.R. Carpenter, AMERICAN SpiralWeld Pipe, Birmingham, Ala.
K.G. Couture, AMERICAN SpiralWeld Pipe , Birmingham, Ala.
D. Dechant, Consultant , Aurora, Colo.
W.B. Geyer, Steel Plate Fabricators Association, Lake Zurich, Ill.
J.W. Green, Lockwood Andrews & Newnam Inc., Oakbrook Terrace, Ill.
B. Hansen, National Welding Corporation, Midvale, Utah
B.D. Keil, Northwest Pipe Company, Draper, Utah
T. Kennedy, Hanson Pressure Pipe, Dallas, Texas
J.L. Luka, AMERICAN SpiralWeld Pipe , Columbia S.C.
R. Mielke, Northwest Pipe Company, Raleigh, N.C.
J. Olmos, Ameron Water Transmission Group, Rancho Cucamonga, Calif.
S. Rahman, Northwest Pipe Company, Corona, Calif.
G.F. Ruchti, Jr., Consultant, Punta Gorda, Fla.
R.N. Satyarthi, Baker Coupling Company Inc., Los Angeles, Calif.
C.P. Shelley, Victaulic Company, Doraville, Ga.
B. Simpson, AMERICAN SpiralWeld Pipe , Birmingham, Ala.
D.G. Tantalean, Ameron Water Transmission Group, Rancho Cucamonga, Calif.
B.F. Vanderploeg, Northwest Pipe Company, Vancouver, Wash.
G.L. Washburn, Consultant, Easton, Pa.
N. Williams, National Welding Corporation, Midvale, Utah

This revision was reviewed and approved by the Standards Committee on Steel Pipe. The Standards Committee on Steel Pipe had the following personnel at the time of review and approval:

John H. Bambei Jr., *Chairman*
Dennis Dechant, *Vice Chairman*
John L. Luka, *Secretary*

General Interest Members

J.H. Bambei, Jr., Bambei Engineering Services, Arvada, Colo.
W.R. Brunzell, Brunzell Associates Ltd., Skokie, Ill.
R.J. Card, Lockwood Andrews & Newnam Inc, Sugar Hill, Ga.

R.L. Coffey, HDR Engineering Inc., Omaha, Neb.
S.N. Foellmi, Black & Veatch Corporation, Irvine, Calif.
R.L. Gibson, Freese and Nichols Inc., Fort Worth, Texas
*M.D. Gossett,** HDR, Denver, Colo.
*M.B. Horsley,** Horsley Engineering LLC, Overland Park, Kan.
R.A. Kufaas, Norske Corrosion & Inspection Services Ltd., Surrey, BC, Canada
J.L. Mattson, Corrosion Control Technologies, Sandy, Utah
A. Murdock, CH2M Hill, Salt Lake City, Utah
*R. Ortega,** Rafael Ortega, Spring, Texas
E.S. Ralph,† Standards Engineer Liaison, AWWA, Denver, Colo.
A.E. Romer, AECOM, Orange, Calif.
J.R. Snow, MWH Americas Inc., Denver, Colo.
W.R. Whidden, Woolpert, Winter Park, Fla.

Producer

D.W. Angell,† Standards Council Liaison, American Flow Control, Birmingham, Ala.
S.A. Arnaout, Hanson Pressure Pipe, Grand Prairie, Texas
H.H. Bardakjian, Consultant, Glendale, Calif.
D. Dechant, Dechant Infrastructure Service, Aurora, Colo.
*V. DeGrande,** Ameron Water Tranmission Group, Rancho Cucamonga, Calif.
W.B. Geyer, Steel Plate Fabricators Association, Lake Zurich, Ill.
B.D. Keil, Northwest Pipe Company, Draper, Utah
J.L. Luka, American SpiralWeld Pipe Company, Columbia S.C.
*R. Mielke,** Northwest Pipe Company, Raleigh, N.C.
J. Olmos, Ameron Water Transmission Group, Rancho Cucamonga, Calif.
*G.F. Ruchti Jr.,** Consultant, Punta Gorda, Fla.
*B. Simpson,** American Cast Iron Pipe Company, Birmingham, Ala.
C.C. Sundberg, Victaulic, Issaquah, Wash.
D. Walker, Avid Protective Products Ltd/Tnemec Company, Oakville, Ontario, Canada
J.A. Wise, Canus International Sales Inc., Surrey, B.C., Canada

User

G.A. Anderson, New York City Bureau of Water Supply, Little Neck, N.Y.
J.H. Bambei, Jr., Bambei Engineering Services, Arvada, Colo.
B. Cheng, Metro Vancouver, Burnaby, B.C., Canada
M.E. Conner, San Diego County Water Authority, San Diego, Calif.
R.V. Frisz, U.S. Bureau of Reclamation, Denver, Colo.
S. Hattan, Tarrant Regional Water District, Ft. Worth, Texas
*T.J. Jordan,** Metro Water District of Southern California, LaVerne, Calif.
P.K. Karna, Tacoma Water, Tacoma, Wash.
M. McReynolds, Metro Water District of Southern California, Los Angeles, Calif.
*M. Turney,** Denver Water, Denver, Colo.
N.A. Wigner, Los Angeles Department of Water & Power, Los Angeles, Calif.

* Alternate
† Liaison

Chapter 1

History, Uses, and Physical Characteristics of Steel Pipe

HISTORY

Steel pipe has been used for water lines in the United States since the early 1850s. The pipe was first manufactured by rolling steel sheets or plates into shape and riveting the seams. Recognized very early in its development as a significant benefit, steel pipe offered flexibility that allowed variations in the steel sheet thickness being rolled to handle the different pressures based on the pipe's elevation and the hydraulic gradient. Roll-formed pipe with riveted seams was the dominant method of pipe fabrication until the 1930s when the electric welding process replaced the labor-intensive riveted seams.

In consideration of the relatively low tensile strength of steels produced in the second half of the nineteenth century and the inefficiencies of cold-riveted seams and riveted or drive stovepipe joints, engineers set the allowable design stress at 10,000 psi. As riveted-pipe fabrication methods improved through the early part of the twentieth century, concurrently higher strength steels were being produced. As a result, allowable design stresses progressed in this period from 10,000 psi to 12,500 psi, to 13,750 psi, and finally to 15,000 psi, in all cases maintaining a safety factor of 4 to the steel's tensile strength. Allowable design stresses were adjusted as necessary to account for the inefficiency of the riveted seam. The pipe was produced in diameters ranging from 4 in. through 144 in. and in thicknesses from 16 gauge to 1.5 in. Fabrication methods consisted of single-, double-, triple-, and quadruple-riveted seams, varying in efficiency from 45 percent to 70 percent, depending on the design.

Lockbar pipe, introduced in 1905, had nearly supplanted riveted pipe by 1930. Fabrication involved milling 30-ft-long plates to a width approximately equal to half the intended circumference, cold forming the longitudinal edges, and rolling the plates into

30-ft-long half-circle troughs. The lockbar was a specially configured H-shaped bar that was applied to the mating edges of two 30-ft troughs and clamped into position to form a full-circle pipe section. Lockbar pipe had notable advantages over riveted pipe: It had only one or two straight seams and no round seams. The straight seams were considered to be 100 percent efficient, in that the seam developed the full strength of the pipe wall, as compared to the 45 percent to 70 percent efficiency for riveted seams. Manufactured in sizes from 20 in. through 74 in., from plate ranging in thicknesses from 3⁄16 in. to ½ in., lockbar played an increasingly greater role in the market until the advent of automatic electric welding in the mid-1920s.

The period beginning circa 1930 saw a very abrupt reduction in the use of both riveted-seam and lockbar pipe manufacturing methods. These methods were replaced by seams that were fused together using electric-fusion welding. Pipe produced using electric-fusion welding was advantageous because the plate could be prewelded into a single flat sheet that could be fed into the three-roll forming machine to form a cylinder with only a single longitudinal seam to weld. This resulted in faster production, minimal weld-seam protrusion, and 100 percent welded-seam efficiency. The fabricators of fusion-welded seam pipe followed similar initial production sequences as for lockbar; first rolling two long half sections, then using electric-fusion welding, joining the two long pipe-halves into a single section. Also developed in the 1930s was the pipe roll forming method that is a U-ing and O-ing process producing a longitudinal weld or fused seam. Through this decade and into the 1940s, 30-ft to 40-ft-long pipe cylinders were being formed from plate.

The helical process, more commonly referred to as the *spiral-weld forming process,* for fabricating welded seam steel pipe was also developed in the early 1930s and was first used extensively to produce steel pipe in diameters from 4 in. through 36 in. This method was typically more efficient to manufacture and also offered lower weld seam stress than longitudinal welded pipe. Welding was performed using the electric-fusion method. After World War II, US manufacturers adapted German spiral weld–seam technology and developed new equipment capable of forming spiral weld seam steel pipe to diameters in excess of 144 in.

The development of the spiral-weld forming process coincided chronologically with the option developed by the steel industry to roll or coil steel sheet and plate. Steel in coil form allows modern day spiral weld forming equipment and roll-forming equipment to be very efficient in maximizing production. Present day steel mill capacities for coil allow for steel thicknesses up to 1 in. and widths up to 100 in., with mechanical properties up to 100-ksi yield strength.

The welding renaissance of the 1930s brought confidence in the design and use of steel pipe with welded seams and joints. In the prewelding era, it had been common practice to design steel pipe using a safety factor of 4 based on the tensile strength. The performance of the welded seams proved to be so significantly better than riveted joints that a change in design parameters was adopted. Pipeline designers and users no longer needed high safety factors to compensate for inefficient seams and joints. The design methodology would be changed to reflect the use of an allowable design stress of 50 percent of the material's yield strength.

USES

Steel water pipe meeting the requirements of appropriate ANSI/AWWA standards has many applications, some of which follow:

- Aqueducts
- Supply lines

Figure 1-1 Steel pipe penstock on bridge

- Transmission mains
- Distribution mains
- Penstocks
- Horizontal directional drilling
- Tunneled casing pipe
- Treatment-plant piping
- Self supporting spans
- Force mains
- Circulating-water lines
- Underwater crossings, intakes, and outfalls
- Relining and sliplining

General data and project details on some of the notable steel pipeline projects are readily available on numerous Web sites. See Figure 1-1 for an example of a steel pipe penstock on a bridge.

CHEMISTRY, CASTING, AND HEAT TREATMENT

General

The steel industry produces very high quality steels that demand accurate control of chemistry and precise control of the casting and rolling process. These steels, available in sheet, plate, and coil, meet or exceed the requirements of the ASTM standards listed in

ANSI/AWWA C200, Steel Water Pipe, 6 In. (150 mm) and Larger (latest edition), for use in steel water pipe. ASTM steel standards in ANSI/AWWA C200 allow for grades with yield strengths from 30 ksi to 100 ksi without significant changes in chemistry. ANSI/AWWA C200 utilizes grades from the ASTM standards up to about 55-ksi-minimum specified yield strength for ease of manufacturing and welding. By adding small amounts of carbon and manganese or various other metals called microalloying, the strength and other properties of these steels are modified.

Properties and chemical composition of steels listed in ANSI/AWWA C200 are governed by the applicable ASTM standards and are also a function of the processes used to transform the base metal into a shape, and, when appropriate, by controlling the heat during the steel rolling process. The effects of these parameters on the properties of steels are discussed in this section.

Chemical Composition

In general, steel is a mixture of iron and carbon with varying amounts of other elements—primarily manganese, phosphorus, sulfur, and silicon. These and other elements are present or added in various combinations to achieve specific characteristics and physical properties of the finished steel. The effects of the commonly used chemical elements on the properties of hot-rolled and heat-treated carbon and alloy steels are presented in Table 1-1. Additionally, the effects of carbon, manganese, sulfur, silicon, and aluminum will be discussed.

Carbon is the principal hardening element in steel. Incremental addition of carbon increases the hardness and tensile strength of the steel. Carbon has a moderate tendency to segregate, and an excessive amount of carbon can cause a decrease in ductility, toughness, and weldability.

Manganese increases the hardness and strength of steels but to a lesser degree than carbon. Manganese combines with sulfur to form manganese sulfides, therefore decreasing the harmful effects of sulfur.

Sulfur is generally considered an undesirable element except when machinability is an important consideration. Sulfur adversely affects surface quality, has a strong tendency to segregate, and decreases ductility, toughness, and weldability.

Silicon and aluminum are the principal deoxidizers used in the manufacture of carbon and alloy steels. Aluminum is also used to control and refine grain size. The terms used to describe the degree to which these two elements deoxidize the steel are *killed steel* or *semikilled steel*. Killed steels have a very low oxygen level, while semikilled steels have indications of slightly higher levels of oxygen.

Casting

Historically, the steel-making process involved pouring molten steel into a series of molds to form castings known as *ingots*. The ingots were removed from the molds, reheated, and then rolled into products with square or rectangular cross sections. This hot-rolling operation elongated the ingots and produced semifinished products known as *blooms*, *slabs*, or *billets*. Typically, ingots exhibited some degree of nonuniformity of chemical composition known as *segregation*. This chemical segregation was associated with yield losses and processing inefficiencies.

Most modern day steel producers use the continuous casting process to avoid the inherent detrimental characteristics that resulted from the cooling and solidification of the molten steel in the ingot mold. Continuous casting is a process where the molten steel is poured at a controlled rate directly from the ladle through a water-cooled mold to form a continuous slab. The cross section of the water-cooled mold will be dimensioned so as to correspond to that of the desired slab. This steel-making process bypasses the operations

Table 1-1 Effects of alloying elements

Alloying Element	Effects
Aluminum (Al)	Used to deoxidize or "kill" molten steel
	Refines grain structure
Boron (B)	Small amounts (0.005%) can be used to tie up nitrogen and soften steel
	Used only in aluminum-killed steels and where titanium is added to tie up nitrogen
	Most effective at low carbon levels, but there are a number of medium carbon steels in use today that employ boron for hardenability
Carbon (C)	Principal hardening element in steel
	Increases strength and hardness
	Decreases ductility, toughness, and weldability
	Moderate tendency to segregate
Chromium (Cr)	Increases strength
	Increases atmospheric corrosion resistance
Copper (Cu)	Primary contributor to atmospheric corrosion resistance
	Decreases weldability
Manganese (Mn)	Increases strength
	Controls harmful effects of sulfur
Nickel (Ni)	Increases strength and toughness
Nitrogen (N)	Increases strength and hardness
	Decreases ductility and toughness
Phosphorus (P)	Increases strength and hardness
	Decreases ductility and toughness
	Considered an impurity but sometimes added for atmospheric corrosion resistance
Silicon (Si)	Used to deoxidize or "kill" molten steel
Sulfur (S)	Considered undesirable except for machinability
	Decreases ductility, toughness, and weldability
	Adversely affects surface quality
	Strong tendency to segregate
Titanium (Ti)	In small amounts, it ties up nitrogen to improve toughness, and in greater amounts it can strengthen steel
Vanadium (V) and Columbium (Nb)	Small additions increase strength
	Often referred to as *microalloying elements*

between molten steel and the semifinished product that are inherent in making steel products from ingots. As the molten metal begins to solidify along the walls of the water-cooled mold, it forms a shell that permits the gradual withdrawal of the strand product from the bottom of the die into a water-spray chamber where solidification is completed. The solidified strand is cut to length and then reheated and rolled into finished products, as in the conventional ingot process. Continuous casting produces a smaller size and higher cooling rate for the strand, resulting in less segregation and greater uniformity in composition and properties than for ingot products.

Killed and Semikilled Steels

The primary reaction involved in most steel-making processes is the combination of carbon and oxygen to form carbon monoxide gas. The solubility of this and other gases dissolved in the steel decreases as the molten metal cools to the solidification temperature range. Excess gases are expelled from the metal and, unless controlled, continue to evolve during solidification. The oxygen available for the reaction can be eliminated and the gaseous evolution inhibited by deoxidizing the molten steel using additions of silicon or aluminum or both.

Steels that are deoxidized do not evolve any gases and are called *killed steels* because they lie quietly in the mold. Killed steels are less segregated and contain negligible porosity when compared to semikilled steels. Consequently, killed-steel products exhibit a higher degree of uniformity in composition and properties than do semikilled steel products.

Heat Treatment for Steels

Steels respond to a variety of heat treatment methods that produce desirable characteristics. These heat treatment methods can be divided into slow cooling treatment and rapid cooling treatment. Slow cooling treatment decreases hardness, can increase toughness, and promotes uniformity of structure. Slow cooling includes the processes of annealing, normalizing, and stress relieving. Rapid cooling treatment increases strength, hardness, and toughness, and includes the processes of quenching and tempering. Heat treatments of base metal are generally mill options or ASTM requirements, and are generally performed on plates rather than coils.

Annealing. Annealing consists of heating steels to a predetermined temperature followed by slow cooling. The temperature, the rates of heating and cooling, and the amount of time the metal is held at temperature depend on the composition, shape, and size of the steel product being treated and the desired properties. Usually steels are annealed to remove stresses, induce softness, increase ductility, increase toughness depending on the parameters of the process, produce a given microstructure, increase uniformity of microstructure, improve machinability, or to facilitate cold forming.

Normalizing. Normalizing consists of heating steels to between 1,650°F and 1,700°F followed by slow cooling in air. This heat treatment is commonly used to refine the grain size, improve uniformity of microstructure, and improve ductility and fracture toughness.

Stress Relieving. Stress relieving of carbon steels consists of heating steels to between 1,000°F and 1,200°F and holding for the appropriate amount of time to equalize the temperature throughout the piece followed by slow cooling. The stress-relieving temperature for quenched and tempered steels must be maintained below the tempering temperature for the product. Stress relieving is used to relieve internal stresses induced by welding, normalizing, cold working, cutting, quenching, and machining. It is not intended to alter the microstructure or the mechanical properties significantly.

Quenching and Tempering. Quenching and tempering consist of heating and holding steels at the appropriate austenizing temperature (about 1,650°F) for a significant amount of time to produce a desired change in microstructure, then quenching by immersion in a suitable medium (water for bridge steels). After quenching, the steel is tempered by reheating to an appropriate temperature, usually between 800°F and 1,200°F, holding for a specified time at that temperature, and cooling under suitable conditions to obtain the desired properties. Quenching and tempering increase the strength and improve the toughness of the steel.

Controlled Rolling. Controlled rolling is a thermomechanical treatment performed at the rolling mill. It tailors the time-temperature-deformation process by controlling the rolling parameters. The parameters of primary importance are (1) the temperature at the start of controlled rolling in the finished strand after the roughing mill reduction; (2) the percentage reduction from the start of controlled rolling to the final plate thickness; and (3) the plate finishing temperature.

Hot-rolled plates are deformed as quickly as possible at temperatures above about 1,900°F to take advantage of the workability of the steel at high temperatures. In contrast, controlled rolling incorporates a hold or delay time to allow the partially rolled slab to reach the desired temperature before the start of final rolling. Controlled rolling involves deformation at temperatures ranging between 1,500°F and 1,800°F as recrystallization ceases to occur below this temperature range. Because rolling deformation at these low temperatures increases the mill loads significantly, controlled rolling is usually restricted to less than 2-in.-thick plates. Controlled rolling increases the strength, refines the grain size, improves the toughness, and may eliminate the need for normalizing.

Controlled Finishing-Temperature Rolling. Controlled finishing-temperature rolling is a less severe practice than controlled rolling and is aimed primarily at improving notch toughness of plates up to 2½-in. thick. The finishing temperatures in this practice (about 1,600°F) are on the lower end of those required for controlled rolling. However, because heavier plates are involved than in controlled rolling, mill delays are still required to reach the desired finishing temperatures. By controlling the finishing temperature, fine grain size and improved notch toughness can be obtained.

MECHANICAL CHARACTERISTICS

The commercial success of steel as an engineered material stems from the ability to provide a wide spectrum of mechanical properties. Steel offers a balance of strength, ductility, fracture resistance, and weldability. The design engineer should understand the importance of each of these properties, how they interact, and the correct methods of incorporating them into a final design.

Ductility and Yield Strength

Solid materials can be divided into two classes: ductile and brittle. Engineering practice treats these two classes differently because they behave differently under load. A ductile material exhibits a marked plastic deformation or flow at a fairly definite stress level (yield point or yield strength) and shows a considerable total elongation, stretch, or plastic deformation before failure. With a brittle material, the plastic deformation is not well defined, and the ultimate elongation before failure is small. Steels, as listed in ANSI/AWWA C200, are typical of the ductile class materials used for steel water pipe.

Ductility of steel is measured as an elongation, or stretch, under a tension load in a tensile-testing machine. Elongation is a measurement of change in length under the load and is expressed as a percentage of the original gauge length of the test specimen.

Ductility allows comparatively thin-walled steel pipe to perform satisfactorily, even when the vertical diameter is decreased 2 to 5 percent by external earth pressures, provided the true required strength has been incorporated in the design. Additionally, ductility allows steel pipe with theoretically high localized stresses at connection points of flanges, saddles, supports, and joint-harness lugs to continue to perform satisfactorily.

Designers who determine stress using formulas based on Hooke's law find that the calculated results do not reflect the integrity exhibited by the structures discussed in this manual. These discrepancies occur because the conventional formulas apply only up to a certain stress level and not beyond (stress-based design). Many otherwise safe structures and parts of structures contain calculated stresses above this level (strain-based design). A full understanding of the performance of such structures requires that the designer empirically examines the actual behavior of steel as it is loaded from zero to the fracture point.

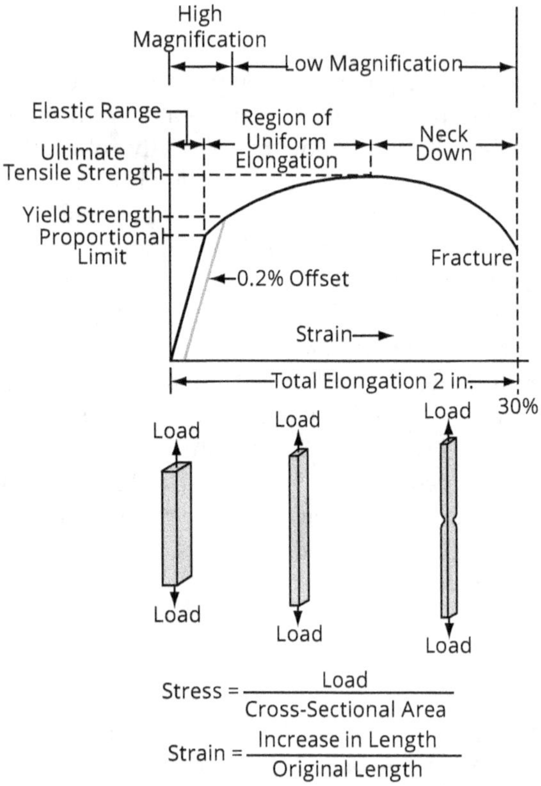

The change in shape of the test piece of steel, which occurred during the test, is shown by the bars drawn under the curve.

Figure 1-2 Stress-strain curve for steel

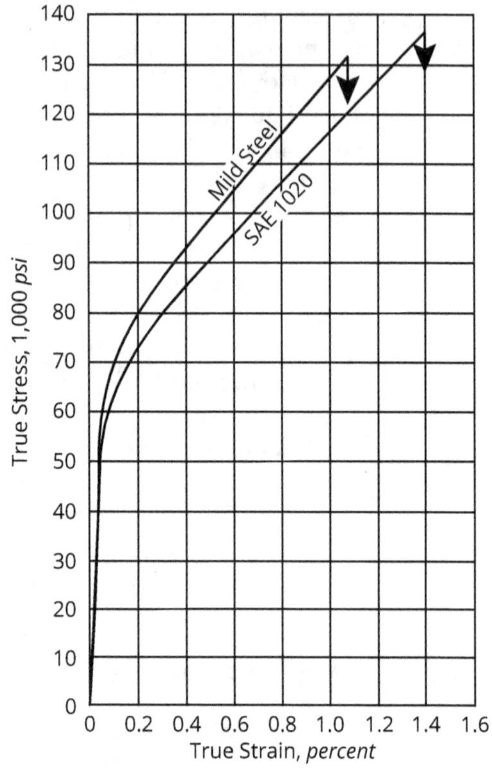

Unlike conventional stress–strain curves, both true stress and true strain have been calculated for the curves shown.

Figure 1-3 True stress-strain for steel

The physical properties of steel (yield strength and ultimate tensile strength) used as the basis for design and purchase specifications are determined from tension tests made on standard specimens pulled in a tensile-testing machine. The strength of ductile materials, in terms of design, is defined by the yield strength as measured by the lower yield point, where one exists, or by the ASTM International offset yield stress, where a yield point does not exist. For steel typically used in water pipe, the yield strength is defined by the material specification as the stress determined by the 0.5 percent extension-under-load method, or the 0.2 percent offset method. The yield strength determined by the 0.2 percent offset method is most commonly used. Based on the 0.2 percent offset method, the value of the yield strength is defined as the stress represented by the intersection of the stress-strain curve and a line, beginning at the 0.002 value on the strain axis, drawn parallel to the elastic portion of the stress-strain curve. Such a line is shown in Figure 1-2. The yield strength of steel is considered the same for either tension or compression loads.

Stress and Strain

In engineering, stress is a value obtained by dividing a load by an area. Strain is a length change per unit of length. The relation between stress and strain, as shown on a stress-strain diagram, is of basic importance to the designer.

A stress-strain diagram for any given material is a graph showing the stress that occurs when the material is subjected to a given strain. For example, a bar of steel is pulled in a tensile-testing machine with suitable instrumentation for measuring the load and indicating the dimensional changes. While the bar is under load, it stretches. The change in length under load per unit of length is called *strain* or *unit strain*; it is usually expressed as percentage elongation or, in stress analysis, microinches (μin.) per inch, where 1 μin. = 0.000001 in. (For metric units, strain is defined as μmm/mm or μm/m.) The values of strain are plotted along the horizontal axis of the stress-strain diagram. For purposes of plotting, the load is converted into units of stress (pounds per square inch) by dividing the load in pounds by the original cross-sectional area of the bar in square inches. The values of stress are plotted along the vertical axis of the diagram. The result is a conventional stress-strain diagram.

Because the stress plotted on the conventional stress-strain diagram is obtained by dividing the load by the original cross-sectional area of the bar, the stress appears to reach a peak and then diminish as the load increases. However, if the stress is calculated by dividing the load by the actual cross-sectional area of the bar as it decreases in cross section under increasing load, it is found that the true stress never decreases. Figure 1-3 is a stress-strain diagram on which both true stress and true strain have been plotted. Because conventional stress-strain diagrams are used commercially, only conventional diagrams are used for the remainder of this discussion.

Figure 1-2 shows various parts of a pure-tension stress-strain curve for steel such as that used in steel water pipe. The change in shape of the test piece during the test is indicated by the bars drawn under the curve. As the bar stretches, the cross section decreases in area up to the maximum tensile strength, at which point local reduction of area (necking in) takes place.

Many types of steel used in construction have stress-strain diagrams of the general form shown in Figure 1-2; whereas many other types used structurally and for machine parts have much higher yield and ultimate strengths, with reduced ductility. Still other useful engineered steels are quite brittle. In general, low-ductility steels must be used at relatively low strains, even though they may have high strength.

The ascending line on the left side of the graph in Figure 1-2 is straight or nearly straight and has a recognizable slope with respect to the vertical axis. The break in the slope of the curve is rather sudden. For this type of curve, the point where the first deviation from a straight line occurs marks the proportional limit of the steel. The yield strength is defined as a slightly higher stress level as discussed previously. Most engineering formulas involving stress calculation presuppose a loading such that working stresses will be well below the proportional limit.

Stresses and strains that fall below the proportional limit—such as those that fall on the straight portion of the ascending line—are said to be in the elastic range. Steel structures loaded to create stresses or strains within the elastic range return to their original shape when the load is removed. Exceptions may occur with certain kinds and conditions of loading not usually encountered in steel water pipe installations. Within the elastic range, stress increases in direct proportion to strain.

The modulus of elasticity (Young's modulus) is defined as the slope of the ascending straight portion of the stress-strain diagram. The modulus of elasticity of steel is about 30,000,000 psi, which means that for each increment of load that creates a strain or stretch of 1 μin./in. of length, a stress of 30 psi is imposed on the steel cross section (30,000,000 x 0.000001 = 30).

Immediately above the proportional limit lies a portion of the stress-strain curve that is termed the *plastic range of the material*. Typical stress-strain curves with the elastic range and the initial portion of the plastic range are shown in Figures 1-4 and 1-5 for two grades of carbon steel used for water pipe. Electric-resistance strain gauges provide a means of studying both the elastic and plastic regions of the curve. These and associated instruments allow minute examination of the shape of the curve in a manner not possible before development of these instruments.

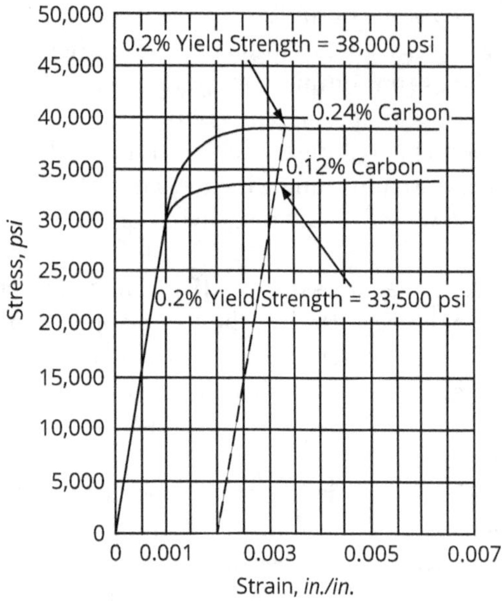

The curves show the elastic-plastic range for two grades of carbon steel.

Figure 1-4 Stress-strain curves for carbon steel

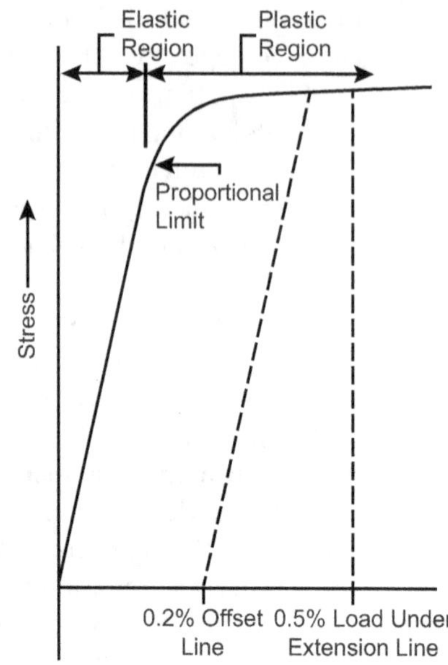

Shown are the elastic and plastic portions of a stress–strain curve. Note: the 0.2% and 0.5% lines intersect at a specific location that may not be located on the stress–strain curve.

Figure 1-5 Plastic and elastic strains

The plastic range is important to the designer. Analysis of this range was necessary, for example, to determine and explain the successful performance of thin steel flanges on thin steel pipe (Barnard 1950). Designs that load steel to within the plastic range are safe only for certain types of apparatus, structures, or parts of structures. For example, designing within this range is safe for the hinge points or yield hinges in steel ring flanges on steel pipe; for hinge points in structures where local yielding or relaxation of stress must occur; and for bending in the wall of pipe under external earth pressure in trenches or under high fills. Such areas can generally involve secondary stresses, which will be discussed in the following section. It is not safe to rely on performance within this plastic range to handle principal tension stress in the walls of pipe or pressure vessels or to rely on such performance in other situations where the accompanying deformation is uncontrolled or cannot be tolerated.

Figure 1-6 shows graphically how a completely fictitious stress is determined by a formula based on Hooke's law, if the total strain is multiplied by the modulus of elasticity. The actual stress (Figure 1-7) is determined using only the elastic strain with the modulus of elasticity, but neglects what actually occurs to the steel in the plastic range.

Stress in Design

Stress can be generally categorized as either principal or secondary. Although both types of stress can be present in a structure at the same time, the driving mechanism for, and a structure's response to, each differ significantly. A principal stress results from applied loads and is necessary to maintain the laws of equilibrium of a structure. If the level of a principal stress substantially exceeds the yield strength, a structure's deformation will continue toward failure. Therefore, a principal stress is not considered self-limiting. In the case of steel pipe, longitudinal and circumferential stresses resulting from internal pressure are examples of principal stresses. In contrast, secondary stress is developed when the deformation of a component due to applied loads is restrained by other components.

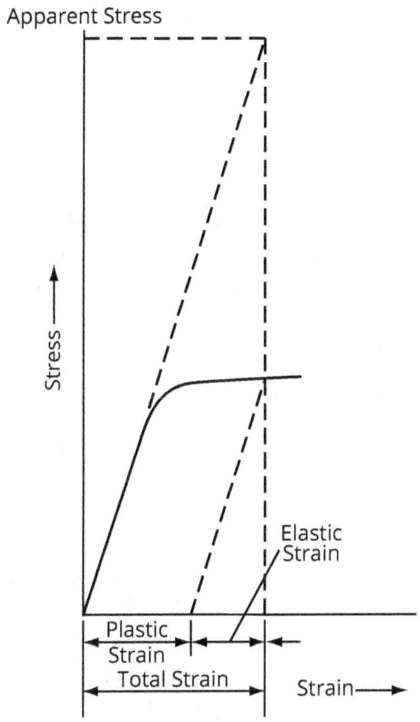

If the total strain is multiplied by the modulus of elasticity, the stress determined by use of a formula based on Hooke's law is fictitious.

Figure 1-6 Actual and apparent stresses

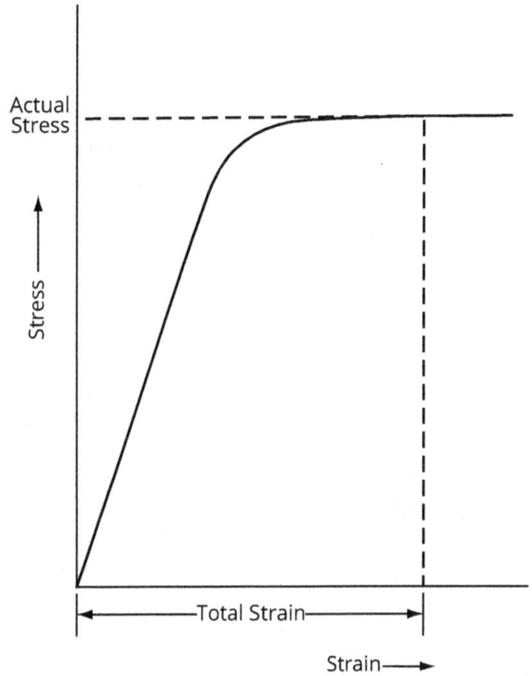

When the total measured strain is known, the actual stress can be determined by use of the stress–strain curve.

Figure 1-7 Determination of actual stress

Secondary stresses are considered self-limiting in that they are strain driven, not load driven; localized yielding absorbs the driving strain, which "relaxes" or redistributes the secondary stresses to lower levels without causing failure. Once the developed strain has been absorbed by the localized yielding, the driving mechanism for further deformation no longer exists. In the case of steel pipe, shell-bending stresses at hinge points such as flange connections, ring attachments, or other gross structural discontinuities, as well as induced thermal stress, are examples of secondary stresses.

Strain in Design

Analysis of a structure becomes more complete when considering strain as well as stress. For example, it is known that apparent stresses calculated using classic formulas based on the theory of elasticity are erroneous at hinge-point stress levels. The magnitude of this error near the yield-strength stress is demonstrated in the next paragraph, where the classically calculated result is compared with the measured performance.

By definition, the yield-strength load of a steel specimen is that load that causes a 0.5 percent extension of the gauge length or 0.2 percent offset from the linear elastic line. In the elastic range, a stress of 30 psi is imposed on the cross-sectional area for each microinch-per-inch increase in length under load. Because a load extension of 0.5 percent corresponds to 5,000 μin/in., the calculated yield-strength stress is 5,000 x 30 = 150,000 psi. The measured yield-strength stress, however, is approximately 30,000–35,000 psi or about one-fourth the calculated stress.

Similarly varied results between strain and stress analyses occur when the performance of steel, at its yield strength, is compared to the performance at its ultimate strength. There is a great difference in strain between the 0.2 percent offset yield strength of low- or medium-carbon steel and the specified ultimate strength at 30 percent elongation. This difference has a crucial bearing on design safety. The specified yield strength corresponds to a strain of about 2,000 μin/in. To pass a specification requirement of 30 percent elongation, the strain at ultimate strength must be no less than 0.3 in./in. or 300,000 μin/in.. The ratio of strain at ultimate strength to strain at yield strength, therefore, is 300,000:2,000 or 150:1. On a stress basis, assuming an ultimate tensile strength of 60,000 psi from the stress-strain diagram, the ratio of ultimate strength to yield strength is 60,000:30,000 or only 2:1.

Steels, such as those used in waterworks pipe, show nearly linear stress-strain diagrams up to the proportional limit, after which strains of 10 to 20 times the elastic-yield strain occur with no increase in actual load. Tests on bolt behavior under tension substantiate this effect (Bethlehem Steel Co. 1946). The ability of bolts to hold securely and safely when they are drawn into the region of the yield, especially under vibration conditions, is easily explained by the strain concept but not by the stress concept. The bolts act somewhat like extremely stiff springs at the yield-strength level.

ANALYSIS BASED ON STRAIN

In some structures and in many welded assemblies, conditions permit the initial adjustment of strain to working load but limit the action automatically either because of the nature of the loading or because of the mechanics of the assembly. Examples are, respectively, pipe under deep earth loads and steel flanges on steel pipe. In these instances, bending stresses may be in the region of yield, but deformation is limited.

In bending, there are three distinguishable phases that a structure passes through when being loaded from zero to failure. In the first phase, all fibers undergo strain less than the proportional limit in a uniaxial stress field. In this phase, a structure will act in a completely elastic fashion, to which the classic laws of stress and strain are applicable.

In the second phase, some of the fibers undergo strain greater than the proportional or elastic limit of the material in a uniaxial stress field; however, a more predominant portion of the fibers undergo strain less than the proportional limit, so that the structure still acts in an essentially elastic manner. The classic formulas for stress do not apply but the strains can be adequately defined in this phase.

In the third phase, the fiber strains are predominantly greater than the elastic limit of the material in a uniaxial stress field. Under these conditions, the structure as a whole no longer acts in an elastic manner.

An experimental determination of strain characteristics in bending and tension was made on medium-carbon steel (<0.25 percent carbon) similar to that required by ANSI/AWWA C200. Results are shown in Figure 1-8. Note that the proportional-limit strains in bending are 1.52 times those in tension for the same material. Moreover, the specimen in bending showed fully elastic behavior at a strain of 1,750 μin/in., which corresponds to a calculated stress of 52,500 psi (1,750 x 30 = 52,500) using the modulus of elasticity. The specimens were taken from material having an actual yield of 39,000 psi. Therefore, this steel could be loaded in bending to produce strains up to 1,750 μin/in. and still possess full elastic behavior.

Steel ring flanges made of plate and fillet welded to pipe with a comparatively thin wall have been used successfully for many years in water service. Calculations were made to determine the strain that would occur in the pipe wall adjacent to the flanges. The flanges ranged from 4 in. through 96 in. in diameter. Table 1-2 shows the results.

Note that from the table, in practice, the limiting strain was always below the 1940s' recognized yield-strength strain of 5,000 μin/in. but did approach it closely in at least one instance. All of these flanges are sufficiently satisfactory, however, to warrant their continued use by designers.

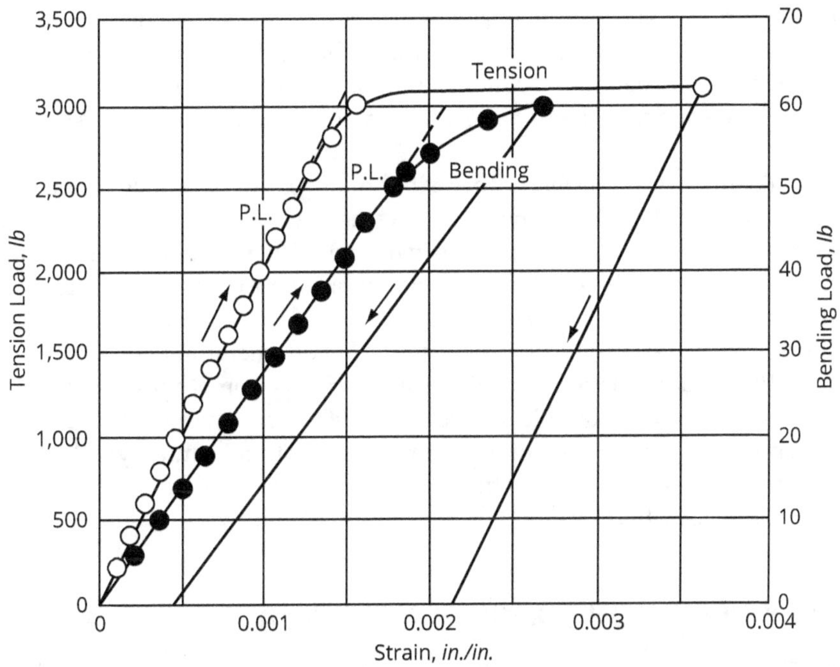

The proportional limit (P.L.) strains in bending are 1.52 times those in tension for the same material.

Figure 1-8 Experimental determination of strain characteristics

Table 1-2 Maximum strain in pipe wall developed in practice

	Operating Pressure		Maximum Strain	
Standard Flange	*psi*	*kPa*	*μin./in.*	*(mm/m)*
A	75	(517.1)	1,550–3,900	(1.55–3.90)
	150	(1,034.2)	2,200–4,650	(2.20–4.65)
B	150	(1,034.2)	1,100–3,850	(1.10–3.85)

Source: Barnard, R.E., Design of Steel Ring Flanges for Water Works Service—A Progress Report. Jour. AWWA, 42:10:931 (Oct. 1950).

Designing a structure on the basis of ultimate load capacity from test data rather than entirely on allowable stress is a return to an empirical point of view, a point of view that early engineers accepted in the absence of knowledge of the mathematics and statistics necessary to calculate stresses. The recent development of mathematical processes for stress analysis has, in some instances, overemphasized the importance of stress and underemphasized the importance of the overall strength of a structure.

DUCTILITY IN DESIGN

The plastic, or ductile, behavior of steel in welded assemblies may be especially important. Current design practice allows the stress at certain points in a steel structure to go beyond the elastic range. For many years, in buildings and in bridges, specifications have allowed the designer to use average or nominal stresses because of bending, shear, and bearing, resulting in local yielding around pins and rivets and at other points. This local yield, which redistributes both load and stress, is caused by stress concentrations that are neglected in the simple design formulas. Plastic action is and has been depended on to ensure the safety of steel structures. Experience has shown that these average or nominal maximum stresses form a satisfactory basis for design. During the manufacturing process, the steel in steel pipe has been forced beyond its yield strength many times, and the same thing may happen during installation. Similar yielding can be permitted after installation by design, provided the resulting deformation has no adverse effect on the function of the structure.

Basing design solely on approximations for real stress does not always produce safe results. The collapse of some structures has been traced to a trigger action of neglected points of high stress concentrations in materials that are not ductile at these points. Ductile materials may fail in a brittle fashion if subjected to overload in three planes at the same time. Careful attention to such conditions will result in safer design and will eliminate grossly over designed structures that waste both material and money.

Plastic deformation, especially at key points, sometimes is the real measure of structural strength. For example, a crack, once started, may be propagated by almost infinite stress, because at the bottom of the crack the material cannot yield a finite amount in virtually zero distance. In a ductile material, the crack will continue until the splitting load is resisted elsewhere.

EFFECTS OF COLD WORKING ON STRENGTH AND DUCTILITY

During pipe fabrication, the steel plates or sheets are often formed into the desired shape at room temperatures. Such cold-forming operations obviously cause inelastic deformation because the steel retains its formed shape. To illustrate the general effects of such deformation on strength and ductility, the elemental behavior of a carbon-steel tension specimen subjected to plastic deformation and subsequent reloading will be discussed. The behavior of actual cold-formed plates may be much more complex.

As illustrated in Figure 1-9, if a steel specimen of plate material is unloaded after being stressed into either the plastic or strain-hardening range, the unloading curve will follow a path parallel to the elastic portion of the stress-strain curve, and a residual strain or permanent set will remain after the load is removed.

If the specimen is promptly reloaded, it will follow the unloading curve to the stress-strain curve of the virgin (unstrained) material. If the amount of plastic deformation is less than that required for the onset of strain hardening, the yield strength of the plastically deformed steel will be approximately the same as that of the virgin material. However, if the amount of plastic deformation is sufficient to cause strain hardening, the yield strength of the steel will be increased. In either case, the tensile strength will remain the same, but the ductility measured from the point of reloading will be decreased. As indicated in Figure 1-9, the decrease in ductility is approximately equal to the amount of inelastic prestrain.

A steel specimen that has been strained into the strain-hardening range, unloaded, and allowed to age for several days at room temperature (or for a much shorter time at a moderately elevated temperature) will tend to follow the path indicated in Figure 1-10 during reloading (Dieter 1961). This phenomenon, known as *strain aging*, has the effect of increasing yield and tensile strength while decreasing ductility (Chajes et al. 1963).

The effects of cold work on the strength and ductility of the structural steels can be eliminated largely by thermal stress relief, or annealing. Such treatment is not always possible; fortunately, it is not often necessary.

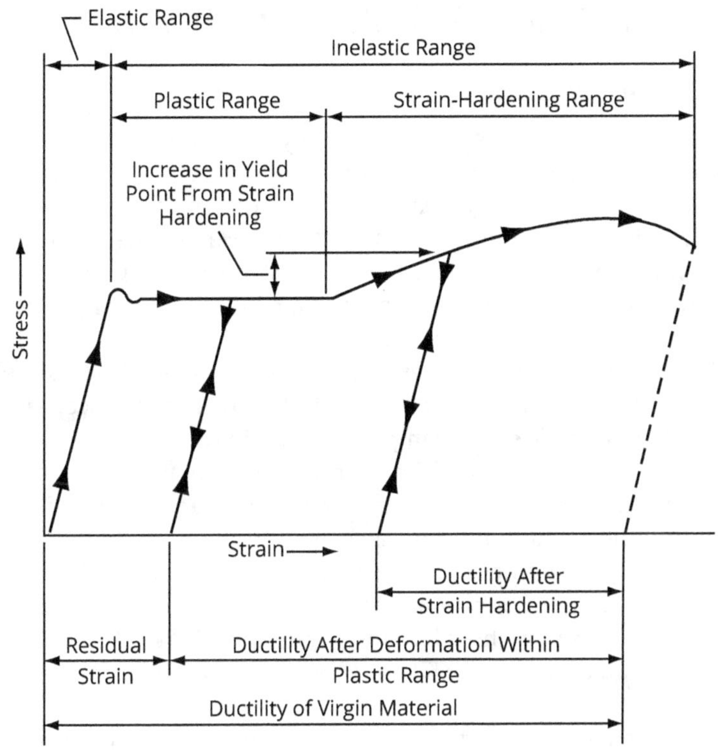

Note: Diagram is schematic and not to scale.

Source: Brockenbrough and Johnston 1981.

Figure 1-9 Effects of strain hardening

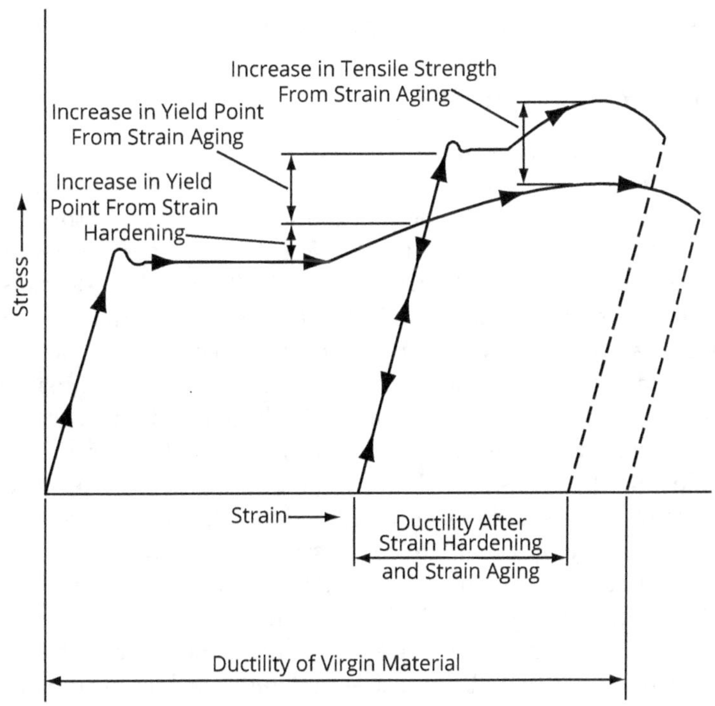

Note: Diagram is schematic and not to scale.

Source: Brockenbrough and Johnston 1981.

Figure 1-10 Effects of strain aging

BRITTLE FRACTURE CONSIDERATIONS IN STRUCTURAL DESIGN

General Considerations

As temperature decreases, there generally is an increase in the yield strength, tensile strength, modulus of elasticity, and fatigue strength of the plate steels. In contrast, the ductility of these steels, as measured by reduction in area or by elongation under load, decreases with decreasing temperatures. Furthermore, there is a temperature below which a structural steel that is subjected to tensile stresses may fracture by cleavage with little or no plastic deformation, rather than by shear, which is usually preceded by a considerable amount of plastic deformation or yielding.*

Fracture that occurs by cleavage at a nominal tensile stress below the yield stress is referred to as *brittle fracture*. Generally, a brittle fracture can occur when there is an adverse combination of tensile stress, temperature strain rate, and geometrical discontinuity (such as a notch). Other design and fabrication factors may also have an important influence. Because of the interrelation of these effects, the exact combination of stress, temperature, notch, and other conditions that cause brittle fracture in a given structure cannot be readily calculated. Preventing brittle fracture often consists mainly of avoiding conditions that tend to cause brittle fracture and selecting steel appropriate for the application. These factors are discussed in the following paragraphs. Parker (1957), Lightner and Vanderbeck (1956), Rolfe and Barsom (1977), and Barsom (1993) have described the subject in much more detail.

Fracture mechanics offer a more direct approach for prediction of crack propagation. For this analysis, it is assumed that an internal imperfection forming a crack is present in the structure. By linear-elastic stress analysis and laboratory tests on precracked specimens, the applied stress causing rapid crack propagation is related to the size of the imperfection. Fracture mechanics has become increasingly useful in developing a fracture-control plan and establishing, on a rational basis, the interrelated requirements of material selection, design stress level, fabrication, and inspection requirements (Barsom 1993).

Conditions Causing Brittle Fracture

Plastic deformation occurs only in the presence of shear stresses. Shear stresses are always present in a uniaxial or a biaxial state of stress. However, in a triaxial state of stress, the maximum shear stress approaches zero as the principal stresses approach a common value. As a result, under equal triaxial tensile stresses, failure occurs by cleavage rather than by shear. Consequently, triaxial tensile stresses tend to cause brittle fracture and should be avoided. As discussed in the following material, a triaxial state of stress can result from a uniaxial loading when notches or geometrical discontinuities are present.

If a transversely notched bar is subjected to a longitudinal tensile force, the stress concentration effect of the notch causes high longitudinal tensile stresses at the apex of the notch and lower longitudinal stresses in adjacent material. The lateral contraction in the width and thickness direction of the highly stressed material at the apex of the notch is restrained by the smaller lateral contraction of the lower stressed material. Therefore, in addition to the longitudinal tensile stresses, tensile stresses are created in the width and thickness directions, so that a triaxial state of stress is present near the apex of the notch.

The effect of a geometrical discontinuity in a structure is generally similar to, although not necessarily as severe as, the effect of the notch in the bar. Examples of geometrical discontinuities include poor design details (such as abrupt changes in cross section,

---

* Shear and cleavage are used in the metallurgical sense (macroscopically) to denote different fracture mechanisms. Parker (1957), as well as most elementary textbooks on metallurgy, discussed these mechanisms.

attachment welds on components in tension, and square-cornered cutouts) and fabrication flaws (such as weld cracks, undercuts, arc strikes, and scars from chipping hammers).

Increased strain rates tend to increase the possibility of brittle behavior. Therefore, structures that are loaded at fast rates are more susceptible to brittle fracture. However, a rapid strain rate or impact load is not a required condition for a brittle fracture.

Cold work and the strain aging that normally follows generally increase the likelihood of brittle fractures. This behavior is usually attributed to a reduction in ductility. The effect of cold work occurring in cold-forming operations can be minimized by selecting a generous forming radius, therefore limiting the amount of strain. The amount of strain that can be tolerated depends on both the steel and the application. A more severe but quite localized type of cold work occurs at sheared edges, but this effect can be essentially eliminated by machining or grinding the edges after shearing. Severe hammer blows may also produce enough cold work to locally reduce the toughness of the steel.

When tensile residual stresses are present, such as those resulting from welding, they increase any applied tensile stress, resulting in the actual tensile stress in the member being greater than the applied stress. Consequently, the likelihood of brittle fracture in a structure that contains high residual stresses may be minimized by a postweld heat treatment. The decision to use a postweld heat treatment should be made with assurance that the anticipated benefits are needed and will be realized, and that possible harmful effects can be tolerated. Many modern steels for welded construction are designed for use in the less costly as-welded condition when possible. The soundness and mechanical properties of welded joints in some steels may be adversely affected by a postweld heat treatment.

Welding may also contribute to brittle fracture by introducing notches and flaws into a structure and changing the microstructure of the base metal. Such detrimental effects can be minimized by properly designing welds, by selecting their appropriate location, and by using good welding practice. The proper electrode must be selected so that the weld metal will be as resistant to brittle fracture as the base metal.

Charpy V-Notch Impact Test

Some steels will sustain more adverse temperature, notching, and loading conditions without fracture than other steels. Numerous tests have been developed to evaluate and assign a numerical value determining the relative susceptibility of steels to brittle fracture. Each of these tests can establish with certainty only the relative susceptibility to brittle fracture under the particular conditions in the test; however, some tests provide a meaningful guide to the relative performance of steels in structures subjected to severe temperature and stress conditions. The most commonly used rating test, the Charpy V-notch impact test, is described in this section, and the interpretation of its results is discussed briefly.

The Charpy V-notch impact test specifically evaluates notch toughness—the resistance to fracture in the presence of a notch—and is widely used as a guide to the performance of steels in structures susceptible to brittle fracture. In this test, a small rectangular bar with a V-shaped notch of specified size at its midlength is supported at its ends as a beam and fractured by a blow from a swinging pendulum. The energy required to fracture the specimen (which can be calculated from the height to which the pendulum raises after breaking the specimen) or the appearance of the fracture surface is determined for a range of temperatures. The appearance of the fracture surface is usually expressed as the percentage of the surface that appears to have fractured by shear as indicated by a fibrous appearance. A shiny or crystalline appearance is associated with a cleavage fracture.

These data are used to plot curves of energy (see Figure 1-11) or percentage of shear fracture as a function of temperature. For most ferritic steels, the energy and percentage of shear fracture decrease from relatively high values to relatively low values with decreasing temperature. The temperature near the lower end of the energy-temperature curve, at which

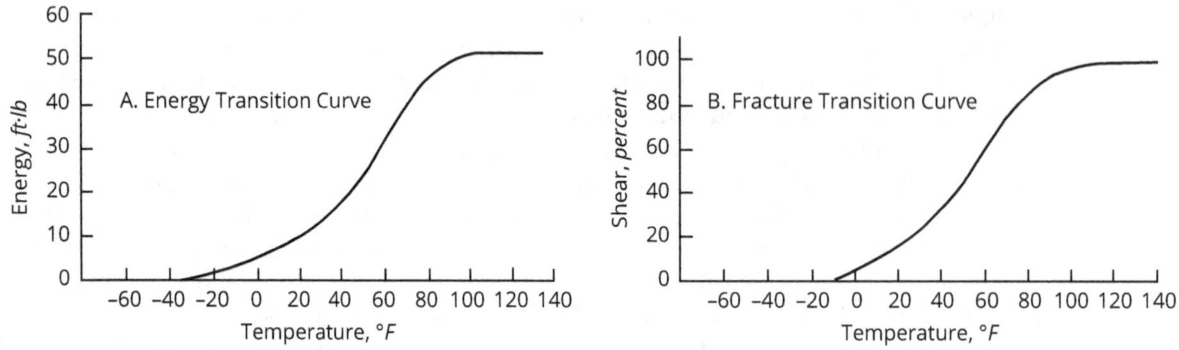

Note: Curves are for carbon steel and are taken from the Welding Research Council (1957).

Source: Brockenbrough and Johnston 1981.

Figure 1-11 Transition curves obtained from Charpy V-notch impact tests

a selected value of energy is absorbed (often 15 ft-lb), is called the *ductility transition temperature*. The temperature at which the percentage of shear fracture decreases to 50 percent is often called the *fracture-appearance transition temperature* or *fracture transition temperature*. Both transition temperatures provide a rating of the brittle fracture resistance of various steels; the lower the transition temperature, the better the resistance to brittle fracture. The ductility transition temperature and the fracture transition temperature depend on many parameters (such as composition, thickness, and thermomechanical processing) and, therefore, can vary significantly for a given grade of steel.

Steel Selection

Requirements for notch toughness of steels used for specific applications can be determined through correlations with service performance. Fracture mechanics, when applied in conjunction with a thorough study of material properties, design, fabrication, inspection, erection, and service conditions, has been beneficial. In general, where a given steel has been used successfully for an extensive period in a given application, brittle fracture is not likely to occur in similar applications unless unusual temperature, notch, or stress conditions are present. Nevertheless, it is always desirable to avoid or minimize the previously cited adverse conditions that increase the susceptibility to brittle fracture.

REFERENCES

American National Standards Institute/American Water Works Association (ANSI/AWWA) C200, Steel Water Pipe, 6 In. (150 mm) and Larger. Latest edition. Denver, CO: American Water Works Association.

Barnard, R.E. 1950. Design of Steel Ring Flanges for Water Works Service—A Progress Report. *Jour. AWWA*, 42(10):931.

Barsom, J.M. 1993. *Welded Steel Water Pipe Fracture Toughness and Structural Performance.* Bull. 11-3. Cincinnati, OH: SPFA Steel Water Pipe.

Bethlehem Steel Co. 1946. *Bolt Tests—Tension Applied by Tightening Nut Versus Pure Tension.* Bethlehem, PA: Bethlehem Steel Company (unpublished).

Brockenbrough, R.L., and B.G. Johnston. 1981. *USS Steel Design Manual.* ADUSS 27-3400-04. Pittsburgh, PA: US Steel Corporation.

Chajes, A., S.J. Britvec, and C. Winter. 1963. Effects of Cold-Straining on Structural Sheet Steels. *Jour. of the Structural Div.*, Proc., ASCE, 89, No. ST2.

Dieter, G.E., Jr. 1961. *Mechanical Metallurgy*. New York: McGraw-Hill Book Company.

Lightner; M.W., and R.W. Vanderbeck. 1956. *Factors Involved in Brittle Fracture Regional Technical Meetings*. Washington, DC: American Iron and Steel institute.

Parker, E.R. 1957. *Brittle Behavior of Engineering Structures*. New York: John Wiley & Sons.

Rolfe, S.T., and J.M. Barsom. 1977. *Fracture and Fatigue Control in Structures—Applications of Fracture Mechanics*. Englewood Cliffs, NJ: Prentice Hall.

Welding Research Council. 1957. *Control of Steel Construction to Avoid Brittle Failure*. New York: Welding Research Council.

The following references are not cited in the text.

Farr, J.R., and M.H. Jawad. 2006. *Guidebook for the Design of ASME Section VIII Pressure Vessels*, 3rd Ed. New York: ASME Press.

Neal, B.G. 1956. *The Plastic Methods of Structural Analysis*. New York: John Wiley & Sons.

Sommer, B. 1982. Spiral-Weld Pipe Meets High-Pressure Needs. *Oil and Gas Journal* (Feb.):106–116.

This page intentionally blank.

Chapter 2

Steel Pipe Manufacture and Testing

MANUFACTURE

The manufacture of steel pipe for the water industry and other water system facilities is covered by ANSI/AWWA C200, Steel Water Pipe, 6 In. (150 mm) and Larger. Although some seamless pipe is acceptable under ANSI/AWWA C200, large-diameter pipe is generally fabricated from steel coil or plate.

Electric-fusion welding and electric-resistance welding are the most common methods used to convert flatrolled steel bars, plates, sheets, and strips into tubular products. For large-diameter steel pipe electric-fusion-welded spiral-seam pipe is the most common product.

Electric-Fusion-Welded (EFW) Pipe

Electric-fusion welding incorporates a filler metal and heat to fabricate the pipe seam. Pipe produced with this process can have spiral or straight seams.

Spiral-Seam Pipe. Spiral-seam pipe is made from coiled steel by a continuous process (Figure 2-1). An automatic machine unrolls the coil, prepares the edges for welding, and spirally forms the strip into a cylindrical shape and welds the inside and outside seam automatically. The pipe is cut to the desired length by an automatic cutoff device.

Straight-Seam Pipe. Straight-seam pipe is made from plate with edges planed parallel to each other and square with the ends. Curving the plate edges with crimping rolls, where necessary, is the first step of the forming process. This step is followed by presses that form the plate first into a U-shaped trough and then into a full O-shaped tube (Figure 2-2). The O-shaped tube is then fed into a longitudinal seamwelding machine. Straight-seam pipe may also be produced on pyramid-style rollers in shorter sections that are then usually welded together to form a longer length pipe (Figure 2-3).

Pipe manufactured using the EFW process is limited only by size capabilities of individual pipe manufacturers.

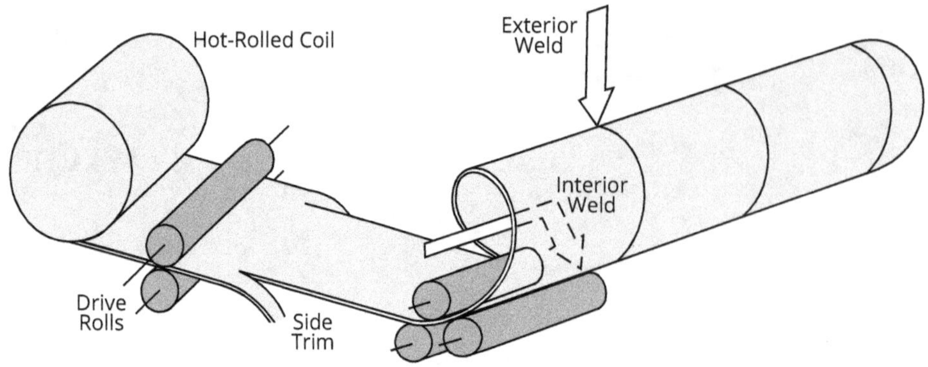

Figure 2-1 Schematic diagram of process for making spiral-seam pipe

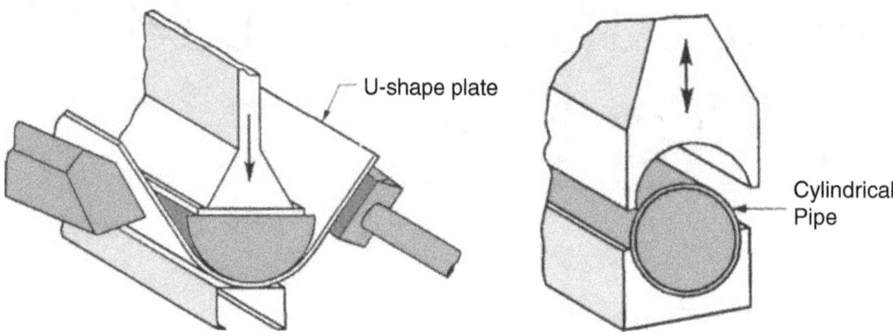

Figure 2-2 U-ing and O-ing straight-seam double-fusion-welded pipe

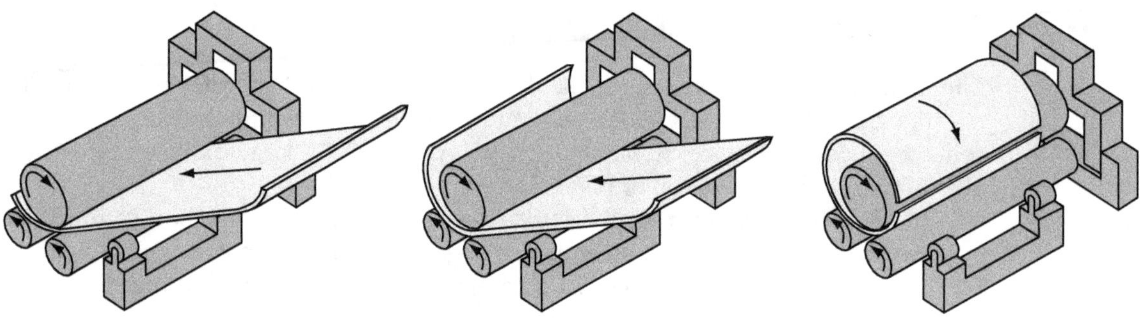

Figure 2-3 Schematic diagram for making plate pipe

Electric-Resistance Welded (ERW) Pipe

Electric-resistance welding (ERW) is performed without filler metals (Figures 2-4 through 2-7). The flat strip, with edges previously trimmed to provide a clean, even surface for welding, is formed progressively into a tubular shape as it travels through a series of rolls. This process is typical for small-diameter pipe, with the pipe diameter limited by the width of the sheet or coil. The material is cold formed. Welding is then effected by the application of heat and pressure. The welding heat for the tubular edges is generated by resistance to the flow of an electric current, which can be introduced through electrodes or by induction. Pressure rolls force the heated edges together to effect the weld. The squeezing action of the pressure rolls forming the weld causes some of the hot weld metal to extrude from the joint, forming a bead of weld flash both inside and outside the pipe. The flash is normally trimmed within tolerance limits while it is still hot from welding using mechanical cutting tools contoured to the shape of the pipe.

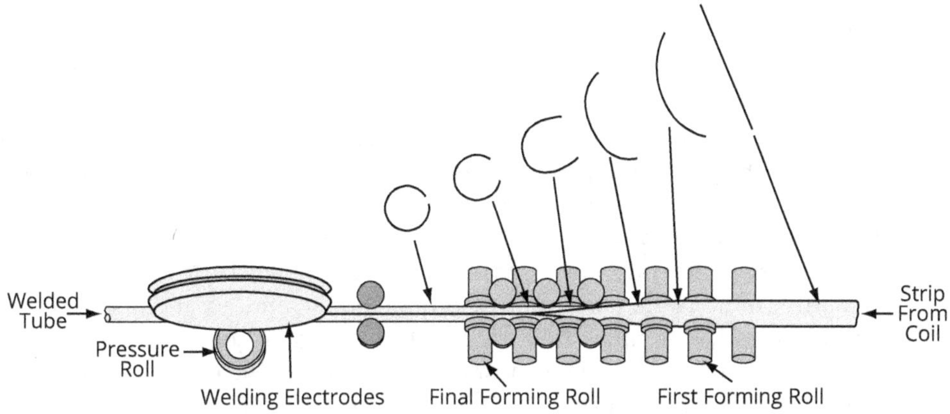

Source: American Iron and Steel Institute 1982.

Figure 2-4 Schematic representation of the sequence of operations performed by a typical machine for making electric-resistance-welded tubes from steel strip

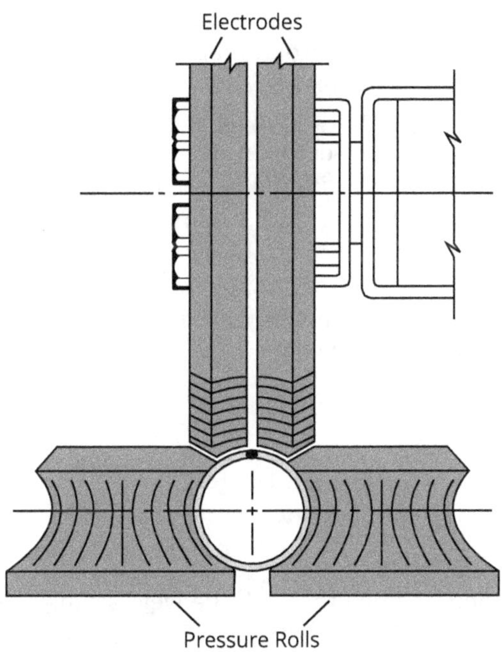

Application of pressure by rolls on both sides and beneath the electrodes forces the heated tube edges together to form a weld.

Source: American Iron and Steel Institute 1982.

Figure 2-5 Cross section through weld point

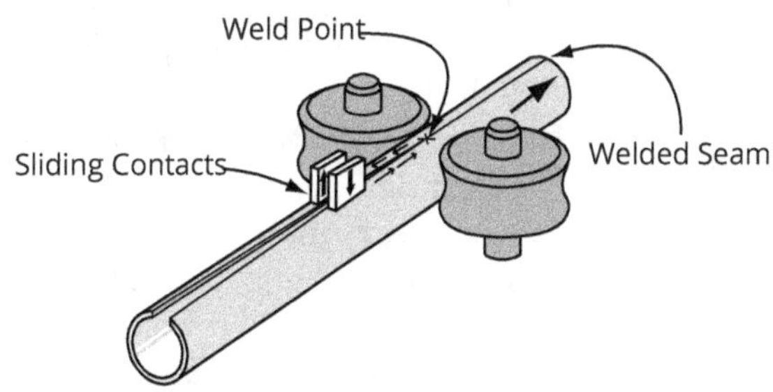

The current enters the tube via sliding contacts and flows along Vee edges to and from weld point.

Source: American Iron and Steel Institute 1982.

Figure 2-6 Electric-resistance welding using high-frequency welding current

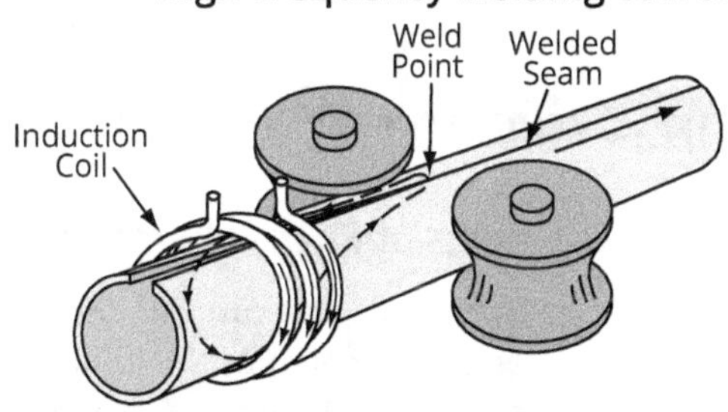

Eddy current flows around the back of the tube and along the edges to and from the weld point.

Source: American Iron and Steel Institute 1982.

Figure 2-7 Electric-resistance welding by induction using high-frequency welding current

MATERIALS

The various services for which steel pipe is used require a variety of chemical compositions to produce the necessary characteristics. The chemical compositions established in the ANSI/AWWA steel pipe standards are appropriate for the needs of water utility applications. However, there are other steel materials that may be equally suitable.

TESTING—COIL AND PLATE

Heat Analysis

Heat analysis is the chemical analysis representative of the heat of steel. This analysis is reported to the purchaser. Analysis results are determined by testing for specific elements, using a test sample obtained from the first or middle part of the heat during the pouring of the steel from the ladle.

Most steel-melting operations obtain more than one ladle test sample from each heat; often three or more are taken, representing the first, middle, and last portions of the heat. The first or middle samples are used in determining the ladle analysis because experience has shown that these locations most closely represent the chemical analysis of the entire heat. The additional samples are used for a survey of uniformity and for control purposes.

Product Analysis

Product analysis, as used in the steel industry, means analysis of the steel after it has been rolled or forged into semifinished or finished forms. Such an analysis is made either to verify the average composition of the heat, to verify the composition of a lot as represented by the heat analysis, or to determine variations in the composition of a heat or lot. Product analysis of known heats is justified only where a high degree of uniformity of composition is essential—for example, on material that is to be heat treated. Such analysis should rarely be necessary for water pipe except to identify or confirm the assumed analysis of plates or pipe that have lost identity. The results of analyses representing different locations in the same piece, or taken from different pieces of a lot, may differ from each other and from the heat analysis because of segregation and analytical reproducibility. These permissible variations from the specified ranges or limits have been established in the applicable specification or by common practice. The variations are a natural phenomenon that must be recognized by inspectors. The methods of analysis commonly used are in accordance with ASTM A751, those approved by the National Institute of Standards and Technology, or others of equivalent accuracy.

TESTING—FORMED PIPE

Tests of Physical Properties

Standard pipe production tests include tension and bend tests of the weld seam as well as a macroetch of the joint weld to verify complete joint penetration (CJP). Other testing, such as Charpy V-notch testing, of the steel material or weld seam may be required by the purchaser. The methods of testing the physical properties of steel pipe are based on ASTM A370. The required test frequencies and values of the physical properties are contained in ANSI/AWWA C200, Steel Water Pipe, 6 In. (150 mm) and Larger.

Hydrostatic Test of Straight Pipe

Straight lengths of pressure pipe are customarily subjected to an internal hydrostatic pressure test. This operation is conducted as a part of the regular mill inspection procedure to detect defects. It is not intended to bear a direct relationship to bursting pressures, working pressures, axial loads, or design data, although test pressures sometimes influence higher future allowable design pressures. ANSI/AWWA C200 contains a formula for determining the required hydrostatic test pressure.

Hydrostatic tests are performed at the pressure required by the standard during manufacture of the pipe. The producer, on request, customarily furnishes a certificate confirming such testing.

Specials and Fittings

Specials and fittings may be produced from previously hydrostatically tested straight sections of pipe or may be manufactured independently either by precutting plate and rolling into round or conical sections or by first rolling and then cutting to form the necessary configuration. While hydrostatic testing is generally required of straight lengths of pipe, it is usually not required of specials and fittings for economic reasons. Regardless of the method of fabricating specials and fittings, all nonhydrostatically tested seams should be tested by nondestructive means.

Tests of Dimensional Properties

The pipe and fitting dimensions, wall thickness, straightness, and out-of-roundness are checked as part of the normal manufacturing procedure. Such dimensions are subject to the tolerances prescribed in the appropriate standards or specifications.

REFERENCES

American Iron and Steel Institute. 1982. Carbon Steel Pipe, Structural Tubing, Line Pipe and Oil Country Tubular Goods. In *Steel Products Manual*. Washington, DC: American Iron Steel Institute.

ANSI/AWWA C200, Steel Water Pipe, 6 In. (150 mm) and Larger. Latest edition. Denver, CO: American Water Works Association.

ASTM International (ASTM) A370, Standard Test Methods and Definitions for Mechanical Testing of Steel Products. Philadelphia: ASTM International.

ASTM A751, Standard Test Methods, Practices, and Terminology for Chemical Analysis of Steel Products. Philadelphia: ASTM International.

Chapter 3

Hydraulics of Pipelines, Water Hammer, and Pressure Surge

This chapter concentrates on the flow of water in transmission conduits. It is not intended to cover flows through the complicated networks of distribution systems. Because this manual is a guide to practice rather than a textbook, historical and theoretical development of the many hydraulic flow formulas has been omitted, as has discussion of universal or rational formulas.

Water hammer is the result of a change in flow velocity in a closed conduit causing elastic waves to travel upstream and downstream from the point of origin. The elastic waves, in turn, cause increases or decreases in pressure as they travel along the line, and these pressure changes are variously referred to as *water hammer, surge,* or *transient pressure.*

The phenomenon of water hammer is extremely complex, and no attempt will be made to cover the subject in-depth in this manual. Only the fundamentals of elastic-wave theory and specific data pertaining to the properties of steel pipe will be discussed. For a more detailed understanding of water hammer, the references listed at the end of this chapter should be consulted.

Hydraulics of pipelines will be discussed first in this chapter, followed by a discussion of water hammer and pressure surge.

HYDRAULIC FORMULAS

Hydraulic Symbols

The definitions of hydraulic symbols used in this chapter are as follows:

A = area of pipe, ft^2

C = Hazen-Williams coefficient

d = inside diameter of pipe, ft (used in Figure 3-2)
d_n = nominal diameter of pipe, in.
D = inside diameter of pipe, ft
e = absolute roughness, ft
f = Darcy friction factor
g = acceleration due to gravity, 32.2 ft/sec^2
h_L = frictional head loss in pipe length L, ft
H = head loss in 1,000 units of length of pipe, ft/1,000 ft
K = resistance coefficient
K_s = Scobey constant
L = length of pipe, ft
n = Manning coefficient
P = pressure, lb/in.^{2}
Q = flow (discharge), ft^3/sec
r = hydraulic radius of pipe, ft; $r = D/4$ for laminar flow
R_e = Reynolds number
$s = \frac{h_L}{L} = \frac{H}{1,000}$ = slope of hydraulic gradient
ν = kinematic viscosity of the fluid, ft^2/sec
V = mean velocity, ft/sec

The Hazen-Williams Formula

Probably the most popular formula in current use among waterworks engineers is the Hazen-Williams formula. This formula, developed and published in the early 1900s, is

$$V = 1.318Cr^{0.63}s^{0.54} \quad \text{(Eq 3-1)}$$

The head loss h_L may be calculated from

$$h_L = \frac{4.72Q^{1.852}L}{C^{1.852}D^{4.87}} \quad \text{(Eq 3-2)}$$

Tests have shown that the value of the Hazen-Williams coefficient C is dependent not only on the surface roughness of the pipe interior but also on the diameter of the pipe. Flow measurements indicate that for pipe with smooth interior linings in good condition, the average value of $C = 140 + 0.17d_n$.

However, for design purposes for new pipe the following values are suggested:

Diameter, *in.*	C Value
16 to 48	140
54 to 108	145
114 and larger	150

A graphical solution of the Hazen-Williams formula for $C = 140$ is presented in Figure 3-1 for pipe sizes 6 in. through 144 in. The multiplying factors in Table 3-1 provide a convenient means of changing the flow capacities shown in Figure 3-1 to the flows for other values of C.

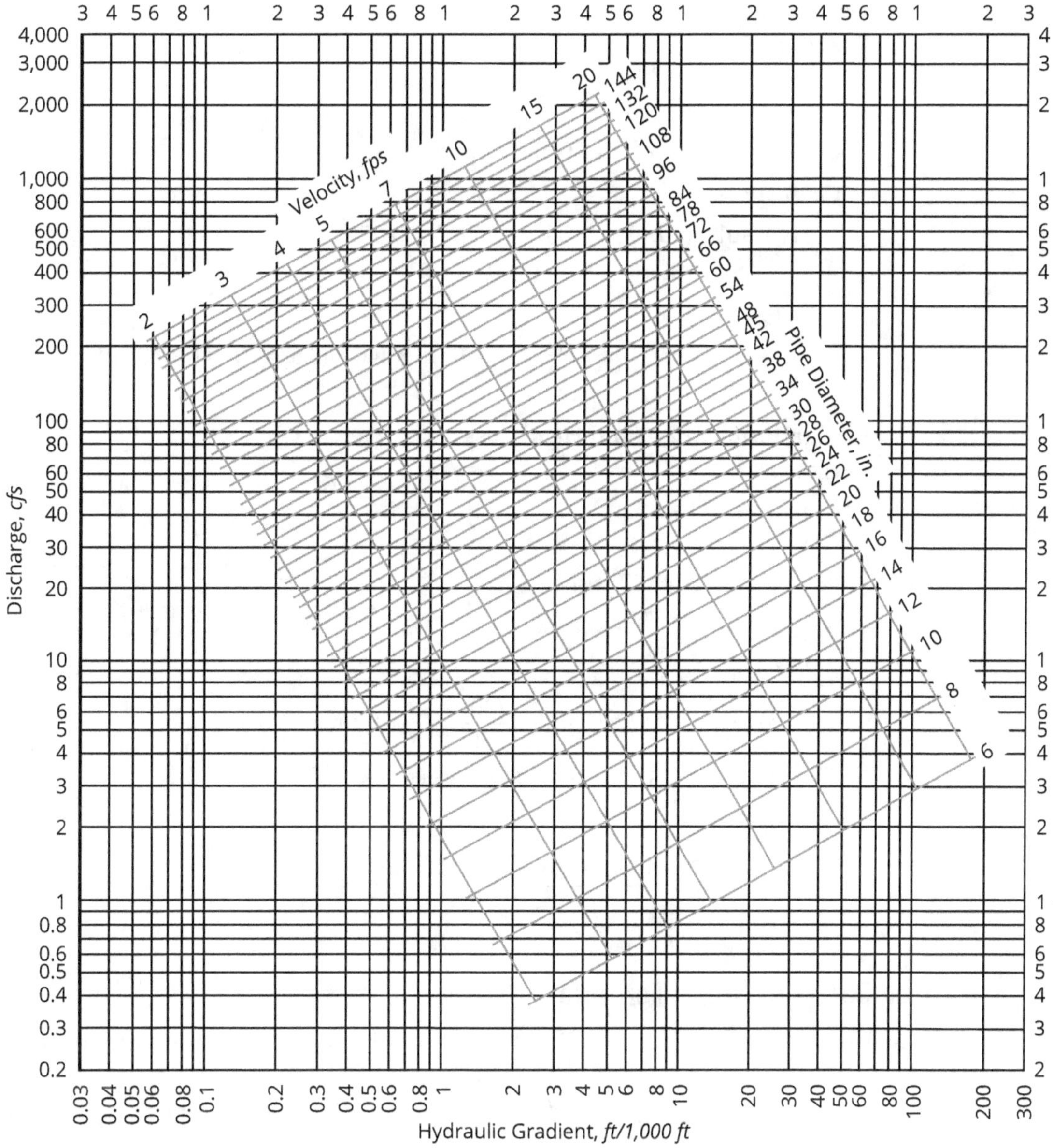

Source: Barnard 1948.

Figure 3-1 Solution of the Hazen-Williams formula (based on $V = 1.318Cr^{0.63}s^{0.54}$ for $C = 140$)

Table 3-1 Multiplying factors corresponding to various values of C in Hazen-Williams formula*

	Values of C								
	160	155	150	145	140	130	120	110	100
					Base C = 140				
Relative discharge and velocity for given loss of head	1.143	1.107	1.071	1.036	1.000	0.929	0.857	0.786	0.714
Relative loss of head for given discharge	0.781	0.828	0.880	0.937	1.000	1.147	1.330	1.563	1.865

Source: Some data drawn from Barnard 1948.

*Data for use with Figure 3-1.

Design Problem 1—Find Velocity

Assume:

D = 30 in.

C = 140

H = 3 ft in 1,000 ft of pipe

Calculations:

Hydraulic radius: $r = 30/4 = 7.5$ in. = 0.625 ft

Slope: $s = H/1{,}000 = 0.003$

Velocity:

$$V = 1.318Cr^{0.63}s^{0.54} = 1.318 \times 140 \times (0.625)^{0.63} \times (0.003)^{0.54}$$
$$= 5.95 \text{ ft/sec}$$

Design Problem 2—Find Head Loss

Assume:

D = 30 in.

C = 140

V = 5.67 ft/sec

L = 750 ft

Calculations:

$$\text{Area: } A = \pi D^2/4 = \pi \times \frac{(30)^2}{4} = 706.9 \text{ in.}^2 = 4.91 \text{ ft}^2$$

$$\text{Flow: } Q = 5.67 \times 4.91 = 27.8 \text{ ft}^3\text{/sec}$$

Head loss:

$$h_L = \frac{4.72 \times (27.8)^{1.852} \times 750}{(140)^{1.852} \times (2.5)^{4.87}} = 2.04 \text{ ft}$$

The Darcy-Weisbach Formula

An alternative to the Hazen-Williams formula is the Darcy-Weisbach formula, which is

$$h_L = f\left(\frac{L}{D}\right)\left(\frac{V^2}{2g}\right) \qquad \text{(Eq 3-3)}$$

The Darcy friction factor *(f)* can be determined from the Moody diagram, presented in Figure 3-2. To use this diagram, the Reynolds number R_e and the relative roughness *(e/d)* must first be calculated. Both terms are defined in the following paragraphs.

The Reynolds number is a function of the flow in the pipe and may be calculated as

$$R_e = \frac{dV}{v} \qquad \text{(Eq 3-4)}$$

The kinematic viscosity (v) of water at various temperatures from freezing to boiling is presented in Table 3-2.

The relative roughness *e/d* of a pipe is a function of the absolute roughness *(e)* of the interior surface of the pipe and the inside pipe diameter *(d)*. The recommended design range of *e* is 3.5×10^{-4} to 4.0×10^{-4} ft for cement-mortar lined pipe and has historically been used. Once the Reynolds number and the relative roughness are determined, the Moody diagram (Figure 3-2) can be used to determine values of the Darcy friction factor *(f)*, which may then be used to solve the Darcy-Weisbach formula.

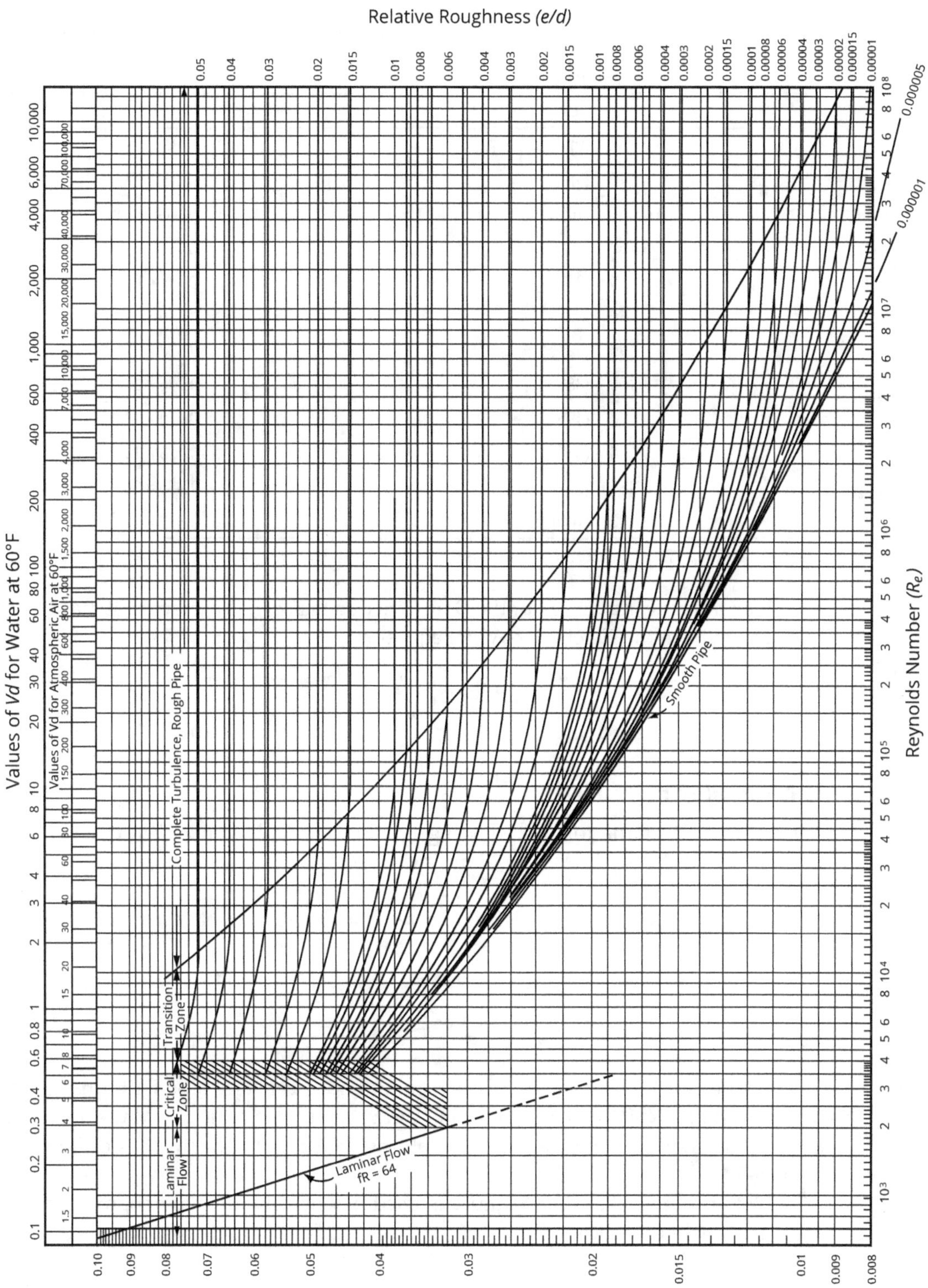

Source: Hydraulic Institute 1954.

Figure 3-2 The Moody diagram for friction in pipe

Table 3-2 Kinematic viscosity of water

Temperature	Kinematic Viscosity (v)
°F (°C)	ft^2/sec
32 (0)	1.93×10^{-5}
40 (4)	1.66×10^{-5}
50 (10)	1.41×10^{-5}
60 (16)	1.22×10^{-5}
70 (21)	1.06×10^{-5}
80 (27)	0.930×10^{-5}
90 (32)	0.823×10^{-5}
100 (38)	0.736×10^{-5}
120 (49)	0.610×10^{-5}
150 (66)	0.476×10^{-5}
180 (82)	0.385×10^{-5}
212 (100)	0.319×10^{-5}

Note: To convert °F to °C, subtract 32 and multiply the result by 5/9.

If an analytical solution for the Darcy friction factor *(f)* is preferred, it may be obtained by iteration from the Colebrook-White equation:

$$\frac{1}{\sqrt{f}} = -2\log_{10}\left[\frac{e}{3.7d} + \frac{2.51}{R_e\sqrt{f}}\right] \quad \text{(Eq 3-5)}$$

The Manning Formula

The Manning formula for pipe flowing full is commonly utilized to determine flow in partially filled gravity lines. The Manning formula is

$$V = \frac{0.59}{n} D^{0.667} s^{0.5} \quad \text{(Eq 3-6)}$$

or

$$h_L = 2.87n^2 \frac{LV^2}{D^{1.33}} \quad \text{(Eq 3-7)}$$

For design, an n value of 0.011 is recommended for steel pipe with linings conforming to current ANSI/AWWA standards. A graphical solution to the Manning formula for $n = 0.011$ is shown in Figure 3-3. Multiplying factors for other values of n are given in Table 3-3.

Note that the Manning formula is used extensively for open channel with limited pressure pipeline applications.

The Scobey Formula

The Scobey formula, used perhaps more commonly in irrigation work than in the waterworks industry, is

$$V = \frac{D^{0.58} h_L{}^{0.526}}{K_s{}^{0.526}} \quad \text{(Eq 3-8)}$$

or for determining head loss:

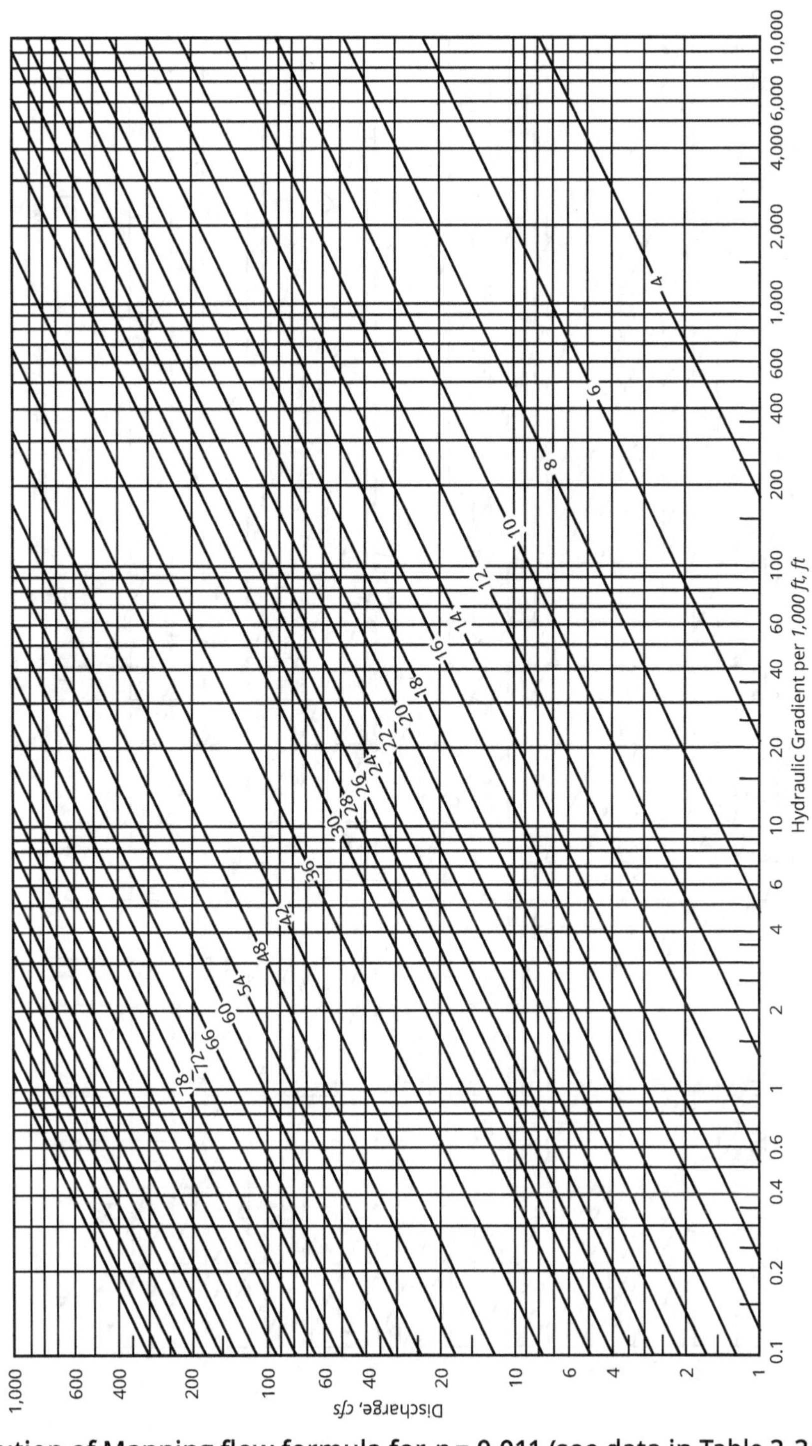

Figure 3-3 Solution of Manning flow formula for *n* = 0.011 (see data in Table 3-3 for other *n* values)

Table 3-3 Multiplying factors for friction coefficient values—base *n* = 0.011*

	n value				
	0.009	0.010	0.011	0.012	0.013
Relative discharge	1.222	1.100	1.000	0.917	0.846

*Data for use with Figure 3-3.

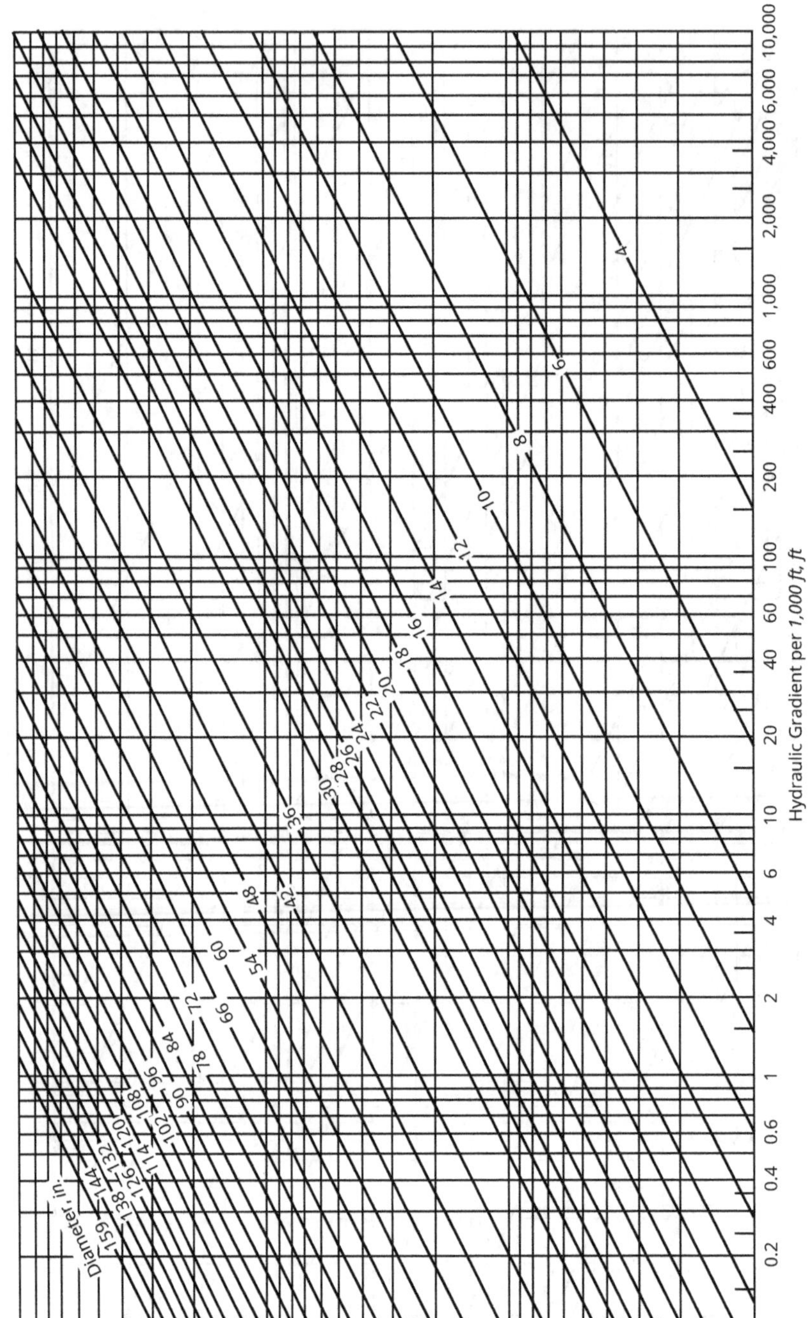

Figure 3-4 Solution of Scobey flow formula for K_s = 0.36 (see data in Table 3-4 for other K_s values)

Table 3-4 Multiplying factors for friction coefficient values—base K_s = 0.36*

	K_s value				
	0.32	0.34	0.36	0.38	0.40
Relative discharge	1.125	1.059	1.000	0.947	0.900

*Data for use with Figure 3-4.

$$h_L = K_s \frac{V^{1.9}}{D^{1.1}} \qquad \text{(Eq 3-9)}$$

The recommended K_s value for steel pipe with linings conforming to current ANSI/AWWA standards is 0.36. A graphical solution to the Scobey formula for K_s = 0.36 is shown in Figure 3-4. Multiplying factors for other friction coefficients are given in Table 3-4.

Design Problem 3—Find Velocity

Assume:

D = 30 in.
K_s = 0.36
h_L = 3 ft

Calculations:

Velocity:

$$V = \frac{(2.5)^{0.58} \times (3)^{0.526}}{(0.36)^{0.526}} = 5.19 \text{ ft/sec}$$

Design Problem 4—Find Head Loss

Assume:

D = 30 in.
K_s = 0.36
V = 4.89 ft/sec

Calculations:

Head loss:

$$h_L = 0.36 \frac{(4.89)^{1.9}}{(2.5)^{1.1}} = 2.68 \text{ ft}$$

CALCULATIONS

Computations for Flow Through Pipe

The quantity of water passing through any given pipe depends on the head (pressure) producing the flow, the diameter and length of the pipe, the condition of the pipe interior (smooth or rough), the number and abruptness of bends or elbows, and the presence of tees, branches, valves, and other appurtenances in the line.

The pressure head, or pressure, affecting flow may be divided into four parts: velocity head loss, entrance head loss, loss of head through friction, and minor losses caused by elbows, fittings, and valves. Pressure *(P)* is related to pressure head *(H)* by the following formula:

$$P = 0.4335H; \ (P \text{ in lb/in}^2, H \text{ in ft}) \qquad \text{(Eq 3-10)}$$

Velocity Head Loss ($V^2/2g$)

Velocity head loss is defined as the height a body must fall in a vacuum to acquire the velocity at which the water flows in the pipe. This loss is usually considered to be unrecoverable at the outlet. This head loss is calculated by the following formula:

$$h_L = \frac{V^2}{2g} \qquad \text{(Eq 3-11)}$$

Entrance Head Loss

Entrance head loss is the head required to overcome the resistance at the entrance to the pipe; it is usually less than the velocity head loss. When the conditions are not specified, it is ordinarily considered equal to one-half the velocity head loss, assuming a sharp-edge entrance. Head losses for other than sharp-edge entrances may be found in treatises on hydraulics.

Loss of Head Through Friction

Friction head loss may be determined by one of the formulas that have been discussed previously. Data are given in this chapter to aid in solving the formulas.

Minor Losses Caused by Elbows, Fittings, and Valves

In any given pipeline, all losses should be considered so that important factors will not be overlooked. Minor losses should always be recognized when evaluating flow tests. In long pipelines, minor losses caused by bends, fittings, and entrance losses are relatively insignificant and occasionally are ignored. In long lines with low velocities, the sum of velocity head loss and entrance head loss may be relatively insignificant. In short pipelines with high velocities, the sum of minor losses becomes relatively significant and should not be ignored. Ordinary tables and charts showing flow of water in pipe usually give only the friction head loss in straight pipe. In long water lines, this is typically the largest loss.

In the final solution to a flow problem, the sum of all losses must equal the available head, or pressure, producing the flow. The foregoing formulas determine H or V, and the volume of flow Q is found from

$$Q = AV \tag{Eq 3-12}$$

The information contained in Table 3-5 is useful when making hydraulic calculations.

Pressure Equivalents

Pressure is commonly referred to by inches of mercury, feet of head, or simply pounds per square inch (lb/in.^{2} or psi). Following are equations to relate these common terms:

$$P_{(mercury)} = P_{(water)} \times 2.036,\ (P_{(mercury)} \text{ in., } P_{(water)} \text{ lb/in.}^2) \tag{Eq 3-13}$$

$$P_{(mercury)} = H_{(water)} \times 0.8827,\ (P_{(mercury)} \text{ in., } H_{(water)} \text{ ft}) \tag{Eq 3-14}$$

Flow Through Fittings—Equivalent-Length Method

Experiments have shown that the head loss in bends, fittings, and valves is related to flow velocity and pipe diameter similar to that in straight pipe. Consequently, it is possible to determine the length of a theoretical piece of straight pipe in which the head loss caused by friction would be the same as for a specific fitting. Pipeline designers commonly utilize this method of equivalent lengths. By developing the total equivalent length (piping plus bends, fittings, valves, etc.), the total head loss in a piping system can be determined.

The classical equation developed by Darcy-Weisbach for energy loss of flow is shown as Eq 3-3.

The term h_L is the head loss caused by friction in the length of pipe L of inside diameter D for mean velocity V. The friction factor f is a function of pipe roughness, velocity, pipe

Table 3-5 Flow equivalents

mgd	*gpm*	*ft³/sec*	*mgd*	*gpm*	*ft³/sec*
1	694	1.55	36	25,000	55.75
2	1,389	3.10	37	25,694	57.30
3	2,083	4.65	38	26,389	58.85
4	2,778	6.19	39	27,083	60.40
5	3,472	7.74	40	27,778	61.94
6	4,167	9.29	42	29,167	65.04
7	4,861	10.84	44	30,556	68.14
8	5,556	12.39	46	31,944	71.24
9	6,250	13.94	48	33,333	74.33
10	6,944	15.49	50	34,722	77.43
11	7,639	17.03	52	36,111	80.53
12	8,333	18.58	54	37,500	83.63
13	9,028	20.13	56	38,889	86.72
14	9,722	21.68	58	40,278	89.82
15	10,417	23.23	60	41,667	92.92
16	11,111	24.78	62	43,056	96.01
17	11,806	26.33	64	44,444	99.11
18	12,500	27.88	66	45,833	102.21
19	13,194	29.42	68	47,222	105.31
20	13,889	30.97	70	48,611	108.40
21	14,583	32.52	72	50,000	111.50
22	15,278	34.07	74	51,389	114.60
23	15,972	35.62	76	52,778	117.69
24	16,667	37.17	78	54,167	120.79
25	17,361	38.72	80	55,556	123.89
26	18,056	40.26	82	56,944	126.99
27	18,750	41.81	84	58,333	130.08
28	19,444	43.36	86	59,722	133.18
29	20,139	44.91	88	61,111	136.28
30	20,833	46.46	90	62,500	139.38
31	21,528	48.01	92	63,889	142.47
32	22,222	49.56	94	65,278	145.57
33	22,917	51.10	96	66,667	148.67
34	23,611	52.65	98	68,056	151.76
35	24,306	54.20	100	69,444	154.86

diameter, and fluid viscosity. Values for f were developed by Moody (1944) and are shown in Figure 3-2. With known f and L/D, the Darcy-Weisbach formula can be expressed as:

$$h_L = K\left(\frac{V^2}{2g}\right) \quad \text{(Eq 3-15)}$$

In this equation, K is the resistance coefficient. Figure 3-5 shows values for K based on a summary of experimental data.

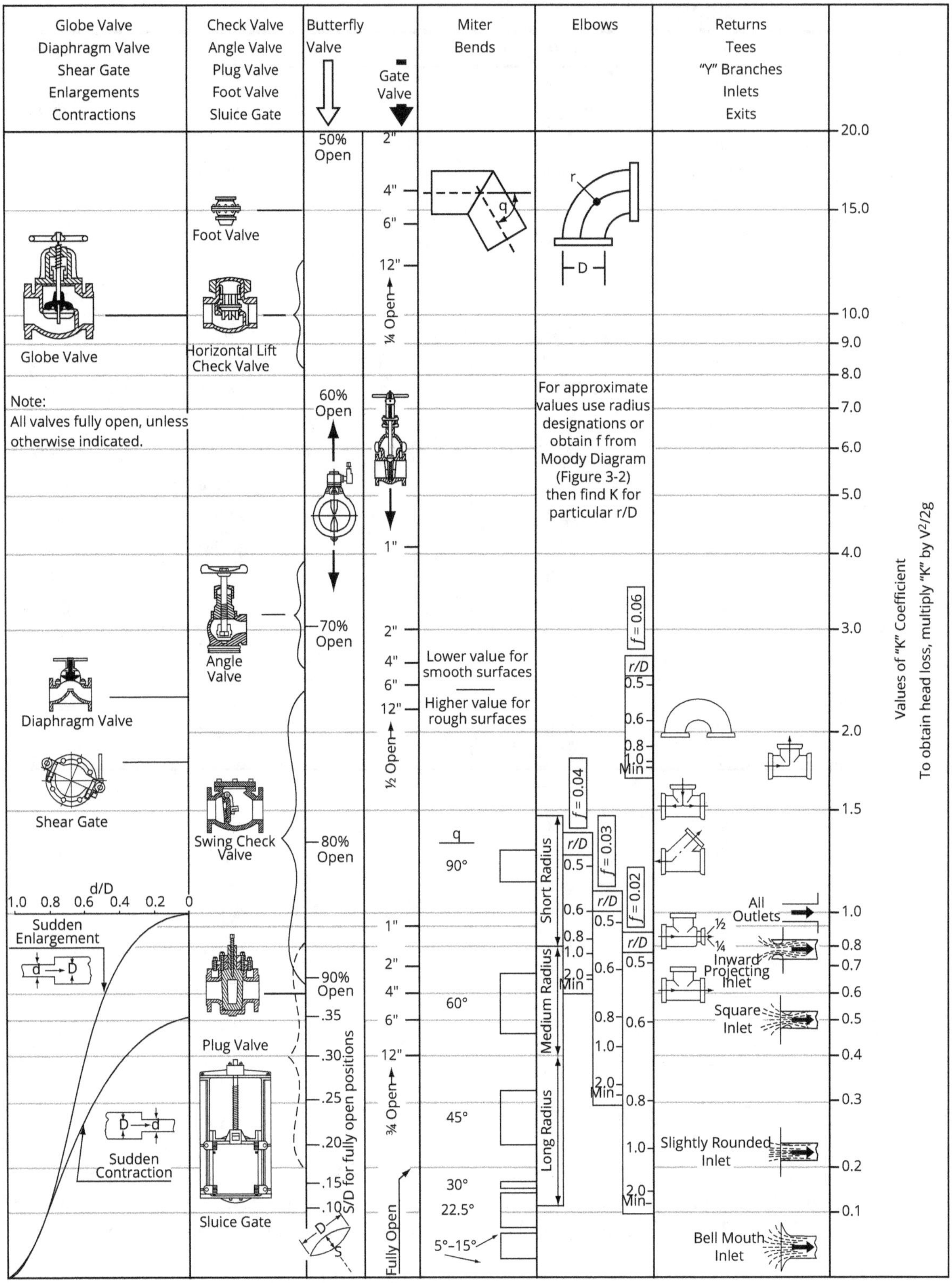

Source: John F. Lenard, President, Lenard Engineering Inc.

Figure 3-5 Resistance coefficients of valves and fittings for fluid flows

Examples to determine head loss h_L for fittings and valves and equivalent pipe lengths using Figure 3-4 are as shown in design problem 5.

Design Problem 5—Find Head Loss Through Fittings

Assume:

D = 6 in.

C = 100

Q = 450 gpm

V = 5.12 ft/sec

Calculations:

Flow: $Q = 450 \times 60 \times 24 = 0.648 \text{ mgd} \times 1.55 = 1 \text{ ft}^3/\text{sec}$

Velocity head: $\dfrac{V^2}{2g} = \dfrac{(5.12)^2}{64.4} = 0.41 \text{ ft}$

a. 6-in. gate valve, fully open:
$K = 0.2$ $\quad h_L = 0.2 \times 0.41 \text{ ft} = 0.08 \text{ ft}$

b. 6-in. swing check valve, 80 percent open:
$K = 1.4$ $\quad h_L = 1.4 \times 0.41 \text{ ft} = 0.57 \text{ ft}$

c. Sudden enlargement from 6 in. to 8 in.: $d/D = 0.75$
$K = 0.18$ $\quad h_L = 0.18 \times 0.41 \text{ ft} = 0.07 \text{ ft}$

d. 6-in. elbow:
$K = 0.5$ $\quad h_L = 0.5 \times 0.41 \text{ ft} = 0.21 \text{ ft}$

Total head loss: 0.93 ft

Solve for the equivalent pipe length of 6-in. pipe using the Hazen-Williams formula with $C = 140$, $h_L = 0.97$ ft as calculated above.

Rewrite Eq 3-2, $h_L = \dfrac{4.72Q^{1.852}L}{C^{1.852}D^{4.87}}$ to solve for length of pipe, $L = \dfrac{h_L C^{1.852} D^{4.87}}{4.72Q^{1.852}}$

$$L = \frac{0.93(140)^{1.852}(0.5)^{4.87}}{4.72(1)^{1.852}}$$

$$L = 63.6 \text{ ft}$$

WATER HAMMER AND PRESSURE SURGE

Basic Relationships

The following fundamental relationships in surge-wave theory determine the magnitude of the pressure rise and its distribution along a conduit. The pressure rise for instantaneous closure is directly proportional to the fluid velocity at cutoff and to the magnitude of the surge wave velocity. It is independent of the length of the conduit. Its value is

$$h = \frac{aV}{g} \qquad \text{(Eq 3-16)}$$

or

$$P = \frac{a\gamma V}{144g} = \left(\frac{a}{g}\right)\left(\frac{\text{sp gr}}{2.3}\right)V \qquad \text{(Eq 3-17)}$$

Where:

$$a = \frac{12}{\sqrt{\frac{\gamma}{g}\left(\frac{1}{k} + \frac{d}{Et}\right)}} \qquad \text{(Eq 3-18)}$$

In the above equations:

h = pressure rise above normal, ft of water
a = wave velocity, ft/sec
V = velocity of flow, ft/sec
g = acceleration due to gravity, 32.2 ft/sec^2
P = pressure rise above normal, lb/in.^{2}
γ = specific weight of fluid, lb/ft^3
sp gr = specific gravity of fluid (water = 1.0)
k = bulk modulus of compressibility of liquid, lb/in.^{2}
d = inside diameter of pipe, in.
E = modulus of elasticity for pipe wall material, lb/in^2
t = thickness of pipe wall, in.
L = length of pipeline, ft
$\frac{2L}{a}$ = critical time of pipeline, sec
T = closing time, sec

For steel pipe, using k = 300,000 lb/in.^{2} and E = 30,000,000 lb/in.^{2}, Eq 3-18 reduces to

$$a = \frac{4{,}660}{\sqrt{1 + \frac{1}{100}\left(\frac{d}{t}\right)}}$$

Figure 3-6 gives values of pressure wave velocity for various pipe materials with *d*/*t* ratios up to 90. For steel pipe, higher ratios are frequently encountered in large sizes and Table 3-6 gives computed values up to *d*/*t* = 288.

When the flow rate is changed in a time greater than zero but less than or equal to the critical time of 2*L*/*a* seconds, the magnitude of the pressure rise is the same as with

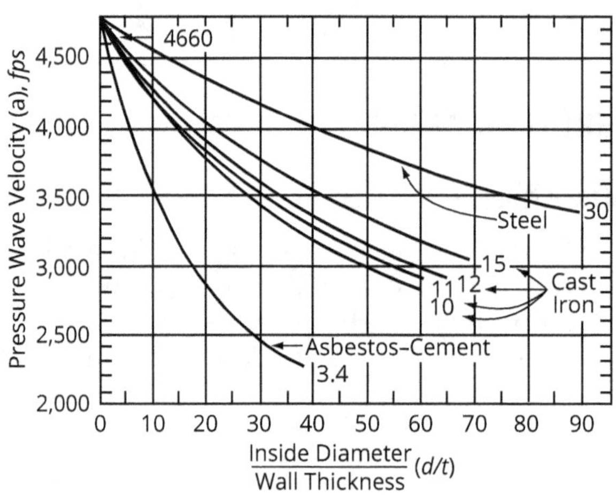

Note: The number at the end of each curve represents the modulus of elasticity (*E*) in 1,000,000-psi units for various pipe materials.

Source: Kerr et al. 1950.

Figure 3-6 Surge wave velocity chart for water

Table 3-6 Velocity of pressure wave for steel pipe

Diameter/Thickness, *d/t*	Wave Velocity, *a* (*ft/sec*)
100	3,300
120	3,140
140	3,010
160	2,890
180	2,780
200	2,690
220	2,605
240	2,530
288	2,365

Note: Wave velocities shown are for bare steel pipe.

instantaneous closure, but the duration of the maximum value decreases as the time of closure approaches $2L/a$ seconds. Under these conditions, the pressure distribution along the pipeline varies as the time of closure varies. The pressure decreases uniformly along the line if closure is in $2L/a$ seconds. The maximum pressure at the control valve exists along the full length of the line with instantaneous closure and for slower rates the pressure travels up the pipe a distance equal to $L - Ta/2$, then decreases uniformly. The surge pressure distribution along the conduit is independent of the profile or ground contour of the line as long as the total pressure remains above the vapor pressure of the fluid.

For valve closing times greater than $2L/a$ seconds, the maximum pressure rise will be the maximum rate of change in flow with respect to time, $\Delta V/\Delta T$. Nonlinear closure rates of a valve can be investigated and the proper valve closing time determined to hold the maximum pressure rise to a desired limiting value. The effect of pumps and quick-closing check valves or control valves can be investigated using a graphical method or by a numerical method using a computer.

The profile of the pipeline leading away from a pumping station may have a major influence on the surge conditions. When high points occur along the line, the surge hydraulic-grade elevation may fall below the pipe profile, causing negative pressures, perhaps as low as the vapor pressure of the fluid. If this occurs, the liquid column may be separated by a zone of vapor for a short time. Parting and rejoining of the liquid column can produce extremely high pressures and may cause failure of the conduit (Richards 1956).

The effect of friction can be accounted for in any surge problem. When friction losses are less than 5 percent of the normal static or working pressure, they can usually be neglected.

Accurate results of a surge analysis depend on knowing the various hydraulic and physical characteristics of the system. The velocity of the pressure wave is a fundamental factor in any surge study, as the surge pressures are directly proportional to its value. This velocity depends on the pipe diameter, wall thickness, material of the pipe wall, and the density and compressibility of the fluid in the pipe.

Determining the physical characteristics of the pipe material is straightforward. Young's modulus for steel lines can be taken at 30,000,000 psi, because it averages between 29,000,000 and 31,000,000 psi, depending on the steel used. If the ratio of diameter to thickness is known, it is necessary to know only the density and the compressibility of the liquid within the pipe to determine the surge wave velocity a.

Within the range of ordinary operating temperatures for water, 32–100°F, and for pressures in the range of 0–1,000 psi, the specific gravity can be assumed to be 1.00. In the same range, the modulus of compressibility, or bulk modulus, has been measured and verified by field tests to be approximately 300,000 psi with a variation of ±3 percent (Kerr et al. 1950).

CHECKLIST FOR PUMPING MAINS

A few factors can be checked to indicate whether surges of serious proportions will occur in any given system, once the physical, hydraulic, and operating characteristics are established. For most transmission mains supplied by motor-driven centrifugal pumps, the following 12 questions will indicate the seriousness of the surge problem:

1. Are there any high spots on the profile of the transmission main where the occurrence of a vacuum can cause a parting of the water column when a pump is shut off?
2. Is the length of the transmission main less than 20 times the head on the pumps (both values expressed in feet)?
3. Is the maximum velocity of flow in the transmission main in excess of 4.0 ft/sec?
4. Is the safety factor of the pipe less than 3 (related to ultimate strength) for normal operating pressures?
5. What is the natural decreasing rate of the water column if the pump is shut off? Will the column come to rest and reverse its direction of flow in less than the critical surge-wave time for the transmission main?
6. Will the check valve close in less than the critical time for the transmission main?
7. Are there any quick-closing automatic valves set to open or close in less than 5 seconds?
8. Will the pump or its driving motor be damaged if allowed to run backward, reaching full speed?
9. Will the pump be tripped off before the discharge valve is fully closed?
10. Will the pump be started with the discharge valve open?
11. Are there booster stations in the system that depend on the operation of the main pumping station?
12. Are there any quick-closing automatic valves used in the pumping system that become inoperative with the failure of the pumping system pressure?

If the answer to any one of these questions is affirmative, there is a strong possibility that serious surges will occur. If the answer to two or more of the questions is affirmative, surges will probably occur with severity in proportion to the number of affirmative answers.

Positive pressure surges are commonly followed by a negative surge or vacuum condition. During the pipeline design phase, consideration must be given to controlling the amount of vacuum in the pipeline, otherwise the pipe must be designed for the anticipated vacuum pressure (see chapters 4 and 5).

GENERAL STUDIES FOR WATER HAMMER CONTROL

Studies of surges can be performed during the design stage. Once the general layout of the system has been completed, the length, diameter, thickness, material, and capacity of the pipe, as well as the type and size of pumps, can be established. The normal operating pressures at various points in the system can be computed and the allowable maximum pressures fixed. Using this information, the margin for water hammer can be determined. The design should then be adjusted to provide either safety factors large enough to withstand such conditions as might be encountered or suitable remedial or control devices. The latter method is usually less costly. It is important to note that there is no single device that will correct all surge difficulties. Only by studying both normal operating conditions and possible emergency conditions can methods be determined to provide proper control.

General recommendations cannot be made on the type, size, and application of surge-control equipment for all systems. Several possible solutions should be considered for any individual installation, and the one selected should give the maximum protection for the least expenditure. Surges can often be reduced substantially by using bypasses around check valves, by cushioning check valves for the last 15–20 percent of the stroke, or by adopting a two-speed rate of valve stroke. Water hammer resulting from power failure to centrifugal pumps can sometimes be held to safe limits by providing flywheels or by allowing the pumps to run backward. Air-inlet valves may be needed, or it may be preferable to use a surge tank, a surge damper, or a hydropneumatic chamber. Under certain operating conditions, no devices will be required to hold the pressure rise within safe limits.

It is essential to coordinate all the elements of a system properly and to ascertain that operating practices conform to safety requirements. As changes take place in the system demands and operating conditions, it may be necessary to review and revise the surge control, particularly if the capacity is increased, additional pumpage or storage is added, or booster stations are planned.

If a complete surge study is made during design and the resulting recommendations are followed, the system should operate without damage. The agreement correlation between the theoretical analyses, properly applied, and the actual tests of installations is extremely close. When a surge study was not performed and dangerous conditions existed, invariably there were serious surges and sometimes costly damage resulted. The time and effort spent on a surge study in advance of the final design are the least expensive means of preventing surges. The elastic-wave theory has been proven in actual practice, and design engineers should take the initiative in performing surge studies and mitigating pressure surges to prevent serious failures.

ALLOWANCE FOR WATER HAMMER

Many conditions have changed since the standard rule-of-thumb empirical allowances for water hammer originated. Automatic stop, check, and throttling valves were not as widely used as they are today. Valve closures measured in seconds and motor-driven centrifugal pumps were practically unknown. New types of pipe have been introduced and used. Consequently, standard allowances for water hammer may not be applied universally to all types of installations. Also, these allowances may not provide full performance under all circumstances. Potential water-hammer problems should be investigated in the design of pumping-station piping, force mains, and long transmission pipelines. Appropriate methods should be provided to reduce water hammer's effect to the minimum that is practicable or economical.

PRESSURE RISE CALCULATIONS

This manual does not cover an analysis of pressure rise in a complicated pipeline. However, basic data are provided for simple problems.

The pressure rise for instantaneous valve closure is given by Eq 3-17. Values of wave velocity are presented in Figure 3-6 for diameter-thickness ratios of 90 and less, and in Table 3-6 for higher ratios.

For solutions to more complex problems, reference can be made to the many publications available (see, for example, Rich 1951, Parmakian 1963, Kinno 1968, and Streeter and Wylie 1967). Computer programs are available that include the effects of pipeline friction and, when properly applied, give accurate results. There are several methods of reducing surges by the addition of devices or revising operating conditions, but these are outside the scope of this manual. Most of the available computer programs evaluate the various methods of reducing or controlling surges. (Streeter and Wylie 1967 described some of these means.)

ECONOMICAL DIAMETER OF PIPE

Hydraulic formulas give the relation between flow rate and head loss in pipes of various diameters and interior surface conditions. When a limited amount of head loss is available, the smallest-diameter pipe that will deliver the required flow is selected. This results in the least construction cost. Where pumping provides head, a part of the cost is for energy to provide head to overcome friction. The cost for energy decreases as pipe diameter increases and friction losses decrease; however, the cost for the pipe increases. The objective is usually to minimize total cost (initial cost, operation, and maintenance) by selecting the pipe diameter and pumping capacity that result in least life-cycle cost. Energy costs may prove to be the most significant cost.

Penstocks

An economic study to determine penstock size generally requires that the annual penstock cost plus the value of power lost in friction be minimal. The annual penstock cost includes amortization of all related construction costs, operation and maintenance costs, and replacement reserve. A precise analytical evaluation, taking all factors into account, may be neither justified nor practical, because all variables in the analysis are subject to varying degrees of uncertainty.

Methods used to determine the economic diameter for steel penstocks and pump lines can be found in the ASCE Manual and Reports on Engineering Practice (MOP) No. 79, *Steel Penstocks* (ASCE 2012), or the *Buried Steel Penstock*, Steel Plate Fabricators Association Engineering Data, Volume 4 (Steel Fabricators Association 2002).

AIR ENTRAPMENT AND RELEASE

Air entrained in flowing water tends to form bubbles at or near the high points in a pipeline. If not removed, such bubbles become serious obstacles to flow. The removal of air will prevent the formation of a hydraulic jump in a pipe at the end of these bubbles. Possible air entrainment and its removal must be considered and remedies applied if needed. The ability of the hydraulic jump to entrain the air and to carry it away by the flowing water has been investigated. Quantitative data have been published (Hall et al. 1943) relating characteristics of the hydraulic jump to the rate of air removal. Removal of air through air valves is discussed in chapter 7.

GOOD PRACTICE

Waterworks engineers should use the hydraulic friction formulas with which they are most familiar and with which they have had experience. Four of the common conventional formulas have been discussed in this chapter. In any particular case, the results calculated using the different conventional formulas could be compared. Engineers should, however, recognize the increasing use of the rational or universal formulas, become familiar with them, and make check calculations using them. A practical coefficient value for the formulas should be conservatively selected and compared or calibrated to actual field measured data for a similar pipeline and water source when these data are available.

The results of flow tests will generally be more useful if they are related to the rational concept of fluid flow. This entails giving more attention to relative surface roughness, water temperature, Reynolds numbers, and an analysis of test results aimed at incorporating them into the fluid-mechanics approach to flow determination.

REFERENCES

Hydraulics

Barnard, R.E. 1948. Design Standards for Steel Water Pipe. *Jour. AWWA*, 40(1):24.

Hall, L.S., A.A. Kalinske, and J.M. Robertson. 1943. Entrainment of Air in Flowing Water—A Symposium. *Trans. ASCE*, 108:1393

Hydraulic Institute. 1954. *Pipe Friction Manual*. 1954. New York: Hydraulic Institute.

Moody, L.F. 1944. *Friction Factors for Pipe Flow*. New York: American Society of Mechanical Engineers.

Steel Fabricators Association. 2002. *Buried Steel Penstocks*, Steel Plate Engineering Data. Vol. 4. New York: AISI.

The following references are not cited in the text:

Aldrich, E.H. 1938. Solution of Transmission Problems of a Water System. *Trans. ASCE*, 103:1759.

American Society of Civil Engineers. 1975. *Pipeline Design for Water and Wastewater*. Report of the Task Committee on Engineering Practice in the Design of Pipelines. New York: American Society of Civil Engineers.

Babbitt, H.E., and J.J. Doland. 1927, 1955. Moody, L.F. 1944. *Water Supply Engineering*. New York: McGraw-Hill Book Co.

Bradley, J.N., and L.R. Thompson. 1951. *Friction Factors of Large Conduits Flowing Full*. Engineering Monograph 7. Denver, CO: US Bureau of Reclamation.

Capen, C.H. 1950. Trends in Coefficients of Large Pressure Pipes. *Jour. AWWA*, 42(9):860.

Crane Co. 1942. *Flow of Fluids Through Valves, Fittings, and Pipe*. 1942. Tech. Paper 409. Chicago: Crane Co.

Crocker, S., ed. 1945. *Piping Handbook*, 4th ed. New York: McGraw-Hill Book Co.

Cross, H. 1936. *Analysis of Flow in Networks of Conduits of Conductors*. Bull. 286. Urbana, IL: Univ. of Illinois, Engrg. Expt. Stn.

Davis, C.V., ed. 1952. *Handbook of Applied Hydraulics*, 2nd ed. New York: McGraw-Hill Book Co.

Farnsworth, G., Jr., and A. Rossano Jr. 1941. Application of the Hardy Cross Method to Distribution System Problems. *Jour. AWWA*, 33(2):224.

Hinds, J. 1937a. Economic Sizes of Pressure Conduits. *Engineering News Record*, 118:443.

Hinds, J. 1937b. Economic Water Conduit Size. *Engineering News Record*, 118:113.

Hinds, J. 1946. Comparison of Formulas for Pipe Flow. *Jour. AWWA*, 38(11):1226.

King, H.W. 1954. *Handbook of Hydraulics*, 4th ed. New York: McGraw-Hill Book Co.

Lischer, V.C. 1948. Determination of Economical Pipe Diameters in Distribution Systems. *Jour. AWWA*, 40(8):849.

Pigott, R.J.S. 1950. Pressure Losses in Tubing, Pipe, and Fittings. *Trans. ASME*, 72:679.

Voetsch, C., and M.H. Fresen. 1938. Economic Diameter of Steel Penstocks. *Trans. ASCE*, 103:89.

White, L. 1913. *Catskill Water Supply of New York, N.Y.* New York: John Wiley & Sons.

Water Hammer and Pressure Surge

American Society of Civil Engineers. 2012. ASCE Manual and Reports on Engineering Practice (MOP) No. 79, *Steel Penstocks*. Reston, VA: ASCE.

Hall, L.S., A.A. Kalinske, and J.M. Robertson. 1943. Entrainment of Air in Flowing Water—A Symposium. *Trans. ASCE*, 108:1393

Kerr, S.L., L.H. Kessler, and M.B. Gamet. 1950. New Method for Bulk Modulus Determination. *Trans. ASME*, 72:1143.

Kinno, H. 1968. Water Hammer Control in Centrifugal Pump System. *ASCE. Jour. Hydraul. Div.*, May.

Parmakian, J. 1963. *Water Hammer Analysis*. New York: Dover Publications.

Rich, G.R. 1951. *Hydraulic Transients* (Engineering Societies Monographs). New York: McGraw-Hill Book Co.

Richards, R.T. 1956. Water Column Separation in Pump Discharge Lines. *Trans. ASME*, 78:1297.

Steel Plate Fabricators Association. 2002. *Buried Steel Penstocks*. Steel Plate Fabricators Association Engineering Data, Volume 4. Lake Zurich, IL: Steel Fabricators Association.

Streeter, V.L., and E.B. Wylie. 1967. *Hydraulic Transients*. New York: McGraw-Hill Book Co.

The following references are not cited in the text:

Alin, A.L. 1944. Penstock Surge Tank at Dennison Hydro Plant. *Civ. Eng.*, 14:296.

Allievi, L. 1925. *Theory of Water Hammer*. New York: American Society of Mechanical Engineers.

Angus, R.W. 1937. Water Hammer in Pipes, Including Those Supplied by Centrifugal Pumps; Graphical Treatment. In *Proc. Inst. Mech. Engrs.*, 136:245.

Angus, R.W. 1939. Water Hammer Pressures in Compound and Branch Pipes. *Trans. ASCE*, 104:340.

ASME. 1949. *Symposium on Water Hammer*. New York: American Society of Mechanical Engineers.

Barnard, R.E. 1948. Design Standards for Steel Water Pipe. *Jour. AWWA*, 40(1):24.

Bennet, R. 1941. Water Hammer Correctives. *Water & Sew. Wks.*, 88:196.

Bergeron, L. 1961. *Water Hammer in Hydraulics and Wave Surge in Electricity*. New York: John Wiley & Sons.

Boerendans, W.L. 1939. Pressure Air Chambers in Centrifugal Pumping. *Jour. AWWA*, 31(11):1865.

Chaudhry, M.H. 1987. *Applied Hydraulic Transients*, 2nd ed. New York: Van Nostrand Reinhold.

Dawson, F.M., and A.A. Kalinske. 1939. Methods of Calculating Water Hammer Pressures. *Jour. AWWA*, 31(11):1835.

Effect of Valve Action on Water Hammer. 1960. *Jour. AWWA*, 52(1):65.

Evans, W.E., and C.C. Crawford. 1916. Charts and Designing Air Chambers for Pump Discharge Lines. In *Proc. ASCE*, 79:57.

Fluid Engineering. 1976. In *Proc. of the Second International Conference on Pressure Surge*. London: British Hydraulic Res. Assoc.

Kerr, S.L. 1948. Practical Aspects of Water Hammer. *Jour. AWWA*, 40(6):599.

Parmakian, J. 1953. Pressure Surges at Large Pump Installations. *Trans. ASME*, 75:995.

Second Symposium on Water Hammer. 1937. *Trans. ASME*, 59:651.

Simin, O. 1904. Water Hammer. (Includes a digest of N. Joukovsky's study.) In *Proc. of the AWWA Annual Conference*. Denver, CO: American Water Works Association.

Standard Allowances for Water Hammer—Panel Discussion. 1952. *Jour. AWWA*, 44(11):977.

Stepanoff, A.J. 1949. Elements of Graphical Solution of Water Hammer Problems in Centrifugal Pump Systems. *Trans. ASME*, 71:515.

Streeter, V.L. 1972. Unsteady Flow Calculations by Numerical Methods. *ASME Jour. of Basic Engrg.*, June.

Surges in Pipe Lines—Oil and Water. 1950. *Trans. ASME*, 72:667.

Thorley, A.R.D. 2004. *Fluid Transients in Pipeline Systems*, 2nd Ed. New York: ASME.

Water Hammer Allowances in Pipe Design. 1958. Committee Report. *Jour. AWWA*, 50(3):340.

Wylie, E.B., and V.L. Streeter 1993. *Fluid Transients in Systems*. New York: Prentice Hall.

Chapter 4

Determination of Pipe Wall Thickness

The primary design consideration for the determination of pipe wall thickness is the pipe's ability to withstand various internal and external pressures. The remaining performance criteria are then checked to verify the adequacy of the overall steel pipe design. The pipe wall thickness design and checks can be accomplished directly by the design engineer, owner, or pipe manufacturer from the performance criteria provided by following the step-by-step process as outlined in this chapter.

Performance criteria and factors that influence the steel pipe design include:

- Pipe diameter
- Yield strength of steel (ANSI/AWWA C200)
- Internal pressures
  1. Working pressure (chapters 3 and 4)
  2. Transient or water-hammer pressures (chapters 3 and 4)
  3. Field-test pressure (chapter 12)
- External loads
  1. Weight of earth cover or dead loads (chapter 5)
  2. Live loads (chapter 5)
  3. Bedding and backfill details (chapter 5)
  4. Uniform radial collapse pressure, atmospheric or hydraulic (chapter 4)
  5. Buckling under buried conditions (chapter 5)
- Joint design (chapter 6)
  1. Thermal force
  2. Poisson's effect
  3. Hydrostatic thrust (pressure × area [PA] force)

- Special loadings
  1. Pipe on saddle supports (chapter 9)
  2. Pipe on ring-girder supports (chapter 9)
  3. Additional stresses (chapter 7)
- Design stress limits for various coatings and linings (chapters 4 and 11)
- Practical handling recommendations (chapter 4)
- Seismic force (appendix D)

INTERNAL PRESSURE

When designing for internal pressure, the pipe wall thickness (0.001-in. increments) should be calculated to limit the circumferential tensile stress or hoop stress to an allowable level. The internal pressures used in design (design pressures) are the actual working pressures and transient pressures or field-test pressures to which the pipe may be subjected during its lifetime.

In a transmission pipeline, the working pressure is based on the elevation difference between the centerline of the pipe and the hydraulic grade line (HGL). For pipelines with inline valves, the maximum static pressure head at any point in the pipe between closed valves is calculated as the elevation difference between the centerline of the pipe and the static HGL at that point. Transient or waterhammer pressures must also be considered as discussed in chapter 3.

In a pump discharge pipeline, the working pressure is based on the elevation difference between the pipe centerline and the HGL created by the pumping operation. Pressure at the outlet and the loss caused by friction should be considered. If it is possible to impose a pressure equal to the shutoff head of the pumps, the pressure is measured between the pipe centerline and the shutoff head HGL. Typical pipeline and hydraulic grade profiles for gravity and pumped flow conditions are shown in Figures 4-1 and 4-2, respectively.

With working, transient, and field-test pressures known, the pipe wall thickness is determined using Eq 4-1:

$$t = \frac{pD_o}{2s} \qquad \text{(Eq 4-1)}$$

Where:

t = pipe wall thickness for the internal pressure, in.

p = internal pressure, psi

D_o = outside diameter of steel pipe cylinder (not including coatings), in.

s = allowable design stress, psi

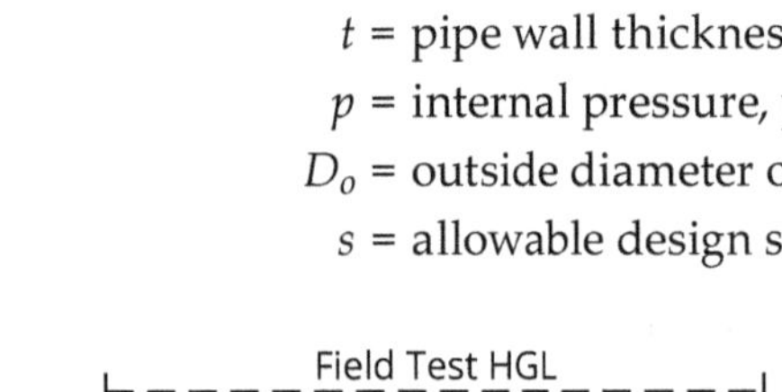

Figure 4-1 Typical pipeline and hydraulic grade profiles for gravity flow

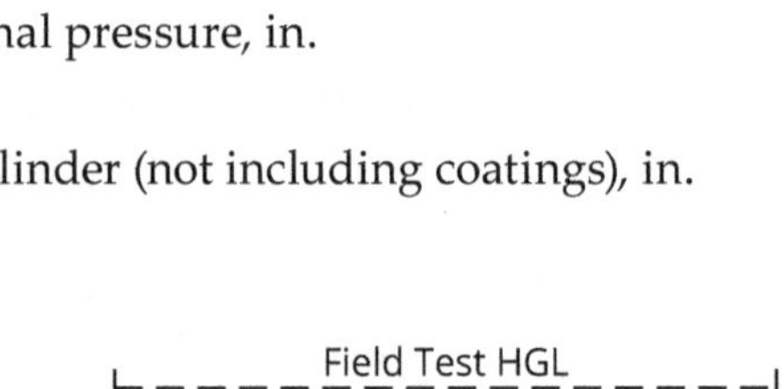

Figure 4-2 Typical pipeline and hydraulic grade profiles for pumped flow

ALLOWABLE STRESS

Tensile Stress and Yield Strength

Working stress is determined using a relationship to the steel's yield strength rather than to its ultimate strength. An allowable design stress equal to 50 percent of the steel specified minimum yield strength for working pressure and 75 percent of the specified minimum yield strength for transient or field-test pressure is typically used for the design of steel water pipe. Using Eq 4-1, these allowable design stresses are considered conservative for typical water-transmission pipelines. ANSI/AWWA C200, Steel Water Pipe, 6 In. (150 mm) and Larger, provides the minimum yield strength for various grades of steel used in the design and manufacture of steel pipe.

Cement-mortar coatings (on exterior of pipe) require the circumferential stress in the steel to be limited for the prevention of detrimental coating damage. A commonly used criterion for cement-mortar coatings is to limit the maximum circumferential stress in the steel cylinder. Values of 50 percent of the specified minimum yield strength of the steel, not exceeding 18,000 psi for working pressure, and 75 percent of the specified minimum yield strength, not exceeding 27,000 psi for transient pressure, are recommended.

Cement-mortar linings (on interior of pipe) perform at higher circumferential steel stress limits than those referenced above for cement-mortar coating. This is primarily caused by water absorption and the resulting expansion of the cement-mortar lining when the pipe is filled with water, and by other dynamics that are inherently different between cement-mortar linings and cement-mortar coatings. When considering stress limits above common industry practice, the pipe manufacturer should be consulted on the applicability of the desired stress level for the particular project conditions.

Flexible coatings and linings do not require similar steel cylinder stress limits and do not limit allowable stresses in the pipe design.

Example. Determine the minimum wall thickness using Eq 4-1, with 48-in. nominal diameter steel pipe with cement-mortar lining, flexible coating, 150-psi working pressure. and 50-psi surge allowance.

D_o = 49.75-in. OD cylinder (assumed OD for design examples)

p_w = 150-psi working pressure

p_s = 50-psi surge allowance

p_t = 200-psi transient pressure (working pressure + surge allowance)

σ_Y = specified minimum yield strength of 40 ksi

Working Pressure Design

s = 50 percent of specified minimum yield strength

$t = 150(49.75)/[(2)(0.5 \times 40{,}000)]$

t = 0.187 in.

Transient Pressure Design

s = 75 percent of specified minimum yield strength

$t = 200(49.75)/[(2)(0.75 \times 40{,}000)]$

t = 0.166 in.

In this example, working pressure is the more severe performance requirement and 0.187 in. is the pipe wall thickness for internal pressure.

HANDLING CHECK

Once the pipe wall thickness for internal pressure has been calculated, a check for handling and transportation thickness recommendations should be made. Nominal plate or sheet thicknesses for handling are often based on one of three equations where D = nominal pipe diameter (in.). They are as follows:

$$t = \frac{D}{288} \text{ (pipe sizes < 54-in. ID)} \qquad \text{(Eq 4-2)}$$

$$t = \frac{D + 20}{400} \text{ (pipe sizes} \geq \text{54-in. ID)} \qquad \text{(Eq 4-3)}$$

$$t = \frac{D}{240} \text{ (for steel pipe with cement-mortar lining and flexible coating)} \qquad \text{(Eq 4-4)}$$

Example. Evaluate the 48-in. nominal-diameter pipe from the previous example for handling.

Using Eq 4-4,

t = 48 in./240

t = 0.200 in.

In this example, the 48-in. pipe with a wall thickness of 0.200 in. is recommended for handling and transportation considerations. A wall thickness of 0.200 in. is greater than that required in the previous example for internal pressure of 0.187 in., so in this case handling thickness governs.

Minimum Wall Thickness

In no case shall the pipe wall thickness be less than 14 gauge (0.0747 in.).

CORROSION ALLOWANCE

Because of the availability of high quality coatings and cathodic protection systems, providing a corrosion allowance is not an economic solution in the waterworks field today and should not be used. The pipe should be designed for the required pipe wall thickness (0.001 in. increments) as determined by the guidelines presented previously in this chapter. Linings, coatings, bonded joints, test stations, and/or cathodic protection may then be selected as required.

EXTERNAL PRESSURE—EXPOSED OR SUBMERGED PIPE

Properly designed exposed or submerged pipe must be capable of resisting external radial forces. Such radial forces may result from outside pressure, either atmospheric (sometimes called *vacuum*) or hydrostatic, both of which are uniform and act radially to generate a collapsing force. The collapsing pressure of the pipe, p_c or p_{cr}, should be greater than the external radial pressure. Buried pipe must be designed to resist earth loading in the pipe trench, and since this is the case, the following equations are not applicable to buried pipe. External load and pressure on buried pipe are discussed in chapter 5.

Atmosphere or Fluid Environments

A general theory of collapse-resistance of a round (i.e., undeflected) steel pipe to a uniform, radially acting force was developed by Timoshenko (Timoshenko 1940). Any unreinforced pipe longer than the critical length can be considered of infinite length, as its collapsing pressure is independent of further increase in length. The original Timoshenko equation for a steel cylinder is

$$p_c = \frac{2E_S}{1 - v_s^2}\left(\frac{t}{D_o}\right)^3 \quad \text{(Eq 4-5)}$$

The following equation, adapted from Timoshenko (Eq 4-5) to include the stiffness of cement-mortar lining and coating, is applied to such pipes:

$$p_c = \frac{2E_S}{1 - v_s^2}\left(\frac{t}{D_o}\right)^3 + \frac{2E_C}{1 - v_c^2}\left(\frac{t_L}{D_I}\right)^3 + \frac{2E_C}{1 - v_c^2}\left(\frac{t_C}{D_c}\right)^3 \quad \text{(Eq 4-6)}$$

Where:

D_o = outside diameter of steel cylinder, in.

D_I = inside diameter of steel cylinder (cement-mortar lining outside diameter), in.

D_c = outside diameter of cement-mortar coating, in.

t = steel thickness, in.

t_L = cement-mortar lining thickness, in.

t_C = cement-mortar coating thickness, in.

E_S = modulus of elasticity for steel, 30,000,000 psi

E_C = modulus of elasticity for cement mortar, 4,000,000 psi

p_c = collapsing pressure, psi

v_s = Poisson's ratio for steel (taken as 0.30)

v_c = Poisson's ratio for cement mortar (taken as 0.25)

NOTE: To consider the effects of ellipticity (out-of-roundness) of the pipe, use Eq 4-7, developed for bare steel pipe but long utilized on cement-mortar lined and coated steel pipe. Eq 4-5 for calculating the collapsing pressure of a round (undeflected) pipe does not include any theoretical safety factor. Calculated collapsing pressure is generally considered an absolute value, hence depending on the use of the equation (e.g., solving for p_c in Eq 4-5 or 4-6 for use in Eq 4-7) and on the design engineer's knowledge and familiarity with the conditions a safety factor may not be applied.

Example. 48-in. nominal diameter, 0.200-in. wall pipe with 0.50-in. cement-mortar lining and flexible coating either exposed to the atmosphere or in fluid such as water.

$$D_o = 49.750\text{-in. outside diameter}$$

$$p_c = \frac{2 \times 30{,}000{,}000}{1 - 0.3^2}\left(\frac{0.2}{49.75}\right)^3 + \frac{2 \times 4{,}000{,}000}{1 - 0.25^2}\left(\frac{0.5}{49.35}\right)^3 + \frac{2 \times 4{,}000{,}000}{1 - 0.25^2}\left(\frac{0}{49.75}\right)^3$$

p_c = steel pipe + cement-mortar lining + cement-mortar coating (0 in this case)

$p_c = 4.3 + 8.9 + 0$

$p_c = 13.2$ psi

Applied Calculations

Circular cylindrical shells under external pressure may fail either by buckling or by yielding. Relatively thin-walled shells fail through instability or buckling under stresses that, on the average, are below the yield strengths of the materials normally encountered in the waterworks field. A number of theoretical and empirical equations have been published to provide for the effect of instability caused by collapsing. They include the equations of Timoshenko (1940), Roark (Young et al. 2012), Love (1888), Stewart (1906), and Bryan (1888).

Timoshenko developed a second-order polynomial equation for the collapsing pressures of circular conduits, which can be applied to pipes. The Timoshenko equation, which accounts for ellipticity and variations in material yield strength, is as follows:

$$(p_{cr})^2 - [\sigma_Y/m + (1 + 6m\Delta x)p_c]p_{cr} + \sigma_Y p_c/m = 0 \qquad \text{(Eq 4-7)}$$

Where:

p_{cr} = critical collapse pressure of a circular conduit with ellipticity (psi)

σ_Y = specified minimum yield strength of material (psi)

m = r_o/t_a

r_o = outside radius of pipe (in.)

t_a = adjusted pipe wall thickness (in.) to account for thickness of lining and coatings = $t + (t_L + t_C)/(E_S/E_C)$

Δx = percent deflection of conduit in decimal form (ellipticity)

p_c = collapse pressure of a perfectly round tube (psi) (see Eq 4-6)

E_S = modulus of elasticity steel (psi)—30,000,000 (psi)

E_C = modulus of elasticity cement mortar (psi)—4,000,000 (psi)

E_S/E_C = modular ratio = 7.5

To account for cement-mortar lining or coating in Eq 4-7, divide the sum of the cement-mortar lining and coating thicknesses by 7.5, the modular ratio. This gives the equivalent steel thickness for the cement mortar that can be added to the cylinder thickness.

Critical collapse pressure p_{cr} can be obtained by application of the quadratic equation to Eq 4-7. The resulting critical collapsing pressure is an absolute design value. The system design engineer and installer should take necessary measures to guard against applied support, appurtenance, or construction loads that would alter the geometry of the conduit to the point that it exceeds the limits of the original design.

Example. From the previous example, a 48-in. nominal diameter with 0.200-in. wall pipe either exposed to the atmosphere or in fluid such as water with 0.50-in. cement-mortar lining and flexible coating, now with 1 percent deflection or ellipticity.

D_o = 49.750-in. OD cylinder

σ_Y = specified minimum yield strength of 40-ksi steel

t = 0.200 in.

t_L = 0.50 in.

Cement-mortar lining equivalent steel thickness

$= t_L/(E_s/E_c)$

$= 0.50/(30{,}000{,}000/4{,}000{,}000)$

$= 0.067$

Adjusted t:

$t_a = 0.200 + 0.067$

$= 0.267$

$m = r_o/t_a = (49.750/2)/0.267 = 93.2$

$\Delta x = 0.01$

p_c = 13.2 psi (see Eq 4-5 example)

$$(p_{cr})^2 - [\sigma_Y/m + (1 + 6m\Delta x)\, p_c]\, p_{cr} + \sigma_Y p_c/m = 0$$

$$(p_{cr})^2 - [40{,}000/93.2 + [1 + 6(93.2)(0.01)]13.2]\, p_{cr} + 40{,}000(13.2)/93.2 = 0$$

$$(p_{cr})^2 - (429.2 + 87.0)p_{cr} + 5{,}665.2 = 0$$

$$(p_{cr})^2 - 516.2(p_{cr}) + 5{,}665.2 = 0$$

$$p_{cr} = \frac{516.2 \pm [(-516.2\,)^2 \; - 4(5{,}665.2\,)]^{1/2}}{2}$$

$$= \frac{516.2 \pm 493.8}{2}$$

= 11.2 psi or 505.0 psi. By inspection, 11.2 psi is the correct value.

GOOD PRACTICE

Steel pipe wall thickness is determined primarily by internal pressures (working, transient, or field test). It is good practice to use 50 percent of the specified minimum yield strength of the steel for working pressure design and 75 percent of the specified minimum yield strength for transient or field-test pressure design. No further safety factors are required in the determination of pipe wall thickness for internal pressure due to the conservative limits of allowable stresses in Eq 4-1. A check for adequacy in resisting external fluid pressures (if applicable), external loads (see chapter 5), and minimum thickness for handling should be performed. A proven AWWA coating and/or lining for corrosion protection as detailed in chapter 11 should be selected since it is neither practical nor economical to add wall thickness for corrosion allowance.

REFERENCES

American National Standards Institute/American Water Works Association (ANSI/AWWA) C200, Steel Water Pipe, 6 In. (150 mm) and Larger. Latest edition. Denver, CO: American Water Works Association.

Bryan, G.H. 1888. Application of the Energy Test to the Collapse of Long Thin Pipe Under External Pressure. *Proc. Cambridge Phil. Soc.*, VI:287–292.

Love, A.E.H. 1888. On the Small Free Vibrations and Deformation of Thin Elastic Shells. *Phil. Trans. Roy. Soc.*, A179:491–546

Stewart, R.T. 1906. Collapsing Pressure of Bessemer Steel Lap-Welded Tubes, Three to Ten Inches in Diameter. *Trans. ASME*, 27:730–822.

Timoshenko, S. 1940. *Strength of Materials*. Part II.D. New York: Van Nostrand Co.

Young, W., R. Budynas, and A. Sadegh. 2012. *Roark's Formulas for Stress and Strain*, 8th ed. New York: McGraw-Hill Companies.

The following references are not cited in the text:

ANSI/AWWA C205, Cement–Mortar Protective Lining and Coating for Steel Water Pipe—4 In. (100 mm) and Larger—Shop Applied. Latest edition. Denver, CO: American Water Works Association.

ANSI/AWWA C602, Cement–Mortar Lining of Water Pipelines in Place—4 In. (100 mm) and Larger. Latest edition. Denver, CO: American Water Works Association.

Parmakian, J. 1982. Minimum Thickness for Handling Pipes, Water Power and Dam Construction. *Water Power and Dam Construction*, June: 18–19.

US Bureau of Reclamation. 1986. *Welded Steel Penstocks*. Engineering Monograph 3. Denver, CO: US Bureau of Reclamation.

Chapter 5

External Loads on Buried Pipe

External loads on buried pipe are generally comprised of the weight of the backfill (earth load) combined with the live load including impact. As described in chapter 4, internal pressure is used to determine pipe wall thickness. Analysis of the effects of external loads and the principles of pipe/soil interaction are used to determine an adequate pipe-soil system for the design wall thickness. To ensure satisfactory performance of the pipe, proper bedding and backfilling installation must also be followed to enable pipe deflection to be controlled. The prudent choice of pipe embedment soil type and the backfill compaction level are the most important factors affecting the performance of the steel pipe-soil system.

EARTH LOAD

The first solution to the problem of soil-induced loads on buried pipe was published by Professor Anson Marston at Iowa State University in 1913. The Marston load theory is simply the weight of the backfill reduced by the friction of the sidefill against the undisturbed trench walls.

The Marston trench load for rigid pipe is defined as

$$W_d = C_d w B_d^2 \qquad \text{(Eq 5-1)}$$

That equation was modified by M.G. Spangler circa 1930 for flexible pipe to

$$W_d = C_d w B_d B_c \qquad \text{(Eq 5-2)}$$

Where:

W_d = trench load on conduit, lb/lin ft of pipe

C_d = load coefficient for ditch conduits (Spangler 1951)

w = unit weight of backfill, lb/ft^3

B_d = horizontal width of trench at top of pipe, ft

B_c = outside diameter of pipe, ft, also = $D_c/12$

D_c = outside diameter of coated pipe, in.

Equations 5-1 and 5-2 have also been used to calculate loads on buried pipe in a narrow trench condition. When these equations are used, care must be taken that the trench width as built does not exceed the design trench width B_d.

In the 1960s, significant testing was done at Utah State University, which concluded that due to the variability and imprecision of soil properties along the alignment of the pipeline and the design engineer's inability to represent soil parameters to the same level of precision as those with other engineering materials such as structural steel or concrete, the conservative prism load should be considered for buried flexible pipe. Simply stated, the prism load is the weight of the column of backfill directly over the pipe without considering soil friction or arching effects and, therefore, is the maximum load that can be imposed by the backfill on flexible pipe over time. The prism load is a conservative design approach, as the actual earth load on a flexible pipe lies somewhere between the Marston/Spangler trench load and the more conservative prism load. Over time the soil load may approach the prism load due to consolidation of the soils.

Prism load:

$$W_c = wH_c \frac{D_c}{12} \qquad \text{(Eq 5-3)}$$

Where:

W_c = prism load = dead load on the conduit, lb/lin ft of pipe

w = unit weight of fill, lb/ft^3

H_c = height of fill above top of pipe, ft

D_c = outside diameter of coated pipe, in.

LIVE LOADS

In addition to supporting dead loads created by soil cover, buried pipelines can also be subjected to superimposed loads: concentrated live loads or distributed live loads. Concentrated live loads are generally caused by truck-wheel loads or railway-car loads. Distributed live loads are caused by surcharges, such as piles of material or temporary structures. The effects of live loads on a pipeline depend on the depth of soil cover over the pipe.

Live-load effects (W_L), when applicable, are added to soil load and are generally based on AASHTO HS-20 truckloads or Cooper E-80 railroad loads as indicated in Table 5-1. These values are given in pounds per square foot and include a 50 percent impact factor. There is no appreciable live-load effect for HS-20 loads when the earth cover exceeds 8 ft or for E-80 loads when the earth cover exceeds 30 ft.

CONSTRUCTION LOADS

During construction operations, it may be necessary for heavy construction equipment to travel over an installed pipe. Before heavy construction equipment is permitted to cross over a pipe, earth fill should be placed over the pipe. A generally accepted minimum elevation for the fill is at least 3 ft over the top of the pipe; however, additional analysis of the heavy equipment load may be necessary. The fill should be of sufficient width to prevent possible lateral displacement of the pipe and should be maintained to ensure rutting does not decrease the effective cover over the pipe and to prevent excessive impact loading on the pipe.

Table 5-1 Live-load effect

Highway HS-20 Loading*		Railroad E-80 Loading*	
Height of Cover	Load†	Height of Cover	Load†
(ft)	*(lb/ft²)*	*(ft)*	*(lb/ft²)*
1	1,800	2	3,800
2	800	5	2,400
3	600	8	1,600
4	400	10	1,100
5	250	12	800
6	200	15	600
7	176	20	300
8	100	30	100

Source: ASTM A796 (ASTM International).

* Neglect live load when less than 100 lb/ft^2; use dead load only.

† Other HS and E loads can be calculated by applying a ratio such as 25/20 to HS-20 for HS-25 loading.

EXTREME EXTERNAL LOADING CONDITIONS

An occasional need to calculate extreme external loading conditions may arise. One example is to determine off-highway loading from heavy construction equipment. A convenient method of solution for such load determination using modified Boussinesq equations is presented by Handy and Spangler (2007).

As an example (using the Newmark coefficient in Table 5-2):

Assume:

Live load from a large dump truck:
Weight on one set of dual wheels = 42,300 lb
Tire pattern is 44 in. × 24 in.
Height of cover, $H_c = 2$ ft

Calculation:

Using Figure 5-1 as reference, calculate:
Tire pattern area or load surface area:
(44) (24) = 1,056 in.^{2} = 7.33 ft^2 (for dual wheels)
Surface pressure:
42,300/7.33 = 5,771 lb/ft^2

The distance from the center to the corner of the surface load is located by (see Figure 5-1)

$$A_T = (1/2)\,(44/12) = 1.83 \text{ ft}$$
$$B_T = (1/2)\,(24/12) = 1.0 \text{ ft}$$

The m and n values can then be found by

$$m = A_T/H_c = 1.83/2 = 0.915 \qquad n = B_T/H_c = 1.0/2 = 0.50$$

Using m and n, the vertical influence coefficient interpolated from Table 5-2 = 0.117.
The product of A_T and B_T is ¼ of the total tire pattern (see Figure 5-1).
Therefore,

$$W_L = (4)(0.117)(5{,}771) = 2{,}701 \text{ lb/ft}^2$$

Table 5-2 Newmark vertical influence coefficients

	$m = A_T/H_c$ or $n = B_T/H_c$																	
m or *n*	0.1	0.2	0.3	0.4	0.5	0.6	0.7	0.8	0.9	1.0	1.2	1.5	2.0	2.5	3.0	5.0	10.0	∞
0.1	0.005	0.009	0.013	0.017	0.020	0.022	0.024	0.026	0.027	0.028	0.029	0.030	0.031	0.031	0.032	0.032	0.032	0.032
0.2	0.009	0.018	0.026	0.033	0.039	0.043	0.047	0.050	0.053	0.055	0.057	0.059	0.061	0.062	0.062	0.062	0.062	0.062
0.3	0.013	0.026	0.037	0.047	0.056	0.063	0.069	0.073	0.077	0.079	0.083	0.086	0.089	0.090	0.090	0.090	0.090	0.090
0.4	0.017	0.033	0.047	0.060	0.071	0.080	0.087	0.093	0.098	0.101	0.106	0.110	0.113	0.115	0.115	0.115	0.115	0.115
0.5	0.020	0.039	0.056	0.071	0.084	0.095	0.103	0.110	0.116	0.120	0.126	0.131	0.135	0.137	0.137	0.137	0.137	0.137
0.6	0.022	0.043	0.063	0.080	0.095	0.107	0.117	0.125	0.131	0.136	0.143	0.149	0.153	0.155	0.156	0.156	0.156	0.156
0.7	0.024	0.047	0.069	0.087	0.103	0.117	0.128	0.137	0.144	0.149	0.157	0.164	0.169	0.170	0.171	0.172	0.172	0.172
0.8	0.026	0.050	0.073	0.093	0.110	0.125	0.137	0.146	0.154	0.160	0.168	0.176	0.181	0.183	0.184	0.185	0.185	0.185
0.9	0.027	0.053	0.077	0.098	0.116	0.131	0.144	0.154	0.162	0.168	0.178	0.186	0.192	0.194	0.195	0.196	0.196	0.196
1.0	0.028	0.055	0.079	0.101	0.120	0.136	0.149	0.160	0.168	0.175	0.185	0.193	0.200	0.202	0.203	0.204	0.205	0.205
1.2	0.029	0.057	0.083	0.106	0.126	0.143	0.157	0.168	0.178	0.185	0.196	0.205	0.212	0.215	0.216	0.217	0.218	0.218
1.5	0.030	0.059	0.086	0.110	0.131	0.149	0.164	0.176	0.186	0.193	0.205	0.215	0.223	0.226	0.228	0.229	0.230	0.230
2.0	0.031	0.061	0.089	0.113	0.135	0.153	0.169	0.181	0.192	0.200	0.212	0.223	0.232	0.236	0.238	0.239	0.240	0.240
2.5	0.031	0.062	0.090	0.115	0.137	0.155	0.170	0.183	0.194	0.202	0.215	0.226	0.236	0.240	0.242	0.244	0.244	0.244
3.0	0.032	0.062	0.090	0.115	0.137	0.156	0.171	0.184	0.195	0.203	0.216	0.228	0.238	0.242	0.244	0.246	0.247	0.247
5.0	0.032	0.062	0.090	0.115	0.137	0.156	0.172	0.185	0.196	0.204	0.217	0.229	0.239	0.244	0.246	0.249	0.249	0.249
10.0	0.032	0.062	0.090	0.115	0.137	0.156	0.172	0.185	0.196	0.205	0.218	0.230	0.240	0.244	0.247	0.249	0.250	0.250
∞	0.032	0.062	0.090	0.115	0.137	0.156	0.172	0.185	0.196	0.205	0.218	0.230	0.240	0.244	0.247	0.249	0.250	0.250

Source: Newmark 1935.

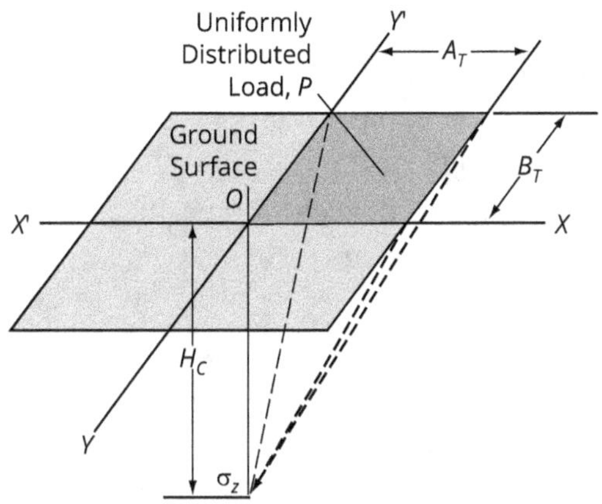

Source: Spangler and Handy 1982.

Figure 5-1 Vertical stress under an imposed area load

PREDICTING DEFLECTION

The Iowa deflection formula was first proposed by M.G. Spangler in 1941. It was later modified by Watkins and Spangler in 1958 and has since been presented in various forms. The Iowa formula was developed as a tool to predict horizontal ring deflection of unpressurized pipe buried in soil. The formula is not the basis for design of a pipeline; it is an estimate or *prediction* of long-term horizontal deflection of unpressurized pipe responding to earth and live loads. It is not applicable to pipe that is pressurized. When pressurized, the internal pressure will tend to reround the pipe. A detailed explanation of the rerounding effect can be found in *Structural Mechanics of Buried Pipes* (Watkins and Anderson 2000).

One of the principal variables in the modified Iowa formula is the modulus of soil reaction E', an empirical representation of soil stiffness in the pipe-soil system. The use of E' in the modified Iowa formula is simplistic and has proven to be reliable over time. The formula also shows that ring deflection is controlled primarily by the soil rather than the pipe; therefore, changing the E' has a much greater effect on controlling deflection than does increasing the pipe stiffness.

As an example, when varying E' values and relative pipe cylinder thicknesses independently, the greater significance of increasing E' versus the impact of increasing cylinder thickness is apparent (see appendix A). In most cases, the pipe stiffness only contributes about 1 to 10 percent of the total resistance to deflection, while the soil contributes 90 to 99 percent. In summary, the benefits of improving the quality of the bedding and backfill as the primary means of controlling pipe deflection are much more cost-effective than increasing the thickness of the steel pipe wall.

In one of its most common historical forms, the modified Iowa formula for predicting deflection is

$$\Delta x = D_l \frac{KWr^3}{EI + 0.061E'r^3} \qquad \text{(Eq 5-4)}$$

This equation in basic terminology is

$$\text{Predicted Deflection} = \frac{\text{Load}}{\text{Pipe Stiffness + Soil Stiffness}}$$

Where:

Δx = predicted horizontal deflection of pipe, in.

D_l = 1.0*

K = bedding constant (0.1)

W = load per unit of pipe length, lb/lin in.

= $[W_c/12 + W_L D_c/144]$

D_c = outside diameter of coated pipe, in.

r = mean radius of the pipe, in.

= $(D_c - t - t_L - t_C) / 2$

E' = modulus of soil reaction of the embedment material, psi

Tables 5-3 and 5-5 show values and nomenclature for various embedment materials.

EI = pipe wall stiffness, lb in.

= $E_S I_S + E_C I_L + E_C I_C$

Where:

E = modulus of elasticity [30,000,000 psi for steel (E_S) and 4,000,000 psi for cement mortar (E_C)]

I = transverse moment of inertia per unit length of individual pipe wall components (for steel cylinder (I_S), for cement-mortar lining (I_L), and for cement-mortar coating (I_C))

= $t^3/12$, in.^{3}

t = pipe cylinder wall thickness, in.

t_L = cement-mortar lining thickness, in.

t_C = cement-mortar coating thickness, in.

Under external loads, the individual elements—the mortar lining, the steel cylinder, and the cement-mortar coating—work together as laminated rings ($E_S I_S + E_C I_L + E_C I_C$). The pipe wall stiffness of these individual elements is additive. Structurally, the combined action of these elements increases the moment of inertia of the pipe section above that of the steel cylinder alone, thus increasing its ability to resist external loads.

In the original Iowa equation, the Marston load (W_d) was used, which partially reduces the earth load by frictional shear forces. If the Marston load W_d is used in place of the prism load W_c in the deflection formula, the use of a D_l greater than 1.0 should be evaluated.

Predicted deflection limits for steel pipe installations are as follows:

- 5 percent of D_c for flexible lining and coating such as liquid epoxy linings and coatings, polyurethane linings and coatings, or tape coatings
- 3 percent of D_c for cement-mortar lining and flexible coating
- 2 percent of D_c for cement-mortar coating

These deflection limits are based on the performance limits of the steel pipe with the various coating options.

A buried steel ring is stable up to a uniform elliptical deflection of approximately 20 percent. Cement-mortar lining will withstand deflections of up to 6 to 10 percent. Cement-mortar coating will withstand deflections of up to 3 to 4 percent. Installed deflection conditions that have stabilized but exceed the predicted deflection limits may still be considered acceptable.

* Deflection lag factor (D_l) accounts for long-term deflection as a result of consolidation or settlement of backfill material at the sides of the pipe. Because the steel pipe is designed for the maximum possible soil load directly over the outside diameter of the pipe when the prism load (W_c) is used, D_l is 1.0.

Table 5-3 Soil stiffness, *E*′, for pipe embedment materials (psi)*

Soil Stiffness Category (SC)	Soil Type†	AASHTO Soil Groups‡	Depth of Cover (ft)	Compaction Level§ 85%	90%	95%	100%
SC1	Clean, coarse-grained soils: SW, SP, GW, GP, or any soil beginning with one of these symbols with 12% or less passing a No. 200 sieve	A1, A3	2–5	700	1,000	1,600	2,500
			5–10	1,000	1,500	2,200	3,300
			10–15	1,050	1,600	2,400	3,600
			15 +	1,100	1,700	2,500	3,800
SC2	Coarse-grained soils with fines: GM, GC, SM, SC, or any soil beginning with one of these symbols more than 12% fines. Sandy or gravelly fine-grained soils: CL, ML (or CL-ML, CL/ML, ML/CL) with more than 25% retained on a No. 200 sieve	A-2-4, A-2-5, A-2-6, or A-4 or A-6 soils with more than 25% retained on a No. 200 sieve	2–5	600	1,000	1,200	1,900
			5–10	900	1,400	1,800	2,700
			10–15	1,000	1,500	2,100	3,200
			15+	1,100	1,600	2,400	3,700
SC3**	Fine-grained soils: CL, ML (or CL-ML, CL/ML, ML/CL) with 25% or less retained on a No. 200 sieve	A-2-7, or A-4 or A-6 soils with 25% or less retained on a No. 200 sieve	2–5	500	700	1,000	1,500
			5–10	600	1,000	1,400	2,000
			10–15	700	1,200	1,600	2,300
			15+	800	1,300	1,800	2,600

NOTE: Soils that have 30% or less retained on the ¾-in. sieve are covered in the standard Proctor test, but the test on some clean gravels can be difficult to complete and the results difficult to interpret. Clean gravel (GP and GW) materials such as "crushed rock" are considered to be the stiffest embedment soils and when consolidated are at least equivalent to 100% standard Proctor compaction for the selection of *E*′ values.

* Derived from Hartley and Duncan 1987.

† ASTM D2487.

‡ AASHTO M145.

§ Standard Proctor densities in accordance with AASHTO T 99 and ASTM D698 are used with this table. See Table 5-4 for comparative soil density tests.

** SC3 soils may require more restricted values for material passing the No. 200 sieve and liquid limit when used with large-diameter (>60 in.) pipe.

Table 5-4 Comparison of standard density tests*

Test	Hammer Weight (lb)	Hammer Drop (in.)	Number of Soil Layers	Blows/Layer	Compactive Energy per Unit Volume ($ft\text{-}lb_f/ft^3$)
Standard Proctor (AASHTO T 99/ ASTM D698)	5.5	12	3	25	12,400
Modified Proctor (AASHTO T 180/ ASTM D1557)	10	18	5	25	56,250

* Natural in-place deposits of soils have densities from 60% to 100% of maximum obtained by the standard AASHTO compaction method. The designer should be sure that the *E*′ value used in design is consistent with this specified degree of compaction and the method of testing that will be used during construction.

For free-draining soils, the relative density should be at least 70 percent as determined by ASTM D4253 and ASTM D4254.

In addition to other considerations, the allowable pipe deflection is also dependent on the type of jointing system being utilized. Contact the pipe manufacturers for additional joint deflection limitations.

Actual percent pipe deflection at a location in an installed pipeline can be determined by measuring the inside diameter along the horizontal axis (D_x) and the vertical axis (D_y). The deflection is then calculated by the following:

Table 5-5 Unified soil classification

Symbol	Description
GW	Well-graded gravels, gravel-sand mixtures, little or no fines
GP	Poorly graded gravels, gravel-sand mixtures, little or no fines
GM	Silty gravels, poorly graded gravel-sand-silt mixtures
GC	Clayey gravels, poorly graded gravel-sand-clay mixtures
SW	Well-graded sands, gravelly sands, little or no fines
SP	Poorly graded sands, gravelly sands, little or no fines
SM	Silty sands, poorly graded sand-silt mixtures
SC	Clayey sands, poorly graded sand-clay mixtures
ML	Inorganic silts and very fine sand, silty or clayey fine sands
CL	Inorganic clays of low to medium plasticity
MH	Inorganic silts, micaceous or diatomaceous fine sandy or silty soils, elastic silts
CH	Inorganic clays of high plasticity, fat clays
OL*	Organic silts and organic silt-clays of low plasticity
OH*	Organic clays of medium to high plasticity
PT*	Peat and other highly organic soils

Source: ASTM Standard D2487.

*Typically not suitable for pipe backfill material.

$$\%\ \text{Deflection} = \left(\frac{D_x - D_y}{D_x + D_y}\right) \times 100 \tag{Eq 5-5}$$

Modulus of soil reaction, E', is a measure of stiffness of the pipe embedment material, which surrounds the pipe as shown in the trench detail (Figure 5-2) and is not generally dependent on soils outside the trench walls. (For poor soils with blow counts of four or less per foot refer to ASCE MOP No. 119, *Buried Flexible Steel Pipe* [ASCE 2009]). E' is a hybrid modulus that has been derived empirically to eliminate the spring constant used in the original Iowa formula. It is the product of the modulus of passive resistance of the pipe zone soil used in Spangler's early derivation and the radius of the pipe. E' increases with depth of cover, is not a fundamental material property, and cannot be measured either in the field or in a geotechnical laboratory using soil samples.

Values of E' were originally determined by measuring deflections of actual installations of metal (corrugated and smooth wall) culvert pipe and then by back-calculating the effective soil reaction.

In the same way, the minimum required E' value can be calculated by rewriting the modified Iowa formula as follows:

$$E' = 16.4\left[\frac{C_\Delta \times W}{r} - \frac{EI}{r^3}\right] \tag{Eq 5-6}$$

Where:

C_Δ = factor to account for predicted pipe deflection limit
W = load per unit of pipe length, lb/lin in.
= $[W_C/12 + W_L D_c/144]$
r = mean radius of the pipe, in.

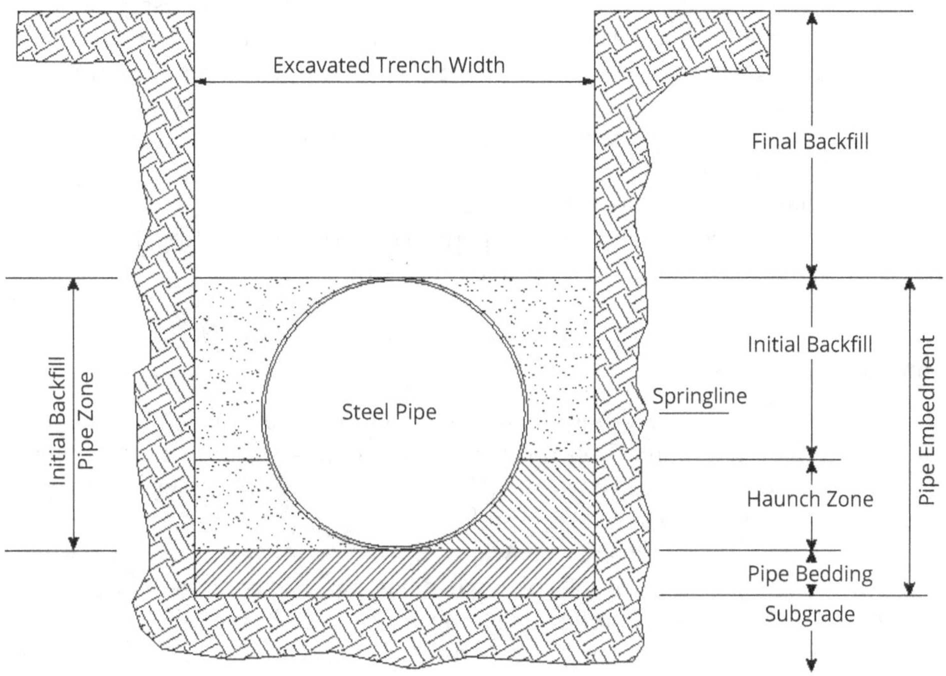

Notes:
—Pipe embedment materials may be SC1, SC2, SC3 or as specified. Materials shall be placed evenly on both sides of pipe and compacted to the density specified by the purchaser.
—Subgrade may need to be replaced or modified if trench bottom material is unacceptable or unstable.
—Trench width shall be adequate to assure elimination of voids in the launch area and/or proper placement and compaction of initial backfill materials.

Source: ANSI/AWWA C604-11.

Figure 5-2 Trench detail

Therefore,

C_Δ = 1 for flexible lining and coating

C_Δ = 1.67 for cement-mortar lining and flexible coating

C_Δ = 2.5 for cement-mortar coating

Using Eq 5-6 and Table 5-3, the most efficient soil type and compaction level of embedment material can be determined at a selected depth of cover.

Example Problem. Is 85 percent compaction of a native fine-grained silt (ML) weighing 110 lb/ft^3 adequate for a 48-in. nominal diameter pipe with a D/t ratio of 240, cement-mortar lined with flexible (liquid or tape) coating, buried with 25 ft of cover?

D_o = 49.75-in. OD cylinder

t = 0.20 in.

t_L = 0.5 in.

C_Δ = 1.67 for cement-mortar lining and flexible coating

Soil stiffness category = SC3

Solving for E' using Eq 5-6,

$$W = (110 \times 25 \times 49.75/12)/12 = 950 \text{ lb/lin in.}$$

$$r^3 = [(49.75 - 0.20)/2]^3 = 15{,}207 \text{ in.}^3$$

$$E_S I_S + E_C I_L = 30{,}000{,}000 \times 0.20^3/12 + 4{,}000{,}000 \times 0.5^3/12 = 20{,}000 + 41{,}667 = 61{,}667 \text{ lb in.}$$

$$E' = 16.4\,[1.67 \times 950/[(49.75 - 0.20)/2] - 61{,}667/15{,}207]$$

Therefore minimum E' = 984 psi

With E' calculated, use Table 5-3 to check that the proposed soil type and compaction level for the application has an E' that is not less than that calculated with Eq 5-6.

From Table 5-3, E' = 800 psi for SC3 soil stiffness, over 15 ft of cover at 85 percent compaction.

When the calculated E' is larger, a higher compaction effort or different soil type should be utilized.

In this case 984 psi > 800 psi; a higher E' is required.

One option is to change to 90 percent compaction resulting in an E' = 1,300 psi, which is acceptable.

CEMENT ENHANCED SOILS

Another option for bedding or haunch material is "flowable fill," which is a slurry that is "poured" into the trench. It is often referred to as *controlled low strength material* (CLSM), *soil-cement*, or *controlled density fill* (CDF). Because the fill is flowable, the pipe is in contact with its embedment and is fully supported.

An additional benefit of flowable fill is its cohesive strength. Flowable fill is capable of supporting additional soil cover when it is confined and, therefore, has greater strength than its unconfined compressive strength. Specifications often call for 50-psi unconfined compressive strength. Greater strength is not warranted. Values of E' for this material of 3,000 to 25,000 psi have been used. Flowable fill is practically noncompressible for typical buried pipe analysis.

Native soil and/or select soil with up to 60 percent fines have been successfully used for flowable fill. Very little Portland cement is needed (one sack per cubic yard) for the mix. Unconfined compressive strength should be in the range of 50 psi to 100 psi, and always under 200 psi to facilitate future excavation. Flow or slump requirements of the material should be based on the type of mix and the application in which it is used.

Care must be taken during placement and curing of the flowable fill to avoid flotation of the pipe and deflection of the ring beyond design limits because of the fluid pressure of flowable fill before it sets. Placement of the flowable fill should be made in lifts as the pipe is held in shape. The usual procedure is to pour flowable fill on one side and watch for it to rise on the other side. Backfill should not be placed over the flowable fill until adequate strength is attained to support the backfill.

As with all compacted pipe zone material, flowable fill will generally transfer the horizontal pressure from pipe ring to trench wall.

TRENCH COMPONENTS

Bedding, haunch area, initial backfill, and final backfill materials can be composed of different materials with varying compaction levels. Subgrade improvement is typically only required when unacceptable material causes an unstable trench bottom. Bedding is typically 2 in. to 6 in. in depth directly under the pipe and is loosely consolidated to provide uniform support for the pipe. The haunch area begins at the bottom of the pipe, and uniform support in this area has the greatest impact on limiting deflection. For more detailed information on pipe embedment and final backfill refer to ANSI/AWWA C604.

SPECIAL CONSIDERATIONS FOR BURIED PIPE

It can become necessary for an analysis of pipe-soil interaction that is beyond the scope of this manual. These conditions may include parallel pipes in a common trench, pipe in parallel trenches, buried pipe on bents, or encased pipe. Methods to address these conditions can be found in ASCE MOP No. 119 (ASCE 2009).

Buckling of Buried Pipe

Pipe embedded in soil may collapse or buckle from elastic instability resulting from loads and deformations. The summation of external loads should be equal to or less than the allowable buckling pressure. The allowable buckling pressure (q_a) based on Moore et al. (1994) may be determined by the following:

$$q_a = \frac{(1.2C_n)(EI)^{0.33}(\varphi_s E'k_v)^{0.67}R_H}{(FS)r_o} \qquad \text{(Eq 5-7)}$$

Where:

q_a = allowable buckling pressure, psi

FS = factor of safety = 2.0

r_o = outside radius of steel cylinder, in.

C_n = scalar calibration factor to account for some nonlinear effects = 0.55

φ_s = factor to account for variability in stiffness of compacted soil; suggested value is 0.9

k_v = modulus correction factor for Poisson's ratio, ν_s, of the soil = $(1 + \nu_s)(1 - 2\nu_s) / (1 - \nu_s)$; in the absence of specific information, it is common to assume $\nu_s = 0.3$ giving $k_v = 0.74$

R_H = correction factor for depth of fill = $11.4/(11 + 2r/H_c)$

H_c = height of ground surface above top of pipe, in.

E' = modulus of soil reaction psi (see Table 5-3)

For determination of buckling loads in normal pipe installations, use the following equation:

$$q_a \geq \frac{\gamma_w}{1{,}728} H_w + R_w \frac{W_c}{12D_o} + P_v \qquad \text{(Eq 5-8)}$$

Where:

H_w = height of water above conduit, in.

γ_w = unit weight of water = 62.4 lb/ft^3

P_v = internal vacuum pressure (psi) = atmospheric pressure (psi) less absolute pressure inside pipe (psia)

R_w = water buoyancy factor = $1 - 0.33(H_w/H_c)$, $0 \leq H_w \leq H_c$

In some situations, live loads should be considered as well. However, simultaneous application of live-load and internal-vacuum transients need not normally be considered. Therefore, if live loads are also considered, the buckling requirement is satisfied by:

$$q_a \geq \frac{\gamma_w}{1{,}728} H_w + R_w \frac{W_c}{12D_o} + \frac{W_L}{144} \qquad \text{(Eq 5-9)}$$

Example Problem. A 48-in. nominal diameter pipe with a 240 *D*/*t*, cement-mortar lined, with flexible (liquid or tape) coating is buried with 25 ft of cover and a water table 10 ft below the ground surface level. The backfill is an SC3 from Table 5-3, compacted to 90 percent of standard Proctor density with a unit weight of 120 lb/ft^3. Check the pipe for buckling with full vacuum.

D_o = 49.75-OD cylinder

t = 0.20 in.

t_L = 0.5 in.

Using Table 5-3,

$$E' = 1{,}300 \text{ psi}$$
$$C_n = 0.55$$
$$\varphi_s = 0.9$$
$$k_v = 0.74$$
$$R_H = 11.4 / (11 + 2r/H_c)$$
$$= 11.4 / [11 + 2 \times (49.75 - 0.2/2) / (25 \times 12)]$$
$$= 1.02$$
$$E_S I_S + E_C I_L = 30{,}000{,}000 \times 0.20^3/12 + 4{,}000{,}000 \times 0.5^3/12 = 20{,}000 + 41{,}667 = 61{,}667 \text{ lb in.}$$
$$R_w = 1 - 0.33\,(h_w/H_C)$$
$$= 1 - 0.33\,[(15 \times 12) / (25 \times 12)]$$
$$= 0.80$$
$$P_v = 14.7 \text{ psi}$$
$$W_L = 0 \text{ (see Table 5-1)}$$
$$W_c = 120 \times 25 \times (49.75/12)$$
$$= 12{,}438 \text{ lb/lin ft}$$
$$q_a = \frac{1.2 \times 0.55 \times 61{,}667^{0.33} \times (0.9 \times 1{,}300 \times 0.74)^{0.67} \times 1.02}{[2 \times (49.75)/2]}$$
$$q_a = 48 \text{ psi}$$

Check for vacuum condition:

$$q_a \geq (62.4/1{,}728) \times (15 \times 12) + 0.80\,[12{,}438 / (12 \times 49.75)] + 14.7$$
$$48 \text{ psi} \geq 38 \text{ psi}$$

The vacuum condition is acceptable.

Check live load condition:

$$q_a \geq (62.4/1728) \times (15 \text{ x } 12) + 0.80\,[12{,}438 / (12 \times 49.75)] + 0 / 144$$
$$48 \text{ psi} \geq 23 \text{ psi}$$

The live load condition is acceptable.

REFERENCES

AASHTO (American Association of State Highway and Transportation Officials). 1991. AASHTO M 145-91, Standard Specification for Classification of Soils and Soil-Aggregate Mixtures for Highway Construction Purposes. Washington DC: AASHTO.

AASHTO. 2010. AASHTO T 99-10. Standard Method of Test for Moisture-Density Relations of Soils Using a 2.5-kg (5.5-lb) Rammer and a 305-mm (12-in.) Drop. Washington, DC: AASHTO.

AASHTO. 2010. AASHTO T 180-10. Standard Method of Test for Moisture-Density Relations of Soils Using a 4.54-kg (10-lb) Rammer and a 457-mm (18-in.) Drop. Washington, DC: AASHTO.

ANSI/AWWA (American National Standards Institute/American Water Works Association) C604, Installation of Buried Steel Water Pipe, 4 In. (100 mm) and Larger. Denver, CO: American National Standards Institute and American Water Works Association.

ASCE (American Society of Civil Engineers). 2009. ASCE Manuals of Reports on Engineering Practice (MOP) No. 119, *Buried Flexible Steel Pipe: Design and Structural Analysis*. Reston, VA: ASCE.

ASTM International. 2012. ASTM D698. Standard Test Methods for Laboratory Compaction Characteristics of Soil Using Standard Effort (12 400 ft-lbf/ft^3 (600 kN-m/m^3)). West Conshohocken, PA: ASTM International.

ASTM International. 2012. ASTM D1557. Standard Test Methods for Laboratory Compaction Characteristics of Soil Using Modified Effort (56,000 ft-lbf/ft^3 (2,700 kN-m/m^3)). West Conshohocken, PA: ASTM International.

ASTM International. 2011. ASTM D2487. Standard Practice for Classification of Soils for Engineering Purposes (Unified Soil Classification System).West Conshohocken, PA: ASTM International.

ASTM International. 2010. ASTM A796/A796M, Standard Practice for Structural Design of Corrugated Steel Pipe, Pipe-Arches, and Arches for Storm and Sanitary Sewers and Other Buried Applications. West Conshohocken, PA: ASTM International.

ASTM International. 2006. ASTM D4253. Standard Test Methods for Maximum Index Density and Unit Weight of Soils Using a Vibratory Table. West Conshohocken, PA: ASTM International.

ASTM International. 2006. ASTM D4254. Standard Test Methods for Minimum Index Density and Unit Weight of Soils and Calculation of Relative Density. West Conshohocken, PA: ASTM International.

Bureau of Reclamation, US Department of the Interior. 1977. REC-ERC-77-1, *Modulus of Soil Reaction (E′) Values for Buried Flexible Pipe*. Denver, CO: US Department of the Interior.

Handy, R.L., and M.G. Spangler. 2007. *Geotechnical Engineering: Soil and Foundation Principles and Practice*, 5th ed. New York: McGraw Hill.

Hartley, J.D., and J.M. Duncan. 1987. *E′* and Its Variation With Depth. *Journal of Transportation*, Division of ASCE, Sept.

Marston, A., and A.O. Adams. 1913. *The Theory of Loads on Pipes in Ditches and Tests of Cement and Clay Drain Tile and Sewer Pipe*, Bulletin 31, Iowa Engineering Experiment Station. Ames, IA: University of Iowa.

Moore, I.D., A. Haggag, and E.T. Selig. 1994. Buckling Strength of Flexible Cylinders With Nonuniform Elastic Support. *Intl. J. Solids Structures*, 31(22):3041–3058.

Newmark, N.M. 1935. *Simplified Computation of Vertical Pressures in Elastic Foundations*. Circ. 24. Engrg. Exp. Sfn. Urbana-Champagne, IL: University of Illinois.

Spangler, M.G. 1941. The Structural Design of Flexible Pipe Culverts. 1941. *Bull. 153*. Ames, IA: Iowa State College.

Spangler, M.G. 1951. *Soil Engineering*. Scranton, PA: International Textbook Company.

Spangler, M.G., and R.L. Handy. 1982. *Soil Engineering,* 4th ed. New York: Harper and Row.

Watkins, R.K., and L.R. Anderson. 2000. *Structural Mechanics of Buried Pipes.* Boca Raton, FL: CRC Press.

Watkins, R.K., and M.G. Spangler. 1958. Some Characteristics of the Modulus of Passive Resistance of Soil: A Study in Similitude. In *Highway Research Board Proc.,* 37:576.

The following references are not cited in the text:

ASTM International. 2010. ASTM C143/C143M, Standard Test Method for Slump of Hydraulic-Cement Concrete, West Conshohocken, PA: ASTM International.

ASTM International. 2013. ASTM C1437. Standard Test Method for Flow of Hydraulic Cement Mortar. West Conshohocken, PA: ASTM International.

Barnard, R.E. 1948. Design Standards for Steel Water Pipe. *Jour. AWWA,* 40:24.

Barnard, R.E. 1955. *Behavior of Flexible Steel Pipe Under Embankments and in Trenches.* Bulletin. Middletown, OH: Armco Drainage and Metal Products.

Barnard, R.E. 1957. Design and Deflection Control of Buried Steel Pipe Supporting Earth Loads and Live Loads. In *Proc. ASTM,* 57:1233.

Luscher, U. 1966. Buckling of Soil Surrounded Tubes. *Jour. Soil Mechanics and Foundations Div.—ASCE,* November.

Marston, A. 1929. The Theory of External Loads on Closed Conduits in the Light of the Latest Experiments. In *Proc. of the Ninth Annual Meeting Highway Res. Board.* Washington, DC: Highway Research Board.

McGrath, T.J., I.D. Moore, E.T. Selig, M.C. Webb, and B. Taleb. 2002. *NCHRP 473—Recommended Specifications for Large-Span Culverts,* National Cooperative Highway Research Program, Report 473. Washington, DC: National Academy Press.

Proctor, R.R. 1933. Design and Construction of Rolled-Earth Dams. *Engineering News Record,* 111:245.

Proctor, R.R. 1948. An Approximate Method for Predicting the Settlement of Foundations and Footings. In *Proc. Second International Conference on Soil Mechanics & Foundation Engr.* The Hague, Netherlands.

Proudfit, D.P. 1963. Performance of Large Diameter Steel Pipe at St. Paul. *Jour. AWWA,* 55(3):303.

Spangler, M.G. 1947. Underground Conduits—an Appraisal of Modern Research. In *Proc. ASCE.* June.

Spangler, M.G. 1948. Underground Conduits—an Appraisal of Modern Research. *Trans. ASCE,* 113:316.

Spangler, M.G. 1951–1952. *Protective Casings for Pipelines.* Engineering Reports 11. Ames, IA: Iowa State College.

Spangler, M.G. 1969. *Soil Engineering*, 4th printing. Scranton, PA: International Textbook Company.

Spangler, M.G., and D.L. Phillips. 1955. *Deflections of Timber-Strutted Corrugated Metal Pipe Culverts Under Earth Fills*. Bull. 102. Highway Research Board; Pub. 350. Washington, DC: National Academy of Sciences—National Research Council.

Terzaghi, K. 1943. *Theoretical Soil Mechanics*. New York: John Wiley and Sons.

Wiggin, T.H., M.L. Enger, and W.J. Schlick. 1939. A Proposed New Method for Determining Barrel Thicknesses of Cast Iron Pipe. *Jour. AWWA*, 31:811.

This page intentionally blank.

Chapter 6

Pipe Joints

Many kinds of joints are used with steel water pipe. Common types are bell-and-spigot rubber gasket joints, field-welded joints (both illustrated in Figure 6-1), flanges, couplings, and expansion joints. When making joint selections, various operational and site-specific conditions should be considered by a designer familiar with the options. Various coatings may be applied to the exposed surfaces of non–field-welded joints at the point of manufacture. The coating selection should consider potential joint interference and certifications required for service, such as NSF certification, e.g., NSF/ANSI Standard 61. Detailed information about all of these joints is available from the manufacturer. The field installation guidelines and practices for most common joint types are described in ANSI/AWWA C604, Installation of Buried Steel Water Pipe—4 In. (100 mm) and Larger.

BELL-AND-SPIGOT JOINT WITH RUBBER GASKET

Several types of rubber-gasket field joints (shown in Figures 6-1A, 6-1B, and 6-1C) have been developed for nonrestrained steel water pipe service. Gasketed joints permit rapid installation in the field and, when properly manufactured and installed, provide a watertight joint that will provide a long service life. The design of the joints allows flexibility in the line, permitting certain angular and longitudinal movement while allowing the joints to remain watertight. The joints are easy to assemble, self-centering, and economical because of the reduced cost of installation.

The rubber gasket should conform to the requirements of ANSI/AWWA C200, Steel Water Pipe, 6 In. (150 mm) and Larger. Bell-and-spigot joints with rubber gaskets should not be used in areas subject to thrust such as at elbows, tees, laterals, wyes, reducers, valves, and dead ends without evaluating appropriate methods to control the thrust condition. Methods to control thrust include substituting welded joints as shown in Figures 6-1D, 6-1E, 6-1F, and 6-1G, or use of thrust blocks, anchors, harnessing, or self-restraining couplings.

Rolled-Groove Rubber Gasket Joints

A rolled-groove rubber gasket joint consists of a rolled spigot shape that is integrally formed on one end of each pipe section, with a bell formed on the adjacent mating end by swedging or expanding as shown in Figure 6-1A. The rubber gasket is compressed when

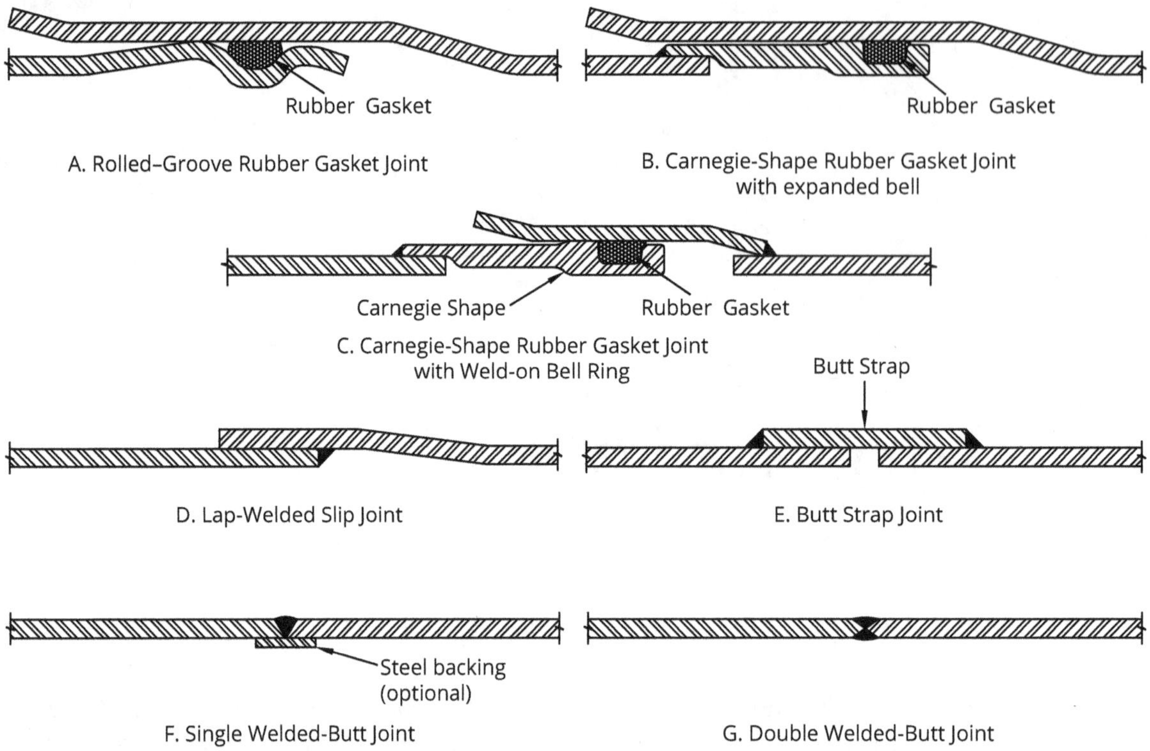

Figure 6-1 Common pipe field joints: Bell-and-spigot rubber gasket joints (A, B, and C) and field-welded joints (D–G)

the spigot-and-bell ends of adjoining pipe sections are engaged, providing a watertight seal.

Sizing and shape of the rubber gasket and the spigot groove are developed by the manufacturer and are dependent on the configuration of the spigot and the compressed volume of the gasket to fill the recess of the groove when the joint is engaged. Swedged bells are formed by forcing one end of a cylinder over a plug die; and expanded bells are formed by means of a hydraulic expander. When the spigot is inserted into the bell, the joint self-centers and the gasket is compressed between the steel surfaces to form a watertight seal. Welds on the inside of the bell and outside of the spigot should be ground flush with the plate surface for a distance not less than the depth of insertion. The design of the groove is such that the gasket is fully confined to the annular space. Water tightness of the joint relies on the compression and the resulting contact pressure of the gasket between the bell and spigot, and is not dependent on water pressure.

Carnegie-Shape Rubber Gasket Joints

There are two types of Carnegie-shape gasket joints: Carnegie-shape gasket joint with expanded bell and Carnegie-shape gasket joint with weld-on bell ring. In the case of the Carnegie gasket joint with an expanded bell, as shown in Figure 6-1B, the bell is formed by swedging or expanding one pipe end and the adjoining pipe end utilizes a weld-on, preformed Carnegie-shape spigot ring. In the case of Figure 6-1C, the Carnegie gasket joint with weld-on bell, both the preformed Carnegie spigot and bell rings are welded on the pipe ends.

The steel bell-and-spigot joint rings are designed and fabricated so when the pipe is laid and jointed, the joint will be self-centering. End rings are formed by joining the ends of one or more pieces of steel using complete-joint-penetration (CJP) butt welding or flash butt welding. Welds on gasket contact surfaces should be smooth and flush with the adjacent surfaces. The joint rings are to be accurately sized by expansion beyond their elastic limits. The rings are attached to the steel cylinder by fillet welds.

The joint rings are so designed that when the pipe is assembled and the joint completed, the gasket will be enclosed on all four sides and confined under compression adequate to ensure watertight performance under the conditions of service. Burrs and sharp edges on the joint-ring surfaces contacting the gasket should be smoothed or blunted.

Field-Welded Joints

Field welding of joints in steel water pipe is a frequently used jointing method that results in fully restrained, permanently sealed joints. The single welded lap joint (shown in Figure 6-1D) is the most common welded joint type. For some special applications, double lap-welded joints can be used. Other welded joint types are shown in Figures 6-1E, 6-1F, and 6-1G. Where welded joints are used, the pipe should be left bare a sufficient distance back from the ends to avoid damaging the protective coatings or linings by the heat produced during welding. If the protective coatings or linings are not held back, they will introduce contaminants during the welding process. The unprotected sections at these joints should be field-coated and/or field-lined after welding.

The welded lap joint is the most common welded joint type used because of its versatility, simple design, ease in forming, ease of assembly, and the resulting watertight joint. When connecting to existing pipes or for closure at structures such as vaults or valves, the pipeline length can be adjusted with a cut to fit pipe and use of a butt strap connection (shown in Figure 6-1E). The welded lap joint or butt strap joint may be welded on the outside only, or if the diameter is of sufficient size to permit access, on the inside only. Under certain special conditions, these joints may be required to be welded on the inside and outside. The butt joints in Figures 6-1F and 6-1G are used for special conditions such as seismic fault crossings or aerial crossings. Specific joint types and corresponding field-weld test options are described in ANSI/AWWA C206, Field Welding of Steel Water Pipe.

The welded lap joint also allows minor changes in alignment during installation by disengaging or pulling one side of the joint. When pulling a joint, the minimum engagement identified in ANSI/AWWA C206 and ANSI/AWWA C604 must be maintained. In addition, mitered bells can provide larger joint deflections and are easily formed by the manufacturer. The maximum angle for a mitered bell angle is described in ANSI/AWWA C208, Dimensions for Fabricated Steel Water Pipe Fittings.

ANSI/AWWA C206 describes the requirements and techniques for satisfactory field welding.

CIRCUMFERENTIAL FILLET WELDS FOR LAP JOINTS

Design of circumferential fillet welds must take into account stress that may develop due to thrust, Poisson's ratio of hoop stress (also referred to as *Poisson's stress*), and thermal loadings. Any circumferential fillet weld that is continuous, properly installed, and properly inspected in accordance with ANSI/AWWA C206 is assumed to be watertight; therefore the weld size should be determined by design considerations.

For a condition where the pipe is fixed or anchored on both sides of a joint, the stress across the joint resulting from thermal expansion and Poisson's ratio must be considered. For a condition where pipe is not fixed or anchored on either side of a joint, the stress across the joint resulting from thrust must also be considered. Since these two conditions cannot occur at the same time, the stresses from each are not additive.

Beam Bending

Beam bending is generally prevented by proper trench and pipe bedding preparation, and is not a design consideration for most buried pipe installations. Should specific conditions dictate the presence of beam bending, inclusion of such in the analysis with any longitudinal stress would be warranted. See chapter 9 for additional information on this subject.

Thrust Condition

Longitudinal thrust in a steel cylinder is transferred to an adjacent cylinder through the circumferential fillet weld at the joint. Longitudinal stress associated with a dead-end thrust condition for any given pressure, p, is only half of the hoop stress, σ_h, associated with the same pressure, p.

For a steel pipe with a single full fillet weld, the conservative approach to recognize longitudinal stress due to thrust is to assume the stress will be transferred through the throat of the fillet weld. At any given pressure, the stress in the weld due to full dead-end thrust will increase over the longitudinal stress in the cylinder by the ratio of the cylinder thickness to that of the fillet weld throat thickness, or $1/0.707 \approx 1.4$. From above, for any hoop stress, σ_h, the resulting longitudinal stress in the cylinder is no more than $0.5\sigma_h$. Accordingly, the resultant stress in the fillet weld is no more than $0.5\sigma_h$ (1.4) = $0.7\sigma_h$. Therefore, stress associated with a dead-end thrust condition does not control fillet weld design.

The analysis of the general condition relationship between longitudinal stress, σ_l due to thrust and hoop stress, for any given pressure, p, is as follows:

$$\sigma_l = \frac{\sigma_h}{2} \qquad \text{(Eq 6-1)}$$

Poisson's Ratio of Hoop Stress and Thermal Stress Condition

Poisson's ratio of hoop stress, σ_v, is load driven, with the load being the internal pressure of the pipe, and it is treated as a primary stress. Poisson's ratio for steel is 0.3. Therefore, Poisson's ratio of hoop stress is limited to 30 percent of the circumferential stress due to internal pressure and by itself will never control the design, even at transient or test pressures, as long as the hoop stress remains in the elastic range of the steel. Although when analyzing circumferential fillet welds, Poisson's effect from hoop stress caused by internal pressure must be added to the thermal stress.

Thermal stress is the result of constraining thermal contraction or expansion in the pipeline. Thermal loading occurs when the pipeline is prevented from moving and experiences temperatures differing from the temperature of the pipe during installation. Increased temperatures place the pipeline in a state of compression, which will relieve tensile stresses from the Poisson's effect. Decreased temperatures, however, create longitudinal tensile stresses that should be analyzed. Thermal stress is a secondary stress, one that is strain driven and not load driven. Thermal stress is dependent only on temperature differential in the steel and can be quantified on a unit basis.

The change in thermal stress, $\Delta\sigma_T$, in a steel cylinder is related to temperature change as follows:

$$\Delta\sigma_T = E_S \varepsilon \Delta T \qquad \text{(Eq 6-2)}$$

Where:

$\Delta\sigma_T$ = change in axial thermal stress, psi

E_S = modulus of elasticity of steel, 30×10^6 psi

ε = coefficient of expansion of steel, 6.5×10^{-6} in./in./°F

ΔT = change in temperature, °F

Substituting the appropriate values for E and ε for steel into Eq 6-2 yields

$$\Delta\sigma_T = 195\Delta T \quad \text{(Eq 6-3)}$$

From Eq 6-3, the change in thermal stress associated with a 1°F change in temperature is

$$\Delta\sigma_T = 195(1°F) = 195 \text{ psi} = 0.195 \text{ ksi}$$

Allowable Stresses

Standard guidelines for the limitation of primary stress such as hoop stress, or longitudinal stress resulting from thrust due to internal pressure, were stated in chapter 4. Secondary stress is considered self-limiting and therefore is not subject to the same design limitations as primary stress. A conservative limitation of secondary stress acting alone is 90 percent of the minimum specified tensile strength of the steel cylinder material (Luka and Ruchti 2008).

Thermal stress is independent of pipe thickness, but the resultant longitudinal force in the cylinder and circumferential fillet weld is a function of each member's respective thickness. To maintain static equilibrium across the joint, the resultant forces in the steel cylinder and the throat of the fillet weld must be equal. To account for the difference in thickness between the steel cylinder and the throat of the fillet weld, the allowable stress in the weld is limited to 0.707 times the allowable stress in the cylinder.

When secondary stress is coupled with primary stress, alternate stress limitations can be applied. The maximum allowable stress for the combination of thermal stress and Poisson's stress, σ_{T+v}, is 90 percent of the minimum specified yield strength of the steel cylinder material, σ_Y, but not exceeding ⅔ of the minimum specified tensile strength, σ_U (Luka and Ruchti 2008), or

$$\sigma_{T+v} = \min[0.9\sigma_Y, 0.67\sigma_U] \quad \text{(Eq 6-4)}$$

These limits are conservative with respect to the typical limits required to achieve elastic shakedown. Example 6-1 details the evaluation of a single full fillet weld.

Example 6-1. A 96-in. diameter pipe (98-in. OD) with a 0.408-in. steel cylinder thickness is installed with pipe temperatures as high as 100°F. The pipe will convey water at temperatures as low as 33°F. The pipe will operate at a pressure of 125 psi with transient pressure equal to 150 percent of the operating pressure. The pipe may be depressurized at service temperature. The pipe is produced in accordance with ANSI/AWWA C200 from steel with a minimum specified yield strength of 42 ksi and a minimum specified tensile strength of 60 ksi. Field welding is to be performed in accordance with ANSI/AWWA C206 using an electrode with strength equal to or greater than that of the steel cylinder material.

a. ***Evaluation of thrust only:*** As shown previously, stress due to longitudinal thrust alone will not control the size of the weld.

b. ***Evaluation of thermal stress only:*** When the pipe is depressurized, there is no Poisson's stress, and the thermal stress alone should be analyzed. Given a minimum specified tensile strength of 60 ksi for the cylinder, the allowable thermal stress in the fillet weld is

$$\sigma_{T+v} = (0.9)\sigma_U(0.707) = (0.9)(60\text{ksi})(0.707) = 38 \text{ ksi}$$

With the allowable stress defined, the maximum sustainable temperature change is then calculated. Recalling that 1°F temperature change yields 0.195-ksi change in thermal stress, the maximum sustainable temperature change for this nonpressurized steel pipe is

$$38 \text{ ksi} / (0.195 \text{ ksi/°F}) = 195°F.$$

The temperature differential is 100°F – 33°F = 67°F, which is less than 195°F; so thermal stress alone will not control the design.

c. ***Evaluation of Poisson's ratio of hoop stress in combination with thermal stress:*** The final evaluation is for fillet weld stress, σ_w, resulting from the combination of Poisson's stress, σ_v, with thermal stress, σ_T. The analysis is performed at the highest expected service pressure—in this case 1.5 times operating pressure—to yield the highest expected Poisson's stress.

$$\sigma_v = 0.3\ \sigma_h = 0.3(pD_o/2T_y) \quad \text{(Eq 6-5)}$$

Where:

σ_h = hoop stress, psi

p = highest service pressure = 1.5 times operating pressure, psi

D_o = pipe steel cylinder outside diameter, in.

T_y = pipe steel cylinder thickness, in.

σ_v = (0.3)(1.5)(125)(98)/[(2)(0.408)]/1,000 ksi

= 6.8 ksi (in the cylinder)

σ_T = 0.195 ksi/°F × 67°F = 13.1 ksi (in the cylinder)

σ_w = $(\sigma_v + \sigma_T)/0.707$ = (6.8 + 13.1)/0.707 = 28.2 ksi (in the weld)

From Eq 6-4, the allowable combined stress, σ_{T+v}, is limited to the smaller of 0.9 σ_Y or ⅔ σ_U. Given the weld metal minimum specified yield strength of 42 ksi and a minimum specified tensile strength of 60 ksi, the individual limits are as follows:

0.9(42) = 37.8 ksi

⅔(60) = 40 ksi.

Therefore, the limiting stress in the fillet weld is 37.8 ksi, which is greater than the calculated stress of 28.2 ksi. In this case, a single full-thickness circumferential fillet weld is adequate.

Resultant Stress

When the calculated axial weld stress exceeds that allowed for a single welded lap joint, the designer must evaluate options available for reducing the stress. Double welding the joint may reduce the weld stress, but the level of reduction is difficult to quantify. Further, the magnitude of reduction in weld stress may not necessarily be justified by the increase in cost associated with the additional weld. The two axial stresses that appear as the best candidates for reduction are Poisson's ratio of hoop stress and thermal stress. Reducing Poisson's ratio of hoop stress can only be accomplished by increasing the cylinder thickness, but doing so is not recommended as it will increase the axial force resulting from any thermal stress. Since thermal stress is generated only as a function of temperature change and is independent of both wall thickness and internal pressure, it is most preferable to mitigate thermal axial stress and thereby reduce the overall axial stress across the weld. See chapter 12, ANSI/AWWA C604, and ANSI/AWWA C206 for methods of minimizing thermal load on a circumferential fillet weld through various installation methods and techniques.

EXPANSION AND CONTRACTION—GENERAL

The coefficient of thermal expansion of steel is 6.5×10^{-6} in./in./°F. The change in length of nonrestrained steel pipe can be determined using

$$\Delta L = (6.5 \times 10^{-6})L(\Delta T) \quad \text{(Eq 6-6)}$$

Where:

ΔL = change in length, in.

L = length between fixed points, in.

ΔT = change in temperature, °F

The expansion or contraction of nonrestrained steel pipe is about ¾ in. per 100 ft of pipe for each 100°F change in temperature.

Expansion and Contraction—Underground

Ordinarily, a buried pipeline under typical operating conditions will not experience significant changes in temperature, and, therefore, the thermal expansion/contraction should be minimal. However, during construction and prior to completion of backfilling, extreme changes in ambient temperatures may cause significant expansion or contraction in the pipe. These extreme temperature changes and the resulting expansion and contraction may be avoided by backfilling the pipe as construction progresses.

For field-welded lines, ANSI/AWWA C206 describes a method that has been used satisfactorily to reduce the thermal stresses resulting from temperature variations occurring during pipeline construction. This method utilizes a special closure lap joint at 400- to 500-ft intervals. These joints usually include a lengthened bell and increased joint lap, which is welded during the coolest part of the day and only after all adjacent pipe has been welded and buried (see chapter 12).

Expansion and Contraction—Aboveground

The expansion and contraction of exposed pipelines with individual pipe sections can be accommodated by anchoring the pipe and utilizing couplings for field joints. Utilizing a coupling joint between anchor points will ordinarily allow enough movement so that expansion or contraction is not cumulative over multiple lengths. When utilizing multiple couplings for this purpose, there are additional considerations to limit movement at each coupling.

Forces caused by expansion and contraction should not be allowed to reach valves, pumps, or other appurtenances that might be damaged by these forces. Appurtenances can be protected by connecting the pipe and appurtenance with an expansion joint or coupling, or by providing anchor rings and thrust blocks of sufficient size and weight to prevent the forces from reaching the appurtenance.

For exposed field-welded lines, expansion joints may be located between anchor points if the pipe at the joint is adequately supported and the pipeline is aligned axially. On slopes, the joint should be placed adjacent to, and on the downhill side of, the anchor point, preferably at a point where the longitudinal bending in the pipe is zero. The coefficient of sliding friction for pipe bearing on supports should be determined. Spacing and positioning of expansion joints should be determined by profile and site-specific requirements. Expansion joints for pipe on bridges should be placed in the same location as the actual bridge structure expansion joints.

FLANGES

Flanged Joints

In water service where it may be necessary to later disassemble the joint for access or maintenance to valves or other components, bolted flanged joints with gaskets are a common practice. This section is intended to provide guidance to the designer for proper installation of flange connections. This bolted connection is made up of three component groups: flanges, fasteners, and a gasket.

Flanges. Of the many possible flange configurations, the most common in water systems is a slip-on type where the flange slides over the end of the pipe and is attached to it with fillet welds as described in ANSI/AWWA C207, Steel Pipe Flanges for Waterworks

Service, Sizes 4 In. Through 144 In. (100 mm Through 3,600 mm). Classes of flanges in ANSI/AWWA C207 are based on the internal pressure and all flanges are of a flat face design.

Flange considerations include:

- *Parallelism and alignment*—for flanges to properly seal, it is imperative that they be parallel and properly aligned with respect to each other. Flanges may not require strict adherence to these requirements, but they should serve as a guideline. Current ASME PCC-1 guidelines on flange alignment address:
  - Centerline alignment
  - Differences in flange gap
  - Bolt holes out of line rotationally
  - Flange gap that cannot be closed at specified torque
    (See ASME PCC-1, appendix E, for additional guidance.)
- *Flatness*—flange surfaces used with soft gasket materials like those in water pipe service should normally not have any excessive waviness or warping that could exceed the gasket's ability to seal.
- *Damages*—scratches or damage to the sealing surfaces may cause leaks that the gasket cannot prevent. Special attention to any scratches that cross the sealing surface from inside to outside should be addressed, and the flange may have to be repaired to eliminate such defects (see ASME PCC-1, appendix E, for additional nonmandatory guidance on surface defects).

Fasteners. The fasteners are the bolts, nuts, and washers. Their job is to create and maintain adequate axial load on the joint so that it will seal under all operating conditions. Bolts or threaded rods (*fully or partially threaded*) are used with nuts and washers or sometimes inserted into a threaded blind hole at one end such as into a valve body. Special considerations include:

- *Grade marking*—care should be taken to ensure that the marked end of the bolt is visible from the tightening side of the joint for inspection purposes.
- *Diameter and thread pitch*—nominal bolt diameter is the distance across the bolt perpendicular to the axis, measured at the outside crest of the threads. Other diameters such as "root diameter" (the distance across the bolt perpendicular to the axis, measured at the thread roots) are often used in such calculations as gasket stress analysis. In most cases nominal diameter is used for torque calculations.

Nuts are designed to have greater load-carrying capacity than the bolts they engage. As a result, the bolt should yield before stripping its own external threads or the internal threads of the nut.

- *Size across flats*—nuts, and therefore sockets, are measured across the flats. For all bolts 1 in. or greater in nominal diameter, the across flats (*A*/*F*) distance = ((bolt diameter × 1.5) + ⅛ in.). The addition of that ⅛ in. is what makes the bolt a "heavy hex." For bolts smaller than 1-in. nominal diameter, the *A*/*F* distance equals the bolt diameter times 1.5. In some cases, clearance around the bolt heads or nuts may require special considerations or tools.

Washers play a role in effective bolting. Washers should be "through-hardened" equivalent to ASTM F436. Hardened washers have the following functions:

- *Protect the flange*—the hardened washer keeps the turning nut from embedding or damaging the flange back.

- *Provide equal friction*—washers give an equally smooth turning surface to the nuts so that equal torque will translate into equal load under all the nuts on the flange irrespective of the flange surface condition.
- *Spread the load*—because of their larger diameter, hardened washers tend to distribute the bolt load over a larger area helping to stiffen the flange and aid in equalizing the load between bolt holes.

The use of split-spring washers, double nuts, lock wiring, and other so-called locking devices is generally ineffective and not necessary except in cases of high vibration and "shear load" cycling. But these conditions are rarely associated with pipe joints. The most important factor in keeping bolts tight is to tighten them properly.

Gaskets. The job of the gasket is to deform due to the bolt load and seal the flanges, achieving and maintaining a watertight seal. Gaskets for water joint use are usually made of some type of compressed fiber sheet or one of several specialized rubber compounds (see ANSI/AWWA C207 for more information).

- *Gasket seating stress*—gaskets have both a minimum and a maximum recommended seating stress. Consult the gasket manufacturer's literature to determine this seating stress range for each gasket type and thickness. To calculate the gasket stress (in psi), divide the sum of all the bolt loads (in pounds) around the flange by the gasket contact area (in square inches). This stress value should fall in the acceptable range. Too much stress or uneven stress may crush the gasket out of the joint and result in a leak.
- *Compression*—the relatively soft materials used in water service typically compress as much as 25 percent of their thickness during assembly.
- *Gasket relaxation*—gasket relaxation is a major cause of bolt load loss and may allow leaks in joints. When gasket materials are first put under load, they initially resist compression. Gaskets tend to flow away from the pressure, thinning and causing a corresponding drop in the bolt load. Each type of gasket has its own relaxation properties. The load loss may be as much as 50 percent, depending on the material. Relaxation generally happens within the first 4 to 6 hours after assembly. It generally does not repeat because the gasket material reaches a stable density and resists further thinning. Compensation, in the form of retightening of bolts in a circular "check pass" to the required torque after at least 4 to 6 hours, is recommended. Controlled tightening to the upper limit of the recommended torque may provide sufficient gasket load to maintain a seal without retightening.

COUPLINGS

Couplings may be used to join pipe sections of all diameters providing a watertight joint with flexibility. There are four categories for couplings used on steel pipe, and each is represented by an applicable ANSI/AWWA standard.

- *Bolted sleeve-type couplings*—ANSI/AWWA C219, Bolted, Sleeve-Type Couplings for Plain-End Pipe
- *Bolted split-sleeve type couplings*—ANSI/AWWA C227, Bolted, Split-Sleeve Restrained and Nonrestrained Couplings for Plain-End Pipe
- *Grooved and shouldered couplings*—ANSI/AWWA C606, Grooved and Shouldered Joints
- *Fabricated mechanical slip-type expansion joint*—ANSI/AWWA C221, Fabricated Steel Mechanical Slip-Type Expansion Joints

Couplings can be used to join pipe lengths securely against internal pressure and vacuum. These couplings can relieve stresses in the pipe such as thermal stress and stress resulting from settlement of the pipe.

Multiple couplings separated by short pipe sections may be used to accommodate differential settlement conditions at structures. A single coupling will not provide for differential settlement at a joint.

Bolted sleeve-type and split-sleeve–type couplings are not intended to be placed in shear, and will not transmit bending moments across a joint when installed and used in accordance with the applicable AWWA standard and the manufacturer's instructions. For coupling joint applications with shear or bending moments, grooved and shouldered coupling joints can be considered when designed, installed, and used in accordance with AWWA standards and the manufacturer's instructions. When excessive shear or moment is expected at the joint, provisions to properly support the pipe should be made.

These couplings are suitable for joining buried or exposed anchored pipes that are laid on curves when the combined deflection does not exceed the maximum capacity of the coupling.

Bolted Sleeve-Type Couplings

Acceptable axial movement in bolted sleeve-type couplings results from shear displacement of the rubber gaskets rather than from sliding of the gaskets on the mating surfaces of the adjacent pipes. Refer to ANSI/AWWA C219 and the manufacturer for installation requirements and performance capabilities when using this type of coupling in an application to accommodate expansion, contraction, and angular deflection. Restraint of a bolted sleeve-type coupling can be achieved through the utilization of an external harness (see chapter 7).

Bolted Split-Sleeve-Type Coupling

The bolted split-sleeve–type coupling is a bolted clamp-type coupling with a body that encloses one or more rubber gaskets and bridging/sealing plates. They may be supplied in one or more segments in order to facilitate handling and/or installation. Refer to ANSI/AWWA C227 and the manufacturer for installation requirements and performance capabilities when using this type of coupling in an application to accommodate expansion, contraction, and angular deflection.

Bolted split-sleeve couplings may offer restraint to the joint through an internal restraint mechanism or restraint may be achieved through the use of an external harness (see chapter 7). Conversely, couplings may be supplied with a slip mechanism that allows the coupling to accommodate pipeline expansion or contraction.

Grooved and Shouldered Couplings

The grooved and shouldered coupling is a bolted, segmental, clamp-type mechanical coupling having a housing that encloses a U-shaped rubber gasket. The housing locks the pipe ends together to prevent end movement yet allows some degree of flexibility and alignment. The rubber gasket is designed to be watertight under either pressure or vacuum service. Both the flexible and rigid coupling joints are fully restrained and can handle full end thrust load when installed and used in accordance with AWWA standards and the manufacturer's instructions.

Ends of pipe must be prepared in accordance with ANSI/AWWA C606 and the manufacturer to accommodate grooved and shouldered coupling installation.

Fabricated Mechanical Slip-Type Expansion Joints

The application of expansion joints with respect to expansion and contraction in a piping system is described in ANSI/AWWA C221. These joints permit movement along the pipeline axis only and are not intended to be put into shear or bending.

When increased axial movement capability is necessary, fabricated mechanical slip-type expansion joints are sometimes made double-ended. Limited-movement features can also be added to both single- and double-ended types. Limiting features are particularly important for double-ended joints. Unless the system or joint has been restrained, the expansion barrel may be subject to pull out.

When installing any type of expansion joint, the initial slip-pipe position should be established with consideration of the expected axial movement of the pipe, the temperature of the pipe when installed, and the pipe length. Pipe should be properly supported and guided to avoid shear or bending at the joint.

INSULATING JOINTS

Long sections of welded steel pipelines may conduct electric currents originating from differences in ground potentials or stray currents. This phenomenon is explained in chapter 10. Where tests indicate the necessity, a long pipeline may be separated into sections or insulated from other parts of a system by electrically insulated joints. This electrical isolation can be provided by installing insulated flanged joints or coupled joints.

Insulated Flanges

Special insulating gaskets, bolts, sleeves, and washers may be used to provide electrical isolation at a flanged joint. These insulating sleeves and washers are made of fabric-reinforced bakelite, micarta, Teflon®, or similar materials that have a long life and good mechanical strength.

The bolts of the insulated flanged joints must be carefully insulated by sleeves and washers. Insulating washers should be used at both ends of the bolts.

It is important that insulating gaskets, sleeves, and washers be installed carefully so that the flanged joint will be properly insulated. After installation of the electrically isolated joint is complete, an electrical resistance test should be performed. The electrical resistance should be at least 10,000 ohms; if the resistance is less, the joint should be inspected for damage, the damage repaired, and the joint retested.

Insulated Couplings

Special insulating sleeves or boots are used to provide electrical isolation between the coupling and the two pipe ends at the joint. These insulating sleeves are made of an elastomer capable of adequately providing electrical isolation at the joint and should extend beyond the coupling. Alternately coupling elements made of Teflon® or similar materials that have a long life and good mechanical strength may be used to isolate the coupling body from the pipe.

Due to the nature of the isolation method used with bolted sleeve-type and split-sleeve type couplings, self-restraining models may not be available, and an alternate method of restraining the pipe may have to be used. Unlike the fasteners used with flanges, the fasteners used with a coupling do not need to be isolated from the coupling; therefore the use of insulating washers or hardware sleeves is not required. In the case of a harness-restrained coupling, the threaded rods will also require electrical isolation.

CONNECTION TO OTHER PIPE MATERIAL

Care must be exercised when connecting dissimilar pipe materials because of the possibility of galvanic corrosion. See chapter 10 for principles of this reaction. When connecting steel pipe to either gray-iron or ductile-iron pipe, to steel-reinforced concrete pipe, or to copper or galvanized pipe, an electrically insulated joint should be considered. Insulating the connection allows the cathodic protection system of both materials to function independently. The insulating joint can be accomplished with an insulating flanged connection or with an insulating coupling connection (see Insulating Joints, this chapter).

Similar precautions are not necessary when connecting to nonmetallic pipe, such as asbestos cement or plastic.

ALTERNATE JOINTS

The joint types provided earlier in this manual represent the most common joint types currently in use for steel water pipelines. However, alternate joints engineered for specific conditions may be utilized.

Valve Connections

Valves are self-contained devices that may not function properly or remain watertight when subjected to substantial external forces. For example, if a valve is rigidly installed in a pipeline, the whole assembly of pipe and valves can be stressed by temperature changes, settlement, and exceptional surface loads. To prevent a valve from being strained, there should be at least one flexible joint located close enough to allow for any anticipated movement. For larger-diameter valves with exceptionally deep cover, refer to guidelines outlined in ANSI/AWWA C504, Rubber-Seated Butterfly Valves, 3 In. (75 mm) Through 72 In. (1,800 mm).

REFERENCES

ANSI/AWWA (American National Standards Institute/American Water Works Association) C200, Steel Water Pipe, 6 In. (150 mm) and Larger. Latest edition. Denver, CO: American Water Works Association.

ANSI/AWWA C206, Field Welding of Steel Water Pipe. Latest edition. Denver, CO: American Water Works Association.

ANSI/AWWA C207, Steel Pipe Flanges for Waterworks Service, Sizes 4 In. Through 144 In. (100 mm Through 3,600 mm). Latest edition. Denver, CO: American Water Works Association.

ANSI/AWWA C208, Dimensions for Fabricated Steel Water Pipe Fittings. Latest edition. Denver, CO: American Water Works Association.

ANSI/AWWA C219, Bolted, Sleeve-Type Couplings for Plain-End Pipe. Latest edition. Denver, CO: American Water Works Association.

ANSI/AWWA C221, Fabricated Steel Mechanical Slip-Type Expansion Joints. Latest edition. Denver, CO: American Water Works Association.

ANSI/AWWA C227, Bolted, Split-Sleeve Restrained and Nonrestrained Couplings for Plain-End Pipe. Latest edition. Denver, CO: American Water Works Association.

ANSI/AWWA C504, Rubber Seated Butterfly Valves, 3 In. (75 mm) Through 72 In. (1,800 mm). Latest edition. Denver, CO: American Water Works Association.

ANSI/AWWA C604, Installation of Buried Steel Water Pipe—4 In. (100 mm) and Larger. Latest edition. Denver, CO: American Water Works Association.

ANSI/AWWA C606, Grooved and Shouldered Joints. Latest edition. Denver, CO: American Water Works Association.

ASME (American Society of Mechanical Engineers). 2010. PCC-1-2010, *Guidelines for Pressure Boundary Bolted Joint Assembly*. New York: ASME.

ASTM International. ASTM F436, Standard Specification for Hardened Steel Washers. Latest edition. West Conshohocken, PA: ASTM International.

Luka, J.L., and G. Ruchti. 2008. Axial Joint Design for Welded Buried Steel Pipe. In *Proc. ASCE Pipeline Conference, Atlanta, GA: Pipelines 2008*. Reston, VA: American Society of Civil Engineers (ASCE).

NSF/ANSI 61. *Drinking Water System Components—Health Effects*. Latest edition. Ann Arbor, MI: NSF International.

This page intentionally blank.

Chapter 7

Fittings Design, Appurtenances, and Miscellaneous Details

Due to the wide range of design possibilities applicable to steel pipe, available welding and fabrication processes provide solutions to most layout configurations through the use of fittings. The design of pipe layouts, especially intricate ones, can in some cases be greatly facilitated by using standardized dimensions for the center-to-face distance or the center-to-end distance of fittings. Reference ANSI/AWWA C208, Dimensions for Fabricated Steel Water Pipe Fittings (latest edition), for dimensions and figures for welded steel pipe fittings, and ANSI/AWWA C200, Steel Water Pipe, 6 In. (150 mm) and Larger (latest edition), for manufacturing requirements for fittings and special joints. In many instances, though, the flexibility that steel pipe fittings afford the design engineer can address an almost infinite range of layout requirements. This chapter will address the design guidelines for various fitting configurations including elbows, miter-cut weld bells, outlets, wyes, reducers, ellipsoidal heads, thrust restraint harness assemblies, and anchor rings. The information provided herein is not intended to prohibit the use of other valid design guidelines presented in recognized design codes or standards.

The standard dimensions of fittings for screwed-joint pipe can be found in the catalogs of many manufacturers. Manufacturers can also provide the dimensions of compression fittings for use on standard plain-end pipe in the smaller sizes.

Definitions for pressure terms used in this chapter are as presented in the glossary of this manual.

DESIGNATION OF FITTINGS

In the context of this manual, fittings are defined as pipes that have aspects other than that of a straight piece of pipe. For instance, a 4-in.-diameter outlet on a 50-ft length of pipe results in the full length of pipe being considered a fitting. Fittings can be as simple as that stated previously or involve any or all of the design configurations presented in this chapter. Fittings are best defined with detailed drawings, as text definitions can lead to ambiguity with respect to the final requirements. Nominal pipe sizes are commonly used when referring to fittings in a general sense, but definition of the specific outside diameter of fittings is critical for most of the design guidelines presented in this chapter and to ensure proper fit when installed. Additionally, the specific steel cylinder thicknesses, reinforcement plate thicknesses, and minimum yield strengths of fitting components are critical. Manufacturing techniques and capabilities vary, and the designer is urged to consult with manufacturers when faced with complicated, involved fittings or combinations of fittings.

MITER END CUTS

Deflection can be accomplished in welded butt joints by miter cutting one or both pipe ends, provided that the maximum radial offset (misalignment) at any point around the resultant joint does not exceed the maximum allowed by the governing specification, standard, or code to which the joint will be welded. A small deflection angle can be accomplished in a welded lap joint using a miter-cut bell end, provided that the following are maintained: bell-and-spigot diameter tolerances, joint formation dimensional requirements, and joint engagement dimensional requirements. To form a miter-cut bell, the pipe end is miter cut and then the bell is expanded square with the face of the miter cut. The limit for the maximum miter-cut angle of a weld bell is a function of design requirements and manufacturing constraints, and cannot be defined explicitly here. Although historically a value of 5° has been a good practice limit, the actual value can be larger or smaller depending on specific design and manufacturing parameters. It is recommended that the designer consult manufacturers regarding the actual limit for specific design requirements. It is standard practice to allow a pulled joint to be combined with a miter-cut weld bell.

ELBOWS

Deflection angles greater than those allowed using miter end cuts can be accomplished using fabricated elbows. Elbows may be designated with a constant diameter or as a reducing elbow with a different diameter on each end. Equations for the specific dimensions of fabricated elbows can be found in ANSI/AWWA C208. Elbows are described by the number of individual pieces that are welded together to yield the finished product. A two-piece elbow is comprised of two mitered sections welded together to form an elbow with one miter joint; a three-piece elbow is comprised of three mitered sections welded together to form an elbow with two mitered joints, and so on. The recommended angular limitations for specific elbow configurations are defined in ANSI/AWWA C208, with the maximum deflection per miter weld being limited to 30°.

In specifying dimensions of an elbow, the designer should consider the hydraulic characteristics, space requirements, manufacturing and shipping constraints, stress considerations, and cost-benefit ratio over the expected life of the pipeline. The optimum radius for a fabricated elbow based on these considerations is 2.5 pipe cylinder outside diameters (D_o). This radius is recommended as a standard for water transmission lines where space requirements permit. For an elbow in plant piping and other locations where space is limited, a radius of less than $2.5D_o$ may be used, provided stress intensification factors are used. ANSI/AWWA C208 defines the radius of an elbow, *R*, as a multiple of the pipe outside diameter, D_o, where $R = n_e D_o$ and n_e is the diameter multiplier for calculating the radius of an elbow. ANSI/AWWA C208 further limits the value of n_e to that which results in the segment length along the inside of the elbow, *S*, being not less than

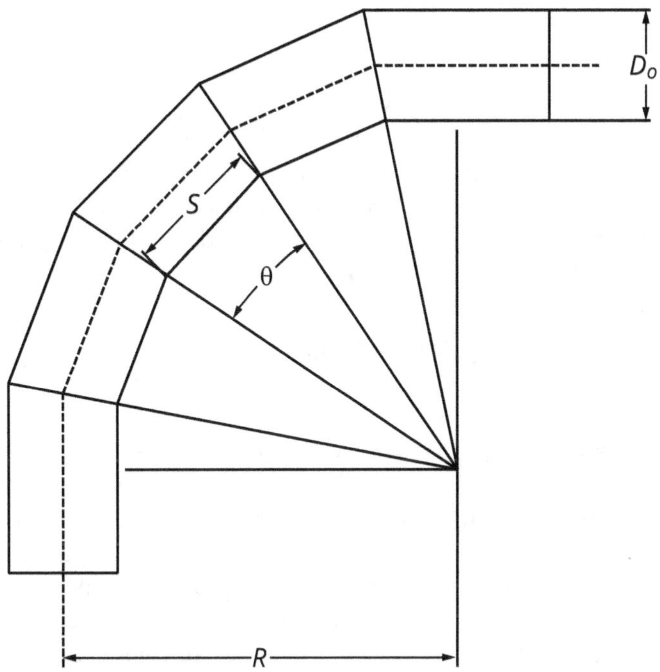

Figure 7-1 Basic elbow dimensions

the greater of 1.5 in. or $6T_y$, where T_y is the steel cylinder thickness (see Figure 7-1). S is calculated using the following formula:

$$S = 2\left(R - \frac{D_o}{2}\right)\tan\left(\frac{\theta}{2}\right) \quad \text{(Eq 7-1)}$$

Where:

S = segment length along inside of elbow, in.

R = radius to centerline of elbow, in.

D_o = outside diameter of pipe, in.

θ = segment deflection angle as shown in Figure 7-1

When minimization of an elbow radius is desired, the minimum S value can be used to calculate the associated value of n_e using the following formula:

$$n_e = \frac{S}{2D_o \tan\left(\frac{\theta}{2}\right)} + \frac{1}{2} \quad \text{(Eq 7-2)}$$

When n_e is greater than or equal to 2.5 and the steel cylinder is subject to longitudinal tensile stress, the stress concentration is minimal and of no concern. When n_e is less than 2.5 or the steel cylinder is not subject to longitudinal stress, such as when thrust blocks are used for thrust restraint, the stress concentration can, depending on the service conditions, yield a larger cylinder thickness. In such cases the steel cylinder thickness shall be calculated using either of the two following formulas:

$$T_y = \frac{pD_o}{2\sigma_A}\left(1 + \frac{D_o}{3R - 1.5D_o}\right) \quad \text{(Eq 7-3)}$$

$$T_y = \frac{pD_o}{S\sigma_A}\left[\frac{S}{2} + \frac{D_o}{3}\tan\left(\frac{\theta}{2}\right)\right] \quad \text{(Eq 7-4)}$$

Where:

T_y = required elbow steel cylinder thickness, in.

p = design internal pressure, psi

σ_A = allowable design stress, psi (The allowable design stress for an elbow is identical to that for design of the mainline cylinder. σ_A is limited to 50 percent of the yield strength of the steel cylinder material when the design internal pressure is equal to working pressure and 75 percent of the yield strength of the steel cylinder when the design internal pressure is equal to the greater of transient pressure and test pressure.)

For n_e = 2.5, $R = 2.5D_o$. Substituting $R = 2.5D_o$ into Eq 7-3 yields the following relationship:

$$T_y = \frac{1.17\,pD_o}{2\sigma_A}, \text{ or for a given } T_y, \text{ the calculated hoop stress, } \sigma_h = \frac{1.17\,pD_o}{2T_y}$$

This relationship shows that for a fully restrained elbow not subject to longitudinal thrust forces, such as when a thrust block is the restraint method used, the hoop stress increases by 17 percent over that of straight pipe adjacent to the elbow. When an elbow is subject to longitudinal thrust forces, though, as in the case of using restrained joints to mitigate thrust forces, cylinder in the elbow develops longitudinal tensile stress. This longitudinal tensile stress is equal to 0.5 times the hoop stress in the adjacent pipe, σ_h, and acts in a direction perpendicular to the hoop stress. Based on this biaxial state of stress, the Hencky-von Mises theory of combined stress can be used to provide a relationship between the effective stress in the elbow cylinder, σ_e, and the hoop stress, σ_h, of the straight pipe adjacent to the elbow.

$$\sigma_e = \sqrt{1.17\sigma_h^2 - 1.17\sigma_h(0.5\sigma_h) + (0.5\sigma_h)^2} = 1.0\sigma_h$$

This relationship shows that for elbows with a radius equal to $2.5D_o$ that are subject to longitudinal tensile stress, as from a thrust force, the wall thickness of the adjacent straight pipe cylinder is acceptable for use in the elbow cylinder.

CALCULATION OF RESULTANT ANGLE OF A COMBINED ANGLE BEND

In many pipe projects, combining a plan and profile deflection in one fitting is necessary. Combined or compound elbows, in which the plane of the bend is neither solely horizontal nor vertical, require certain trigonometric computations. Usually, the plan angle and profile angles are known, and the true angle in the plane of the elbow and the elbow rotations must be determined. The relationship between Δ (the resultant angle of the combined angle fabricated pipe bend), I, O (incoming and outgoing slope angles, respectively), and Δ_p (the plan angle of the combined bend) is as follows:

$$\cos \Delta = \cos I \cos O \cos \Delta_p + \sin I \sin O \qquad \text{(Eq 7-5)}$$

Note: This equation is valid only when the design centerlines for the horizontal and vertical deflections are coincident.

For example, given a 48° plan angle and incoming and outgoing slopes of –2.3° and 6.4°, respectively, the resultant combined angle for the bend is

$$\Delta = \cos^{-1}[\cos(-2.3°) \cos(6.4°) \cos(48°) + \sin(-2.3°) \sin(6.4°)] = 48.70°$$

The process for calculating and locating the field top centerline location on each end of a combined elbow is presented in appendix E.

REDUCERS

Changes in diameter are accomplished using concentric or eccentric cones, generically referred to as *reducers,* placed in the straight section of a pipeline or combined with a mitered elbow, tee, or cross.

The minimum length of a reducer shall be four times the difference in the nominal diameters of each end of the reducer. For reducers of at least the minimum length, the wall thickness of the larger-diameter pipe is adequate for the reducer. Reducers of less than the minimum length should be designed in accordance with ASME Boiler and Pressure Vessel Code, Section VIII, Division 1, where concentric reducers are subject to paragraph UG-32 (g) and appendix 1-5, and eccentric reducers are subject to paragraph UG-36 (g) and appendix 1-5.

REINFORCEMENT OF OUTLETS

Tees, crosses, laterals, wyes, headers, or other fittings that provide means of dividing or uniting flow in pipelines have less resistance to internal pressure than straight pipe of the same wall thickness and size. This is because a portion of the side wall of the pipe in these fittings is removed to allow for the outlet or fitting pipe. Outlet configurations such as full-size crosses, double laterals, and so on may require special design consideration due to their geometric configurations. Figure 7-2 provides a general representation of a single outlet, a double non–full-size outlet, and a full-size cross. The actual configuration of single and double outlets can vary. Figure 7-3 shows some common outlet configurations, but other combinations of configurations not shown in Figure 7-3 are acceptable. For instance, an outlet projecting radially but positioned between a pure radial outlet and a radial-tangential outlet is acceptable subject to the same design guidelines provided herein.

Design pressure, p, for outlet reinforcement shall be equal to the greater of p_w or $p_t/1.5$, where p_w is the working pressure, and p_t is equal to transient or test pressure, whichever is greater. Outlets may be reinforced in various ways for resistance to internal pressure. In some cases, the analysis may reveal that the main pipe and/or outlet pipe cylinder

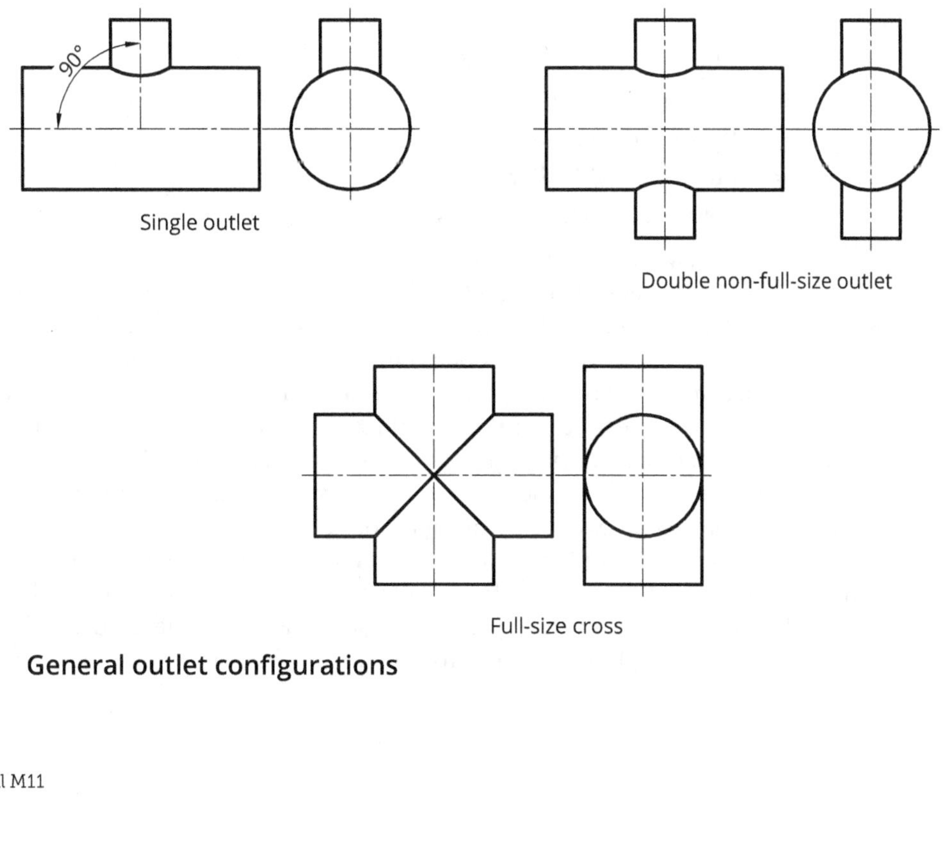

Figure 7-2 General outlet configurations

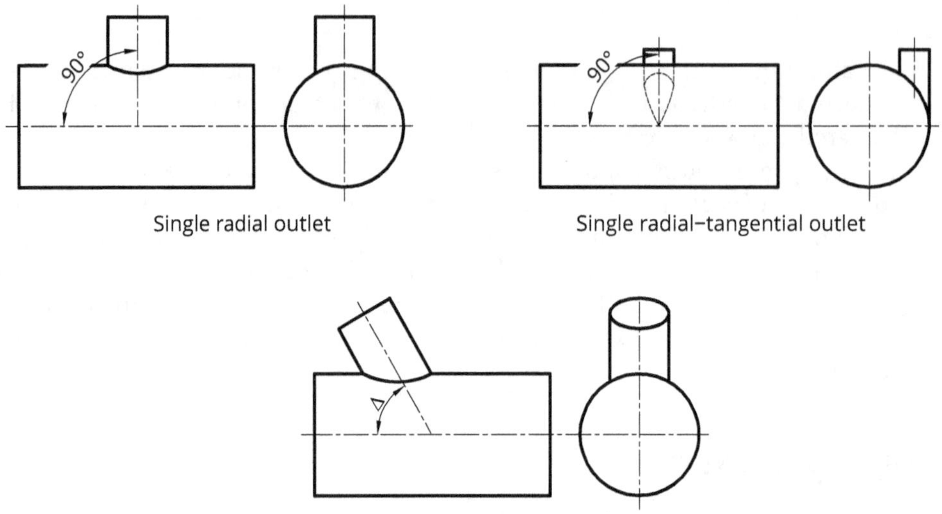

Single radial outlet

Single radial–tangential outlet

Single radial–lateral outlet

Figure 7-3 Common outlet configuration terminology

thicknesses have sufficient excessive material such that no supplemental reinforcement is required. When supplemental reinforcement is required, options for ANSI/AWWA design procedures include increased cylinder thickness, collars, wrappers, and crotch plates. The allowable stress limits used in the reinforcement design, as a percentage of yield strength of the subject material, should not be greater than those used in the design of pipe for hoop stress.

The type of recommended reinforcement can be determined by the magnitude of the pressure-diameter value (PDV) and the criteria listed in Table 7-1. The PDV is calculated as follows:

$$PDV = \frac{Kpd_o^2}{D_o(\sin^2 \Delta)} \quad \text{(Eq 7-6)}$$

Where:

K = multiplier = 1.5 for full-nominal-size cross and 1.0 for all other outlets
p = design pressure, psi
d_o = outlet outside diameter, in.
D_o = main pipe outside diameter, in.
Δ = outlet angle of deflection, degrees

When the magnitude of the PDV is greater than 9,000 (Arch 1980), the recommended reinforcement shall consist of a crotch plate designed in accordance with the method described in the Crotch-Plate Design section of this chapter. When the magnitude of the PDV is less than 9,000, the recommended reinforcement for single outlets and double, non–full-size outlets may be either a collar or wrapper. When the magnitude of the PDV is less than 9,000, the recommended reinforcement for full-size crosses is to increase the cylinder thicknesses of the four adjoining pipes as necessary to yield sufficient excessive material in the cylinders themselves, without the need for supplemental reinforcement. The increased cylinder thickness shall extend a minimum of one diameter from the intersection of the cross on all four adjoining pipes. For non–full-size crosses where the outlets' calculated reinforcements will overlap, the fitting shall be designed as a full-size cross. Although not specifically defined as applicable previously, some designers have successfully designed full-size crosses using the outlet reinforcement process in the previous versions of this

Table 7-1 Recommended reinforcement type based on PDV* and outlet type (see Figure 7-2)

Outlet Type	PDV	*M* Factor	Reinforcement Type†
Single	< 6,000	1.0	ICT or C
Single	6,000–9,000	0.000167 × PDV	ICT or C
Single	> 9,000	Not Applicable	Crotch Plate
Double, Non–full-size	< 6,000	1.0	ICT or C
Double, Non–full-size	6,000–9,000	0.000167 × PDV	ICT or C
Double, Non–full-size	> 9,000	Not Applicable	Crotch Plate
Full-Size Cross	< 6,000	1.0	ICT
Full-Size Cross	6,000–9,000	0.000167 × PDV	ICT
Full-Size Cross	> 9,000	Not Applicable	Crotch Plate

*Reinforcement for double laterals may require additional analyses beyond the criteria discussed in this manual. In addition, to reduce space requirements on certain large-diameter or high-pressure outlets, it may be advantageous to design these fittings without crotch plate reinforcement. Such cases may involve design by other codes, standards, or manuals.
†ICT = increased cylinder thickness; C = collar

manual. Given this historical success, the design procedure presented herein includes design of full-size crosses, with the inclusion of a PDV multiplier, *K*. By including *K*, the effective PDV value in the current design process is held within the historical limit to which successful designs have been provided. When the magnitude of the PDV is such that the recommended reinforcement is not a crotch plate and the subsequent reinforcement calculations yield a theoretical reinforcement area, A_r, less than or equal to the area available, A_a, sufficient material is present in the parent cylinders and no additional reinforcement is required. An effective option for the designer to negate the requirement for a collar or wrapper reinforcement is to thicken the mainline cylinder and/or outlet cylinder to yield an A_a that is greater than A_r. In some instances, design guidelines other than those defined herein may be beneficial to the designer. Alternate design guidelines may result in acceptable reinforcement configurations different from those identified in Table 7-1.

Wrappers may be substituted for collars when following this design procedure. This substitution is especially beneficial to the logistics of the manufacturing process when the d_o/D_o ratio is greater than 0.7. Figure 7-4 shows two outlets; the smaller one with a collar and the larger one with a full circumferential wrapper. When collars or wrappers are manufactured in more than one piece, the welds joining the pieces shall be complete joint-penetration butt welds. Crotch plates may be substituted for collars or wrappers when following this procedure.

The three most common types of outlets are shown in Figure 7-3 and can be generally defined as radial, radial-tangential, and radial-lateral.

A radial outlet is one the centerline of which is perpendicular to, and intersects, the centerline of the main pipe centerline. A radial-tangential outlet is one the centerline of which is perpendicular to the main pipe centerline, but where the outlet outside diameter is effectively tangent to the main pipe outside diameter. A radial-lateral outlet is one the centerline of which is not perpendicular to the main pipe centerline but which does intersect the main pipe centerline.

This design method directs that, for cases where the PDV is less than 9,000, the cross-sectional area of the removed steel at the outlet be replaced in the form of a collar or wrapper. The method further directs that when the PDV is between 6,000 and 9,000, the cross-sectional area of the replaced steel is multiplied by an *M* factor equal to 0.000167 times the PDV. Therefore, from a PDV of 6,000 to 9,000, the resulting cross-sectional area of the replaced steel will be linearly increased up to a maximum of 1.5 times at a PDV of 9,000. Collars and wrappers can be manufactured from plate, sheet, or coil, and the minimum

Figure 7-4 Collar and wrapper

steel thickness shall be 12 gauge (0.1046 in.). Consult the manufacturer regarding common plate thicknesses for a specific design. Figure 7-5 shows the general configuration of an opening for welded steel pipe when collar or wrapper reinforcement is used.

In determining the required steel replacement, credit should be given to any thickness of material in the mainline pipe in excess of that required for internal design pressure, and credit should be given to the similar excess area of the material in the wall of the outlet. The reader is referred to Eq 4-1 for thickness determination based on internal pressure. The design limit of the branch reinforcement in the outlet occurs at a radial distance 2.5 times the thickness of the branch (1) from the surface of the main pipe run when reinforcement is not required or (2) from the top of the collar or wrapper reinforcement. To conservatively simplify the analysis, weld areas are not considered as part of the reinforcement in the design. The overall width of the collar or wrapper, W, is measured parallel to the axis of the pipe at the centerline of the outlet. W should not be less than $d_o/\sin\Delta+3$ in., based on manufacturing logistics, and has a maximum design limit of $2.0d_o/\sin\Delta$. The collar or wrapper design edge width, w, shall be equal on each side of the outlet. Therefore, w should not be less than 1.5 in. and not more than $d_o/(2\sin\Delta)$. Collar edge widths in the circumferential direction should not be less than the design edge width, w. Reinforcement within the w limit in any direction cannot be attributed to the reinforcement requirements of more than one outlet. When initial calculated reinforcement dimensions of adjacent outlets result in overlap between the two reinforcements, the design should be modified as discussed below for limited space areas to the point that overlap is avoided, or one or both of the outlets moved to avoid overlap. The maximum thickness of the collar or wrapper for purposes of design is 2.5 times the mainline cylinder thickness. Reinforcement with edge width or thickness dimension in excess of those maximums noted previously is acceptable, but such excess material shall not be counted as satisfying any portion of the reinforcement design requirements. For areas of limited space, such as in vaults, other areas, or where initial design outlet reinforcement overlaps, options for minimizing or removing the required reinforcement include: (1) increasing the mainline steel cylinder thickness; (2) increasing the outlet steel cylinder thickness; (3) increasing both mainline and outlet steel cylinder

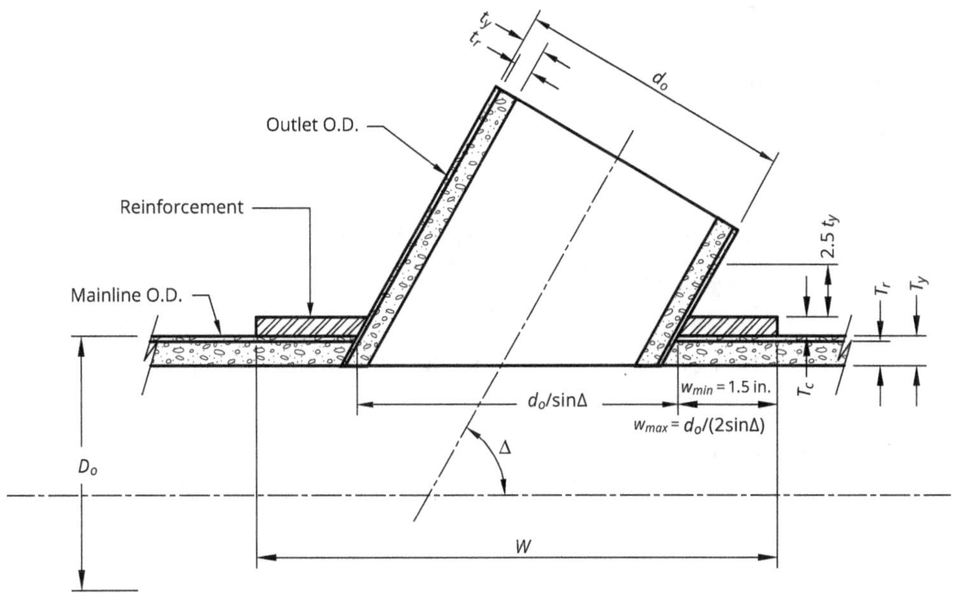

D_o = mainline pipe outside diameter, in.
T_y = mainline cylinder thickness, in.
T_r = required mainline cylinder thickness, in.
d_o = outlet pipe outside diameter, in.
t_y = outlet cylinder thickness, in.
t_r = required outlet cylinder thickness, in.
Δ = outlet deflection angle, degrees
T_c = collar or wrapper thickness, in.
W = overall collar or wrapper width, in.
w = collar or wrapper edge width, in.

Note: Figure does not show the location of necessary welds. See Figures 7-7 and 7-8 for weld definition.

Figure 7-5 Generic sectional view of reinforcement of outlets in welded steel pipe

thicknesses; and (4) using alternate material grades for one or both cylinder thicknesses, reinforcement material, or a combination of any of these subject to the strength reduction factors defined below. Such modification can successfully reduce or remove the need for reinforcement in such limited space areas.

Collars may be oval in shape, or they may be rectangular with rounded corners. The radii at corners should not be less than 4 in. or 20 times the collar thickness (except for collars with a length or width less than 8 in.).

In Figure 7-5, the area $T_y(d_o - 2t_y)/\sin\Delta$ represents the section of the mainline pipe cylinder removed by the opening for the outlet. The hoop tension caused by pressure within the pipe that would be taken by the removed section were it present must be carried by the total areas represented by $2wT_c$ and $5t_y(t_y - t_r)$, or $2.5t_y(t_y - t_r)$ on each side of outlet.

Allowable Stress

The allowable stress shall not exceed 50 percent of the minimum yield strength, σ_Y, of the material at the design pressure. The allowable stress used in calculating the minimum theoretical main pipe and outlet pipe cylinder thicknesses shall be specific to the material for the cylinder being analyzed. To account for varying specified minimum yield strengths between the main pipe cylinder, the outlet pipe cylinder, and the reinforcing material, strength reduction factors, sr_1 and sr_2, shall be used in the analysis. The strength reduction factors are defined as follows:

$$sr_1 = \min[(\text{minimum } \sigma_Y \text{ of outlet pipe})/(\text{minimum } \sigma_Y \text{ of main cylinder}), 1.0]$$

$$sr_2 = \min[(\text{minimum } \sigma_Y \text{ of reinforcement})/(\text{minimum } \sigma_Y \text{ of main cylinder}), 1.0]$$

OUTLET DESIGN EXAMPLES

Example 7-1: Radial Outlet Design

Main pipe cylinder OD	D_o	31.375 in.
Main pipe cylinder thickness	T_y	0.188 in.
Main pipe cylinder material specified minimum yield strength	σ_Y	42 ksi
Outlet pipe cylinder OD	d_o	4.500 in.
Outlet pipe cylinder thickness	t_y	0.237 in.
Outlet pipe cylinder material specified minimum yield strength	σ_Y	35 ksi
Outlet deflection angle	Δ	90°
Working pressure	p_w	150 psi
Reinforcement material specified minimum yield strength	σ_Y	36 ksi

The strength reduction factors are

$$sr_1 = \min[35/42\,,\,1.0] = 0.833$$
$$sr_2 = \min[36/42\,,\,1.0] = 0.857$$

Reinforcement Type

The only pressure condition provided in the example definition is working pressure. Therefore, working pressure becomes the design pressure for the reinforcement analysis.

$$PDV = \frac{Kpd_o^2}{D_o(\sin^2 \Delta)} = \frac{1.0(150)(4.500)^2}{31.375(\sin^2 90°)} = 97$$

Assuming that increased cylinder thickness in not desirable or practical, for PDV < 6,000, select collar reinforcement from Table 7-1.

$$\frac{d_o}{D_o} = \frac{4.500}{31.375} = 0.14$$

Since d_o/D_o is ≤ 0.7, collar reinforcement is appropriate.

Multiplier (*M*-factor)

For PDV < 6,000, $M = 1.0$

Collar Design

Theoretical cylinder thicknesses

Main pipe (T_r)

$$\sigma_{A1} = 0.5(42{,}000) = 21{,}000$$

$$T_r = \frac{pD_o}{2\sigma_{A1}} = \frac{(150)(31.375)}{2(21{,}000)} = 0.112 \text{ in.}$$

Outlet pipe (t_r)

$$\sigma_{A2} = 0.5(35{,}000) = 17{,}500$$

$$t_r = \frac{pd_o}{2\sigma_{A2}} = \frac{(150)(4.500)}{2(17{,}500)} = 0.019 \text{ in.}$$

Theoretical reinforcement area

Theoretical reinforcement area = A_r

$$A_r = M\left[T_r\left(\frac{d_o - 2t_y}{\sin \Delta}\right)\right] = 1.0\left[0.112\left(\frac{4.500 - 2(0.237)}{\sin 90°}\right)\right] = 0.45 \text{ in.}^2$$

Area available as excess T_y and allowable outlet area
Area available = A_a

$$A_a = \frac{(d_o - 2t_y)}{\sin \Delta}(T_y - T_r) + 5t_y(t_y - t_r)s_{r1}$$

$$A_a = \frac{(4.500 - 2(0.237))}{\sin 90°}(0.188 - 0.112) + 5(0.237)(0.237 - 0.019)0.833 = 0.52 \text{ in.}^2$$

Required reinforcement area
Required reinforcement area = A_w

$$A_w = \frac{A_r - A_a}{s_{r2}} = \frac{0.45 - 0.52}{0.857} = -0.08 \text{ in.}^2$$

Since A_w is negative, sufficient inherent reinforcement is available and no supplemental reinforcement is required.

Example 7-2: Radial Outlet Design

Main pipe cylinder OD	D_o	61.875 in.
Main pipe cylinder thickness	T_y	0.258 in.
Main pipe cylinder material specified minimum yield strength	σ_Y	42 ksi
Outlet pipe cylinder OD	d_o	37.875 in.
Outlet pipe cylinder thickness	t_y	0.189 in.
Outlet pipe cylinder material specified minimum yield strength	σ_Y	35 ksi
Outlet deflection angle	Δ	90°
Working pressure	p_w	100 psi
Reinforcement material specified minimum yield strength	σ_Y	36 ksi

The strength reduction factors are

$$sr_1 = \min[35/42\,,\, 1.0] = 0.833$$

$$sr_2 = \min[36/42\,,\, 1.0] = 0.857$$

Reinforcement Type
The only pressure condition provided in the example definition is working pressure. Therefore, working pressure becomes the design pressure for the reinforcement analysis.

$$\text{PDV} = \frac{Kpd_o^2}{D_o(\sin^2 \Delta)} = \frac{1.0(100)(37.875)^2}{61.875(\sin^2 90°)} = 2{,}318$$

Assuming that increased cylinder thickness is not desirable or practical, for PDV < 6,000, select collar reinforcement from Table 7-1.

$$\frac{d_o}{D_o} = \frac{37.875}{61.875} = 0.61$$

Since d_o/D_o is ≤ 0.7, collar reinforcement is appropriate.
Multiplier (*M*-factor)
For PDV < 6,000, M = 1.0
Collar Design
Theoretical cylinder thicknesses
Main pipe (T_r)

$$\sigma_{A1} = 0.5(42{,}000) = 21{,}000$$

$$T_r = \frac{pD_o}{2\sigma_{A1}} = \frac{(100)(61.875)}{2(21,000)} = 0.147 \text{ in.}$$

Outlet pipe (t_r)

$$\sigma_{A2} = 0.5(35,000) = 17,500$$

$$t_r = \frac{pd_o}{2\sigma_{A2}} = \frac{(100)(37.875)}{2(17,500)} = 0.108 \text{ in.}$$

Theoretical reinforcement area

Theoretical reinforcement area = A_r

$$A_r = M\left[T_r\left(\frac{d_o - 2t_y}{\sin \Delta}\right)\right] = 1.0\left[0.147\left(\frac{37.875 - 2(0.189)}{\sin 90°}\right)\right] = 5.51 \text{ in.}^2$$

Area available as excess T_y and allowable outlet area

Area available = A_a

$$A_a = \frac{(d_o - 2t_y)}{\sin \Delta}(T_y - T_r) + 5t_y(t_y - t_r)s_{r1}$$

$$A_a = \frac{(37.875 - 2(0.189))}{\sin 90°}(0.258 - 0.147) + 5(0.189)(0.189 - 0.108)0.833 = 4.23 \text{ in.}^2$$

Required reinforcement area

Required reinforcement area = A_w

$$A_w = \frac{A_r - A_a}{s_{r2}} = \frac{5.51 - 4.23}{0.857} = 1.49 \text{ in.}^2$$

Minimum and maximum reinforcement thicknesses

Minimum reinforcement thickness = T_c

$$w = \frac{d_o}{2 \sin \Delta} = \frac{37.875}{2 \sin 90°} = 18.94 \text{ in.}$$

$$T_c = \frac{A_w}{2w} = \frac{1.49}{2(18.94)} = 0.039 \text{ in.}$$

Therefore, round up to the next commonly available thickness, not less than 12 gauge (0.1046 in.).

$$T_c = 0.1046 \text{ in.}$$

Maximum design reinforcement thickness = T_{max}

Based on the manufacturing logistic of a minimum collar width of 1.50 in., T_{max} is

$$T_{max} = A_w/[2(1.5)] = 1.49/3 = 0.497 \text{ in.}$$

Providing reinforcement thickness in excess of this value is acceptable subject to the 1.50-in. minimum and the following design limitation.

Based on the design parameter of limiting the effective collar thickness to $2.5T_y$,

$$T_{max} = 2.5T_y = 2.5(0.258) = 0.645 \text{ in.}$$

Minimum reinforcement width based on minimum reinforcement thickness

$$w = \frac{A_w}{2T_c} = \frac{1.49}{2(0.1046)} = 7.122 \text{ in.}$$

Minimum allowable reinforcement width verification

$$w_{min} = 1.50 \text{ in.} < 7.122 \text{ in., therefore, use } w = 7.122 \text{ in.}$$

Overall reinforcement width

$$W = 2w + \frac{d_o}{\sin \Delta} = 2(7.122) + \frac{37.875}{\sin 90^\circ} = 52.119 \text{ in.}$$

Common manufacturing practices would round W up to 52.125 for measurement simplicity.

Summary

Use: $T_c = 0.1046$ in. $W = 52\frac{1}{8}$ in.

Alternate Collar Design

Given the range of reinforcement thickness from 0.1046 in. to 0.645 in., an alternate collar thickness of 0.500 in., for example, could be provided as follows:

$$w_{alt} = \frac{A_w}{2T_{alt}} = \frac{1.49}{2(0.500)} = 1.49 \text{ in.}$$

Verify minimum width compliance.

1.49 in. < 1.50 in., therefore, use $w_{alt} = 1.50$ in.

Alternate overall reinforcement width

$$W_{alt} = 2w_{alt} + \frac{d_o}{\sin \Delta} = 2(1.50) + \frac{37.875}{\sin 90^\circ} = 40.875 \text{ in.}$$

Alternate Summary

Use: $T_c = \frac{1}{2}$ in. $W = 40\frac{7}{8}$ in.

Example 7-3: Radial-Tangential Outlet Design

Main pipe cylinder OD	D_o	55.750 in.
Main pipe cylinder thickness	T_y	0.250 in.
Main pipe cylinder material specified minimum yield strength	σ_Y	42 ksi
Outlet pipe cylinder OD	d_o	12.750 in.
Outlet pipe cylinder thickness	t_y	0.250 in.
Outlet pipe cylinder material specified minimum yield strength	σ_Y	35 ksi
Outlet deflection angle	Δ	90°
Working pressure	p_w	150 psi
Field test pressure	p_t	250 psi
Reinforcement material specified minimum yield strength	σ_Y	50 ksi

The strength reduction factors are

$$sr_1 = \min[35/42, 1.0] = 0.833$$

$$sr_2 = \min[50/42, 1.0] = 1.0$$

Reinforcement Type

The field-test pressure of 250 psi is greater than 1.5 times the working pressure and will govern the design. Therefore, the field-test pressure must be reduced by a factor of 1.5 to arrive at the proper design pressure. The design pressure is $p = 250/(1.5) = 167$ psi.

$$\text{PDV} = \frac{Kpd_o^2}{D_o \sin^2 \Delta} = \frac{1.0(167)(12.750)^2}{(55.750)\sin^2(90^\circ)} = 487$$

Assuming that increased cylinder thickness is not desirable or practical, for PDV < 6,000, select collar reinforcement from Table 7-1.

$$\frac{d_o}{D_o} = \frac{12.750}{55.750} = 0.23$$

Since d_o/D_o is ≤ 0.7, collar reinforcement is appropriate.

Multiplier (*M*-factor)

For PDV < 6,000, $M = 1.0$

Reinforcement Design

Theoretical cylinder thicknesses

Main pipe (T_r)

$$\sigma_{A1} = 0.5(42{,}000) = 21{,}000 \text{ psi}$$

$$T_r = \frac{pD_o}{2\sigma_{A1}} = \frac{(167)(55.750)}{2(21{,}000)} = 0.222 \text{ in.}$$

Outlet pipe (t_r)

$$\sigma_{A2} = 0.5(35{,}000) = 17{,}500 \text{ psi}$$

$$t_r = \frac{pd_o}{2\sigma_{A2}} = \frac{(167)(12.750)}{2(17{,}500)} = 0.061 \text{ in.}$$

Theoretical reinforcement area

Theoretical reinforcement area = A_r

$$A_r = M\left[T_r\left(\frac{d_o - 2t_y}{\sin\Delta}\right)\right] = 1.00\left[0.222\left(\frac{12.750 - 2(0.250)}{\sin 90°}\right)\right] = 2.72 \text{ in.}^2$$

Area available as excess T_y and allowable outlet area

Area available = A_a

$$A_a = \frac{(d_o - 2t_y)}{\sin\Delta}(T_y - T_r) + 5t_y(t_y - t_r)s_{r1}$$

$$A_a = \frac{(12.750 - 2(0.250))}{\sin 90°}(0.250 - 0.222) + 5(0.250)(0.250 - 0.061)0.833 = 0.54 \text{ in.}^2$$

Reinforcement area

Reinforcement area = A_w

$$A_w = \frac{A_r - A_a}{s_{r2}} = \frac{2.72 - 0.54}{1.0} = 2.18 \text{ in.}^2$$

Minimum and maximum reinforcement thicknesses

Minimum reinforcement thickness = T_c

$$w = \frac{d_o}{2\sin\Delta} = \frac{12.750}{2\sin 90°} = 6.375 \text{ in.}$$

$$T_c = \frac{A_w}{2w} = \frac{2.18}{2(6.375)} = 0.171 \text{ in.}$$

Therefore, round up to the next commonly available thickness, not less than 12 gauge (0.1046 in.).

$$T_c = 0.188 \text{ in.}$$

Maximum design reinforcement thickness = T_{max}

Based on the manufacturing logistic of a minimum collar width of 1.50 in., T_{max} is

$$T_{max} = A_w/[2(1.5)] = 2.18/3 = 0.727 \text{ in.}$$

Providing reinforcement thickness in excess of this value is acceptable subject to the 1.50-in. minimum and the following design limitation.

Based on the design parameter of limiting the collar effective collar thickness to $2.5T_y$,

$$T_{max} = 2.5T_y = 2.5(0.250) = 0.625 \text{ in.}$$

Minimum reinforcement width based on minimum reinforcement thickness

$$w = \frac{A_w}{2T_c} = \frac{2.18}{2(0.188)} = 5.798 \text{ in.}$$

Minimum allowable reinforcement width verification

w_{min} = 1.50 in. < 5.798 in., therefore, use w = 5.798 in.

Overall reinforcement width

$$W = 2w + \frac{d_o}{\sin \Delta} + 2(5.798) + \frac{12.750}{\sin 90°} = 24.346 \text{ in.}$$

Common manufacturing practices would be to round W up to 24.375 for measurement simplicity.

Summary

Use: T_c = 3⁄16 in. W = 24 3⁄8 in.

Alternate Collar Design

Given the range of reinforcement thickness from 0.188 in. to 0.625 in., an alternate collar thickness of 0.625 in., for example, could be provided as follows:

$$w_{alt} = \frac{A_w}{2T_{alt}} = \frac{2.18}{2(0.625)} = 1.74 \text{ in.}$$

Verify minimum width compliance.

1.74 in. > 1.50 in., therefore, use w_{alt} = 1.74 in.

Alternate overall reinforcement width

$$W_{alt} = 2w_{alt} + \frac{d_o}{\sin \Delta} = 2(1.74) + \frac{12.750}{\sin 90°} = 16.230 \text{ in.}$$

Common manufacturing practices would round W up to 16.250 for measurement simplicity.

Alternate Summary

Use: T_c = 5⁄8 in. W = 16 1⁄4 in.

Example 7-4: Radial-Lateral Outlet Design

Main pipe cylinder OD	D_o	49.750 in.
Main pipe cylinder thickness	T_y	0.219 in.
Main pipe cylinder material specified minimum yield strength	σ_Y	42 ksi
Outlet pipe cylinder OD	d_o	43.750 in.
Outlet pipe cylinder thickness	t_y	0.219 in.
Outlet pipe cylinder material specified minimum yield strength	σ_Y	35 ksi
Outlet deflection angle	Δ	45°

Working pressure	p_w	100 psi
Transient pressure	p_t	135 psi
Reinforcement material specified minimum yield strength	σ_Y	36 ksi

The strength reduction factors are

$$sr_1 = \min[35/42\,,\,1.0] = 0.833$$
$$sr_2 = \min[36/42\,,\,1.0] = 0.857$$

Reinforcement Type

The transient pressure is less than 1.5 times the working pressure. Therefore, the design pressure will equal the working pressure.

$$\text{PDV} = \frac{Kpd_o^2}{D_o \sin^2 \Delta} = \frac{1.0(100)(43.75)^2}{(49.750)\sin^2(45°)} = 7{,}695$$

Assuming that increased cylinder thickness is not desirable or practical, for PDV ≤ 9,000, select collar reinforcement from Table 7-1.

$$\frac{d_o}{D_o} = \frac{43.750}{49.750} = 0.88$$

Since $d_o/D_o > 0.7$, substituting wrapper reinforcement for collar reinforcement may be beneficial to the manufacturing process.

Multiplier (*M*-factor)

For 6,000 < PDV ≤ 9,000

$$M = 0.000167\ \text{PDV} = (0.000167)(7{,}695) = 1.285$$

Therefore, use $M = 1.29$.

Reinforcement Design

Theoretical cylinder thicknesses

Main pipe (T_r)

$$\sigma_{A1} = 0.5(42{,}000) = 21{,}000 \text{ psi}$$

$$T_r = \frac{pD_o}{2\sigma_{A1}} = \frac{(100)(49.750)}{2(21{,}000)} = 0.118 \text{ in.}$$

Outlet pipe (t_r)

$$\sigma_{A2} = 0.5(35{,}000) = 17{,}500 \text{ psi}$$

$$t_r = \frac{pd_o}{2\sigma_{A2}} = \frac{(100)(43.750)}{2(17{,}500)} = 0.125 \text{ in.}$$

Theoretical reinforcement area

Theoretical reinforcement area = A_r

$$A_r = M\left[T_r\left(\frac{d_o - 2t_y}{\sin \Delta}\right)\right] = 1.29\left[0.118\left(\frac{43.750 - 2(0.219)}{\sin 45°}\right)\right] = 9.32 \text{ in.}^2$$

Area available as excess T_y and allowable outlet area

Area available = A_a

$$A_a = \frac{(d_o - 2t_y)}{\sin \Delta}(T_y - T_r) + 5t_y(t_y - t_r)s_{r1}$$

$$A_a = \frac{(43.750 - 2(0.219))}{\sin 45°}(0.219 - 0.118) + 5(0.219)(0.219 - 0.125)0.833 = 6.27 \text{ in.}^2$$

Reinforcement area

Reinforcement area = A_w

$$A_w = \frac{A_r - A_a}{s_{r2}} = \frac{9.32 - 6.27}{0.857} = 3.56 \text{ in.}^2$$

Minimum and maximum reinforcement thicknesses

Minimum reinforcement thickness = T_c

$$w = \frac{d_o}{2 \sin \Delta} = \frac{43.750}{2 \sin 45°} = 30.94 \text{ in.}$$

$$T_c = \frac{A_w}{2w} = \frac{3.56}{2(30.94)} = 0.058 \text{ in.}$$

Therefore, round up to the next commonly available thickness, not less than 12 gauge (0.1046 in.).

$$T_c = 0.1046 \text{ in.}$$

Maximum design reinforcement thickness = T_{max}

Based on the manufacturing logistic of a minimum collar width of 1.50 in., T_{max} is

$$T_{max} = A_w/[2(1.5)] = 3.56/3 = 1.187 \text{ in.}$$

Providing reinforcement thickness in excess of this value is acceptable subject to the 1.50-in. minimum and the following design limitation:

Based on the design parameter of limiting the collar effective collar thickness to $2.5T_y$,

$$T_{max} = 2.5T_y = 2.5(0.219) = 0.548 \text{ in.}$$

Minimum reinforcement width based on minimum reinforcement thickness

$$w = \frac{A_w}{2T_c} = \frac{3.56}{2(0.1046)} = 17.02 \text{ in.}$$

Minimum allowable reinforcement width verification

w_{min} = 1.50 in. < 17.02 in., therefore, use w = 17.02 in.

Overall reinforcement width

$$W = 2w + \frac{d_o}{\sin \Delta} = 2(17.02) + \frac{43.750}{\sin 45°} = 95.912 \text{ in.}$$

Common manufacturing practices would round up W to 96 for measurement simplicity.

Summary

Use: T_c = 0.1046 in. W = 96 in.

Alternate Collar/Wrapper Design

Given the range of reinforcement thickness from 0.1046 in. to 0.548 in., an alternate reinforcement thickness of 0.500 in., for example, could be provided as follows:

$$w_{alt} = \frac{A_w}{2T_{alt}} = \frac{3.56}{2(0.500)} = 3.560 \text{ in.}$$

Verify minimum width compliance.

3.56 in. > 1.50 in., therefore, use w_{alt} = 3.56 in.

Alternate overall collar/wrapper width

$$W_{alt} = 2w_{alt} + \frac{d_o}{\sin \Delta} = 2(3.56) + \frac{43.750}{\sin 45°} = 68.992 \text{ in.}$$

Common manufacturing practices would round up W to 69 for measurement simplicity.

Alternate Summary

Use: $T_c = ½$ in. $W = 69$ in.

Example 7-5: Double-Outlet Design

Main pipe cylinder OD	D_o	55.750 in.
Main pipe cylinder thickness	T_y	0.248 in.
Main pipe cylinder material specified minimum yield strength	σ_Y	42 ksi
Outlet one cylinder OD	d_o	43.750 in.
Outlet one thickness	t_y	0.219 in.
Outlet-one pipe cylinder material specified minimum yield strength	σ_Y	42 ksi
Outlet-one deflection angle	Δ	90°
Outlet-one reinforcement material specified minimum yield strength	σ_Y	36 ksi
Outlet two cylinder OD	d_o	37.500 in.
Outlet two thickness	t_y	0.188 in.
Outlet two pipe cylinder material specified minimum yield strength	σ_Y	35 ksi
Outlet two deflection angle	Δ	75°
Outlet two reinforcement material specified minimum yield strength	σ_Y	36 ksi
Working pressure	p_w	125 psi
Transient pressure	p_t	135 psi
Field-test pressure	p_t	145 psi

The strength reduction factors are

Outlet One	Outlet Two
$sr_1 = \min[42/42\ ,\ 1.0] = 1.0$	$sr_1 = \min[35/42\ ,\ 1.0] = 0.833$
$sr_2 = \min[36/42\ ,\ 1.0] = 0.857$	$sr_2 = \min[36/42\ ,\ 1.0] = 0.857$

Reinforcement Type

The greater of transient and field-test pressures is the field-test pressure. The field-test pressure is less than 1.5 times the working pressure. Therefore, the design pressure will equal the working pressure.

Outlet One / Outlet Two

$$\text{PDV} = \frac{Kpd_o^2}{D_o \sin^2 \Delta} = \frac{1.0(125)(43.75)^2}{(55.750)\sin^2(90°)} = 4{,}292 \qquad = \frac{1.0(125)(37.50)^2}{(55.750)\sin^2(75°)} = 3{,}379$$

Assuming that increased cylinder thickness is not desirable or practical, for PDV < 6,000, select collar reinforcement from Table 7-1.

	Outlet One	Outlet Two
$\dfrac{d_o}{D_o}$	$= \dfrac{43.750}{55.750} = 0.78$	$= \dfrac{37.500}{55.750} = 0.67$

Since $d_o/D_o > 0.7$ for outlet one, substituting wrapper reinforcement for collar reinforcement may be beneficial to the manufacturing process.

Multiplier (*M*-factor)

For PDV < 6,000,

$M = 1.0$ for both outlets

Reinforcement Design

Theoretical cylinder thicknesses

Main pipe (T_r)

$$\sigma_{A1} = 0.5(42{,}000) = 21{,}000 \text{ psi}$$

$$T_r = \frac{pD}{2\sigma_{A1}} = \frac{(125)(55.750)}{2(21{,}000)} = 0.166 \text{ in.}$$

Outlet pipe (t_r)

	Outlet One	Outlet Two
σ_{A2}	$= 0.5(42{,}000) = 21{,}000$ psi	$= 0.5(35{,}000) = 17{,}500$ psi
t_r	$= \dfrac{pd_o}{2\sigma_{A2}} = \dfrac{(125)(43.750)}{2(21{,}000)} = 0.130$ in.	$t_r = \dfrac{pd_o}{2\sigma_{A2}} = \dfrac{(125)(37.500)}{2(17{,}500)} = 0.134$ in.

Theoretical reinforcement area

Theoretical reinforcement area = A_r

Outlet One:

$$A_r = M\left[T_r\left(\frac{d_o - 2t_y}{\sin\Delta}\right)\right] = 1.0\left[0.166\left(\frac{43.750 - 2(0.219)}{\sin 90^\circ}\right)\right] = 7.19 \text{ in.}^2$$

Outlet Two:

$$= 1.0\left[0.166\left(\frac{37.500 - 2(0.188)}{\sin 75^\circ}\right)\right] = 6.38 \text{ in.}^2$$

Area available as excess T_y and allowable outlet area

Area available = A_a

$$A_a = \frac{(d_o - 2t_y)}{\sin\Delta}(T_y - T_r) + 5t_y(t_y - t_r)s_{r1}$$

Outlet One:

$$A_a = \frac{(43.750 - 2(0.219))}{\sin 90^\circ}(0.248 - 0.166) + 5(0.219)(0.219 - 0.130)1.0 = 3.65 \text{ in.}^2$$

Outlet Two:

$$A_a = \frac{(37.500 - 2(0.188))}{\sin 75^\circ}(0.248 - 0.166) + 5(0.188)(0.188 - 0.134)0.833 = 3.19 \text{ in.}^2$$

Reinforcement area

Reinforcement area = A_w

	Outlet One	Outlet Two
	$A_w = \dfrac{A_r - A_a}{s_{r2}} = \dfrac{7.19 - 3.65}{0.857} = 4.13 \text{ in.}^2$	$= \dfrac{6.38 - 3.19}{0.857} = 3.72 \text{ in.}^2$

Minimum and maximum reinforcement thicknesses

Minimum reinforcement thickness = T_c

	Outlet One	Outlet Two
	$w = \dfrac{d_o}{2 \sin \Delta} = \dfrac{43.750}{2 \sin 90°} = 21.88 \text{ in.}$	$= \dfrac{37.500}{2 \sin 75°} = 19.41 \text{ in.}$
	$T_c = \dfrac{A_w}{2w} = \dfrac{4.13}{2(21.88)} = 0.094 \text{ in.}$	$= \dfrac{3.72}{2(19.41)} = 0.096 \text{ in.}$

Therefore, round up to the next commonly available thickness, not less than 12 gauge (0.1046 in.).

$T_c = 3/16$ in. for both outlets

Maximum design reinforcement thickness = T_{max}

Based on the manufacturing logistic of a minimum collar width of 1.50 in., T_{max} is

	Outlet One	Outlet Two
	$T_{max} = A_w/[2(1.5)] = 4.13/3 = 1.377 \text{ in.}$	$= 3.72/3 = 1.240 \text{ in.}$

Providing reinforcement thickness in excess of this value is acceptable subject to the 1.50-in. minimum and the following design limitation.

Based on the design parameter of limiting the collar effective collar thickness to $2.5T_y$,

	Outlet One	Outlet Two
	$T_{max} = 2.5T_y = 2.5(0.248) = 0.620 \text{ in.}$	$= 2.5(0.248) = 0.620 \text{ in.}$

Minimum reinforcement width based on minimum reinforcement thickness

Outlet One:

$$w = \frac{A_w}{2T_c} = \frac{4.13}{2(0.188)} = 10.98 \text{ in.}$$

Outlet Two:

$$w = \frac{A_w}{2T_c} = \frac{3.72}{2(0.188)} = 9.89 \text{ in.}$$

Minimum allowable reinforcement width verification

Outlet One:

$w_{min} = 1.50$ in. < 10.98 in., therefore, use $w = 10.98$ in.

Outlet Two:

$w_{min} = 1.50$ in. < 9.89 in., therefore, use $w = 9.89$ in.

Clearance check

For double-outlet type designs, the geometry needs to include a check for clearance between the two reinforcing elements. As noted previously, outlet reinforcement elements cannot overlap. The check shown herein was accomplished by generating a scale drawing of the resulting geometry; a mathematical check could also be performed. The following scale drawing (Figure 7-6) shows that the two outlets' reinforcing collars do not overlap and the design is geometrically satisfactory.

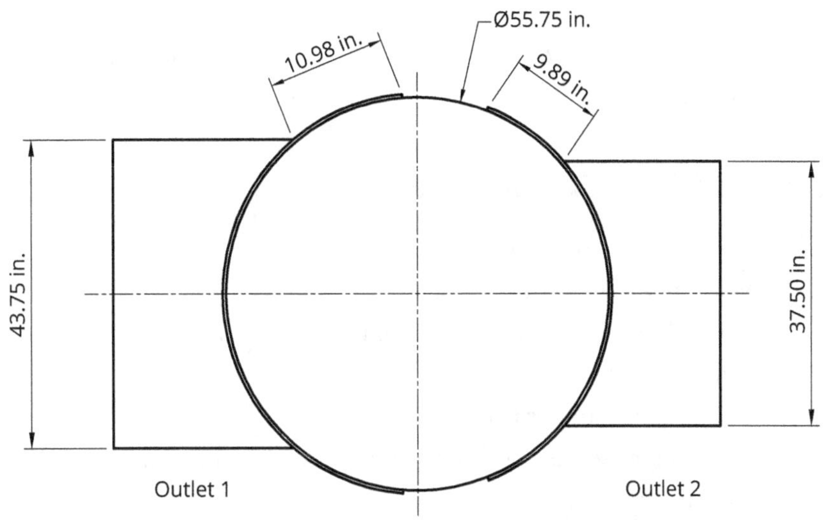

Figure 7-6 Scale drawing of the resulting geometry of double outlet

Overall reinforcement width

Outlet One / Outlet Two

$$W = 2w + \frac{d_o}{\sin \Delta} = 2(10.98) + \frac{43.750}{\sin 90°} = 65.71 \text{ in.} \quad = 2(9.89) + \frac{37.500}{\sin 75°} = 58.60 \text{ in.}$$

Common manufacturing practices would round up W to 65¾ for outlet one and 58⅝ for outlet two for measurement simplicity.

Summary

	Outlet One	Outlet Two
Use:	T_c = 3⁄16 in. W = 65¾ in.	T_c = 3⁄16 in. W = 58⅝ in.

Alternate Collar/Wrapper Design

Given the range of reinforcement thickness from 0.1046 in. to 0.620 in. (outlet one) and 0.620 in. (outlet two), alternate reinforcement thicknesses of 0.500 in. (outlet one) and 0.4375 in. (outlet two), for example, could be provided as follows:

Outlet One / Outlet Two

$$w_{alt} = \frac{A_w}{2T_{alt}} = \frac{4.13}{2(0.500)} = 4.13 \text{ in.} \qquad = \frac{3.72}{2(0.4375)} = 4.25 \text{ in.}$$

Verify minimum width compliance.

Both 4.13 in. and 4.25 in. > 1.50 in., therefore, use w_{alt} = 4.13 in. (outlet one) and 4.25 in. (outlet two).

Alternate overall collar/wrapper width

Outlet One

$$W_{alt} = 2w_{alt} + \frac{d_o}{\sin \Delta} = 2(4.13) + \frac{43.750}{\sin 90°} = 52.01 \text{ in.}$$

Outlet Two

$$= 2(4.25) + \frac{37.500}{\sin 75°} = 47.32 \text{ in.}$$

Common manufacturing practices would round up W to 52.125 for outlet one and 47.375 for outlet two for measurement simplicity.

Alternate summary

	Outlet One	Outlet Two
Use:	T_c = ½ in. W = 52⅛ in.	T_c = 7⁄16 in. W = 47⅜ in.

Example 7-6: Full-Size Cross Design

Main pipe cylinder OD	D_o	61.750 in.
Main pipe cylinder thickness	T_y	0.290 in.
Main pipe cylinder material specified minimum yield strength	σ_Y	35 ksi
Outlet pipe cylinder OD	d_o	61.750 in.
Outlet pipe cylinder thickness	t_y	0.290 in.
Outlet pipe cylinder material specified minimum yield strength	σ_Y	35 ksi
Working pressure	p_w	75 psi
Transient pressure	p_t	100 psi
Field-test pressure	p_t	125 psi

The strength reduction factors are

$$sr_1 = \min[35/35\,,\,1.0] = 1.0$$
$$sr_2 = \min[35/35\,,\,1.0] = 1.0$$

Reinforcement Type

The greater of transient and field test pressures is the field-test pressure. The field-test pressure is more than 1.5 times the working pressure. Therefore, the design pressure will equal the field-test pressure divided by 1.5 = 125/1.5 = 83.3.

$$\text{PDV} = \frac{Kpd_o{}^2}{D_o \sin^2 \Delta} = \frac{1.5(83.3)(61.75)^2}{(61.750)\sin^2(90°)} = 7{,}716$$

From Table 7-1, for PDV ≤ 9,000, wall thickness will be increased as necessary to reinforce all the cylinders of the cross.

Multiplier (*M*-factor)

For 6,000 < PDV ≤ 9,000

M = 0.000167 PDV = (0.000167)(7,716) = 1.289

Therefore, use M = 1.29.

Reinforcement Design

Theoretical cylinder thicknesses

Main pipe (T_r)

$$\sigma_{A1} = 0.5(35{,}000) = 17{,}500 \text{ psi}$$

$$T_r = t_r = \frac{pD_o}{2\sigma_{A1}} = \frac{(83.3)(61.750)}{2(17{,}500)} = 0.147 \text{ in.}$$

Theoretical reinforcement area

Theoretical reinforcement area = A_r

$$A_r = M\left[T_r\left(\frac{d_o - 2t_y}{\sin \Delta}\right)\right] = 1.29\left[0.147\left(\frac{61.750 - 2(0.290)}{\sin 90°}\right)\right] = 11.60 \text{ in.}^2$$

Area available as excess T_y and allowable outlet area

Area available = A_a

$$A_a = \frac{(d_o - 2t_y)}{\sin \Delta}(T_y - T_r) + 5t_y(t_y - t_r)s_{r1}$$

$$A_a = \frac{(61.750 - 2(0.290))}{\sin 90°}(0.290 - 0.147) + 5(0.290)(0.290 - 0.147)1.0 = 8.95 \text{ in.}^2$$

$$A_a < A_r$$

When $A_a < A_r$, additional cylinder thickness is required. Given the symmetry and configuration of a full-size cross, the design procedure requires unilaterally increasing all four cylinder thicknesses in lieu of the addition of reinforcement. Therefore, trial-and-error substitution of increasing cylinder thicknesses is required until a thickness is identified that results in $A_a \geq A_r$.

Try $T_y = 0.375$ in.

Theoretical reinforcement area

Theoretical reinforcement area = A_r

$$A_r = M\left[T_r\left(\frac{d_o - 2t_y}{\sin \Delta}\right)\right] = 1.29\left[0.147\left(\frac{61.750 - 2(0.375)}{\sin 90°}\right)\right] = 11.57 \text{ in.}^2$$

Area available as excess T_y and allowable outlet area

Area available = A_a

$$A_a = \frac{(d_o - 2t_y)}{\sin \Delta}(T_y - T_r) + 5t_y(t_y - t_r)s_{r1}$$

$$A_a = \frac{(61.750 - 2(0.375))}{\sin 90°}(0.375 - 0.147) + 5(0.375)(0.375 - 0.147)1.0 = 14.34 \text{ in.}^2$$

$A_a > A_r$, therefore the full-size cross requires a thickness of ⅜ in. for all four cylinder components. This thickness must extend a distance of 61.75 in. from the intersection point of the cross for each of the four component cylinders of the cross.

OUTLET AND COLLAR/WRAPPER CONNECTION

Outlet connection to a mainline cylinder and reinforcement collar/wrapper connection to an outlet pipe shall be accomplished by a complete joint penetration weld. Configuration alternates and welding requirements for outlet and collar/wrapper connections are identified in Figures 7-7 and 7-8. Configurations shown in Figures 7-7 and 7-8 are representative of outlets attached in the shop and field. Alternate configurations that achieve the intended complete joint penetration and associated fillet welds are acceptable, and performance of the complete joint penetration weld may be made from the inside or outside.

CROTCH PLATE DESIGN FOR OUTLETS AND TRUE WYES

A full treatise on the design of crotch plates for steel pipe, including a nomograph design method, is described in an article prepared by the Department of Water and Power, City of Los Angeles (Swanson et al. 1955). Design using this nomograph method is presented in the following section. Examples are given for single-plate, two-plate, and three-plate designs. Other data on the subject also have been published (Ruud 1964).

CROTCH-PLATE DESIGN

When the PDV exceeds 9,000, crotch-plate reinforcement should be used. Several types of plate reinforcement are illustrated in Figures 7-9 through 7-11. The following section on nomograph use was taken from a published study on crotch-plate (wye-branch) design at Los Angeles (Swanson et al. 1955).

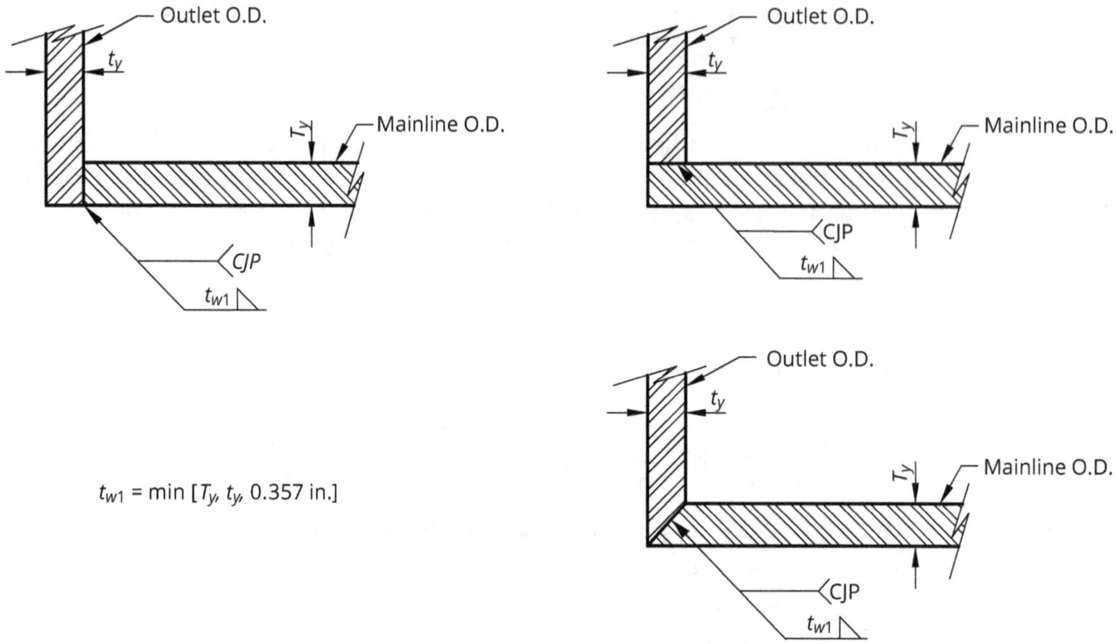

Figure 7-7 Configurations and welding for outlet not requiring reinforcement

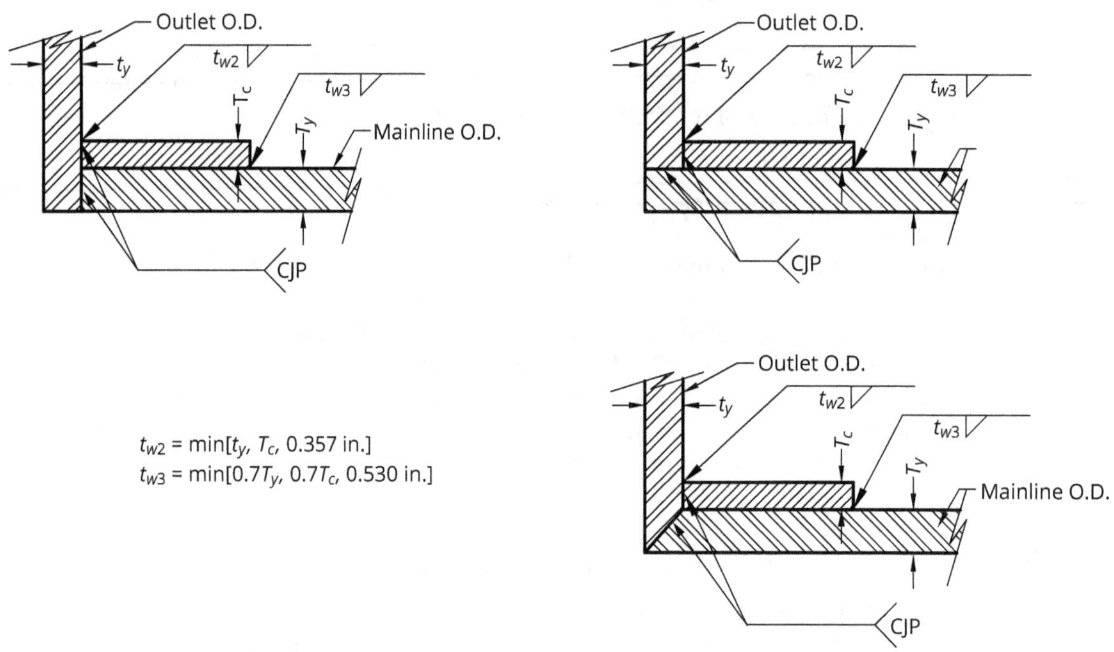

Figure 7-8 Configurations and welding for outlet requiring reinforcement

NOMOGRAPH USE IN RADIAL OUTLET AND WYE-BRANCH DESIGN

The nomograph design, based on design pressure, includes a safety factor that will keep stresses well below the yield point of steel. To maintain this safety factor, the design pressure used to evaluate the initial plate sizes in the nomograph shall be equal to the larger of the field-test pressure, transient pressure, or 1.5 times the working pressure. The minimum yield strength of the steel used to develop this design procedure was 30,000 psi. Stress values are generated from the familiar Mc/I and T/A relations. Both

A single curved plate serves as reinforcement for each branch of this 96-in. × 66-in., 90° included angle wye.

Figure 7-9 One-plate wye

This 15-ft × 15-ft, 90° wye has two crotch plates and one back plate.

Figure 7-10 Three-plate wye

This 126-in. × 126-in., 45° wye section has two plates.

Figure 7-11 Two-plate wye

I and A vary linearly with thickness when rectangular rib cross sections are involved. Therefore, when rectangular rib cross sections are used, if plate with yield strength other than 30,000 psi is used, the thickness can be linearly adjusted to a new value. See Step 1b below.

Step 1a. Lay a straightedge across the nomograph (Figure 7-12) through the appropriate points on the pipe diameter (see Step 2b) and internal-pressure scales; read off the depth of plate from its scale. This reading is the crotch depth for 1-in.-thick plate for a two-plate, 90°, wye-branch pipe. (Thinner plates may be used, provided localized buckling is addressed.) For wye branch sizes in excess of the 120-in. limitation shown in Figure 7-12, the "Diameter of Pipe" scale in the figure can be linearly extrapolated to

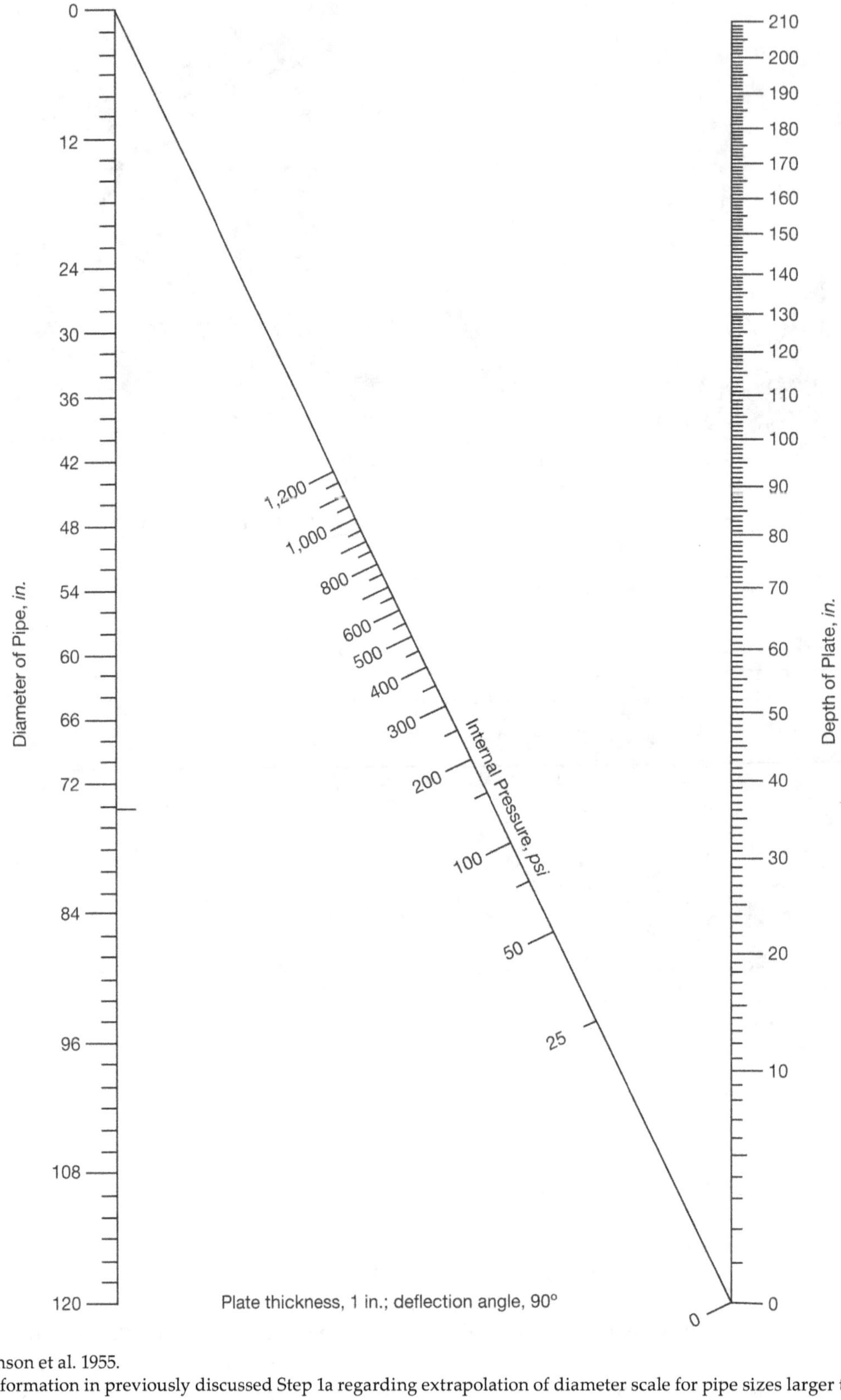

Source: Swanson et al. 1955.

Note: See information in previously discussed Step 1a regarding extrapolation of diameter scale for pipe sizes larger than 120 in.

Figure 7-12 Nomograph for selecting reinforcement plate depths of equal-diameter pipes

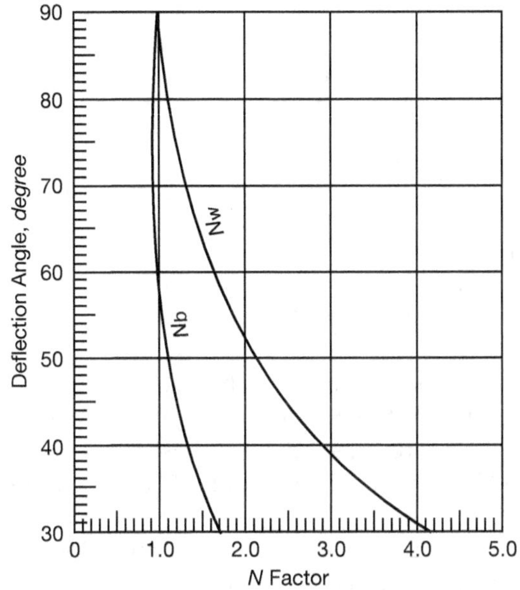

Source: Swanson et al. 1955.

Note: For wyes with deflection angles from 30° to 90°, the N factors obtained from the above curves are applied to the plate depth d, found from the nomograph (Figure 7-12), in accordance with the equations $d_w = N_w d$ and $d_b = N_b d$.

Figure 7-13 *N* factor curves

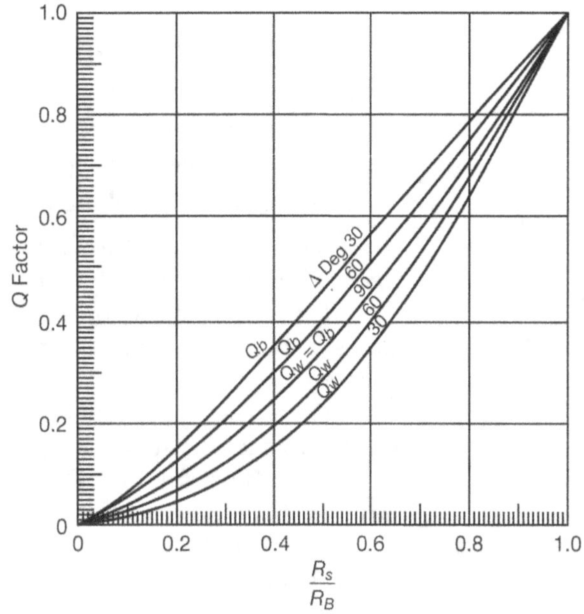

Source: Swanson et al. 1955.

Note: For wyes of unequal diameter, find d_w and d_b for the larger-diameter pipe (from Figures 7-12 and 7-13); then $Q_w d_w = d'_w$ crotch depth of single-plate stiffener; and $Q_b d_b = d'_b$, base depth of a single-plate stiffener.

Figure 7-14 *Q* factor curves

accommodate larger diameters, subject to the following conditions: (1) The line extending from the "Diameter of Pipe" scale through the "Internal Pressure" scale must intersect the "Depth of Plate" scale at a point less than or equal to 210 in.; (2) the design pressure must be less than or equal to the 1,200-psi limit on the "Internal Pressure" scale; and (3) neither the "Depth of Plate" scale nor the "Internal Pressure" scale can be extrapolated beyond the values shown in the figure.

Step 1b. Modification of the crotch-plate thickness when steel with a yield strength greater than 30 ksi is used. If the crotch-plate material has a yield strength greater than 30 ksi, 36 ksi for example, and the crotch plate is of a rectangular cross section, adjust the initial 1-in. plate thickness as follows. The corrected thickness would be 1.0(30)/36 = 0.83-in. The corrected thickness must be verified for compliance to the $30d_w$ limitation identified below in Step 3. If a cross section other than rectangular is used, the 1-in. initial thickness shall not be adjusted.

Step 2a. If the wye branch deflection angle is other than 90°, use the N-factor curve (Figure 7-13) to get the factors that, when multiplied by the depth of plate found in Step 1a or Step 1b, will give the wye depth d_w and the base depth d_b for the new wye branch.

Step 2b. If the wye branch has unequal-diameter pipe, the larger-diameter pipe will have been used in Steps 1a and 2a, and these results should be multiplied by the Q factors found on the single-plate stiffener curves (Figure 7-14) to give d'_w and d'_b. These factors vary with the ratio of the radius of the small pipe to the radius of the large pipe.

Step 3. If the wye depth d_w found so far is greater than 30 times the thickness of the plate, then d_w and d_b should be converted to a greater thickness t using the general equation:

$$d = d_1(t_1/t)^{(0.917-\Delta/360)} \quad \text{(Eq. 7-7)}$$

Where:

d_1 = existing depth of plate, in.

t_1 = existing thickness of plate, in.

d = new depth of plate, in.

t = new thickness of plate selected, in.

Δ = deflection angle of the wye branch, degrees

Step 3a. For mainline cylinder diameters of 60 in. or less, the value of d_w shall not be greater than the mainline cylinder outside diameter.

Step 4. To find the top depth d_t or d'_t, use Figure 7-15, in which d_t or d'_t is plotted against d_b or d'_b. This dimension gives the top and bottom depths of the plate at 90° from the crotch depths.

Step 5. The interior curves follow the cut of the pipe, but the outside crotch radius in both crotches should equal d_t plus the radius of the pipe; or in the single-plate design, d'_t plus the radius of the smaller pipe. Tangents connected between these curves complete the outer shape. Reference Figures 7-17 and 7-23 for clarification of the interior curves and outside plate radii.

The important depths of the reinforcement plates, d_w, d_b, and d_t, can be found from the nomograph. If a curved exterior is desired, a radius equal to the inside pipe radius plus d_t can be used, both for the outside curve of the wye section and for the outside curve of the base section.

CROTCH-PLATE CONNECTIONS

Crotch plates can be connected to the pipe cylinders in either an integral or an external configuration. Integral connection configurations for a two-plate design are shown in Figure 7-16A. External connection configurations for a two-plate design are shown in Figure 7-16B. An external connection configuration provides flexibility in the manufacturing process and a viable option for future upgrade of the pressure rating of a fitting not originally requiring crotch-plate reinforcement. Plate-to-pipe connection configurations for a one-plate design are identical to those shown in Figures 7-16A and 7-16B. Alternate configurations that achieve the intended weld requirements shown below are acceptable. Prior to welding, crotch plates should be inspected for laminations for a minimum distance of two times the plate thickness from the connection locations. When the thinnest material being joined at a connection is greater than 1¼ in., the designer should investigate the applicability of postweld heat treatment (PWHT) as outlined in paragraph UCS-56 of the ASME Code (ASME 2010).

For single curved cold-formed crotch plates with d/D greater than 0.70, consideration should be given to the minimum bend radius of the plate. The manufacturer should be consulted for bend limits in applications where d/D is greater than 0.70, or the designer should consider alternate designs that include tangent cone/cylinder branches. Branches that are tangent to the main cylinder result in straight line intersections that avoid the need for curved crotch plates in favor of planar crotch plates.

General dimensions for a wye-branch and associated crotch-plate reinforcement are shown in Figure 7-17.

Three-Plate Design

The preceding nomograph section described the design of one- and two-plate wye branches without addressing a three-plate design because of its similarity to the two-plate design. Reference Figure 7-18 for a general representation of a three-plate configuration. The function

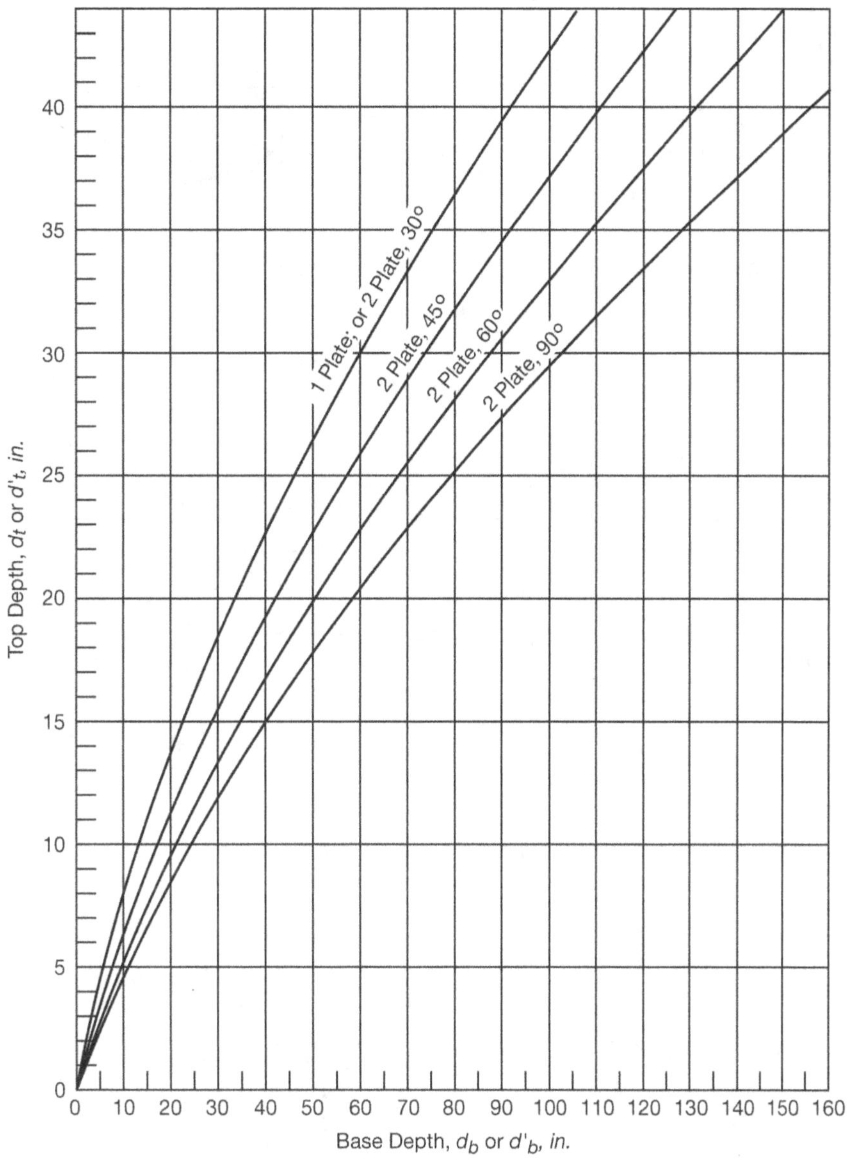

Source: Swanson, et al. 1955.

Note: d'_t and d'_b are for one-plate design dimensions; d_t and d_b are two-plate design dimensions. Two-plate designs are only applicable to size-on-size outlets. One-plate designs are only applicable to outlets with a diameter of a size not equal to the mainline pipe. For one-plate designs with angles from 30° to 90°, use the one-plate curve.

Figure 7-15 Selection of top depth

of the third plate is to act like a clamp by holding down the deflection of the two main plates. In doing so, it accepts part of the stresses of the other plates and permits a smaller design. This decrease in the depths of the two main plates is small enough to make it practical to add a third plate to a two-plate design. The additional plate should be considered a means of reducing the deflection at the junction of the plates. The two factors that dictate the use of a third plate are diameter of pipe and internal pressure. When the diameter is greater than 60 in. (1,500 mm) ID and the internal pressure is greater than 300 psi, a third ring plate can be used advantageously.

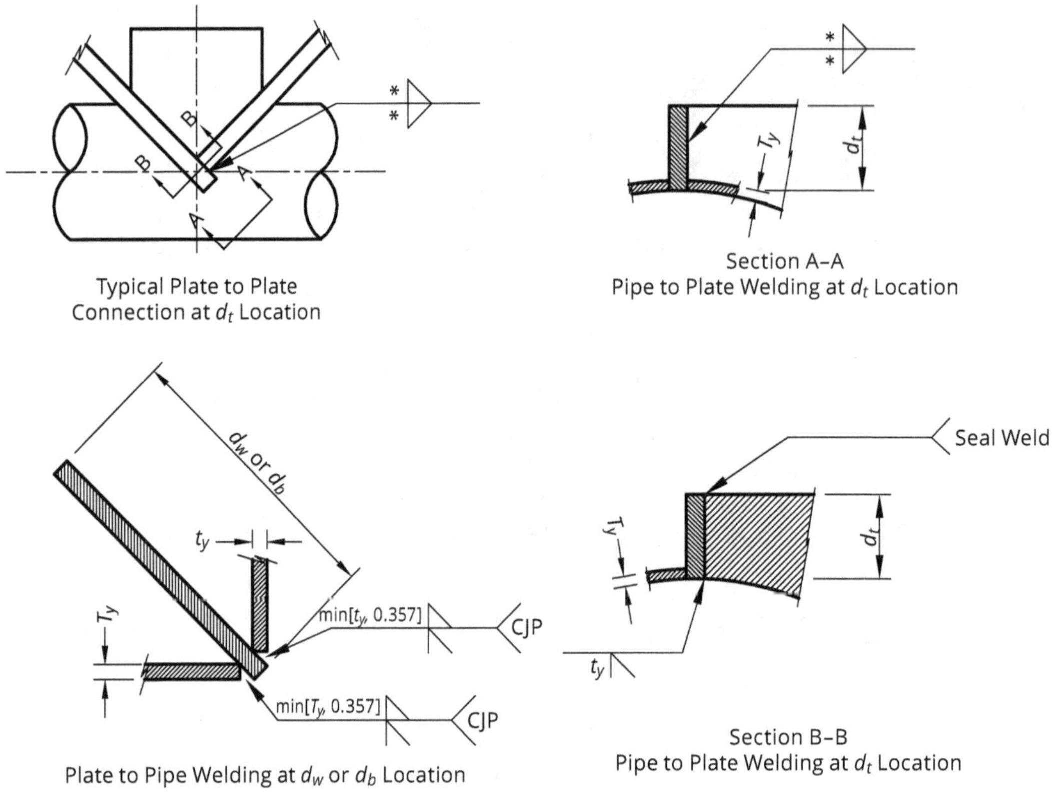

*See Table 7-5 for weld size

Figure 7-16A Two-plate integral crotch-plate connections

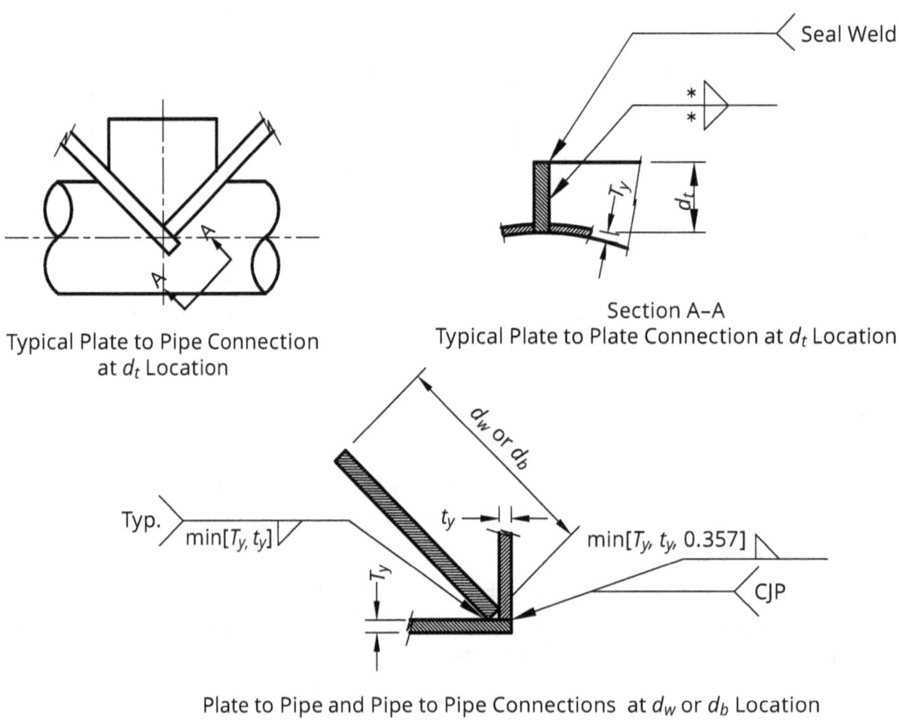

*See Table 7-5 for weld sizes

Figure 7-16B Two-plate external crotch-plate connections

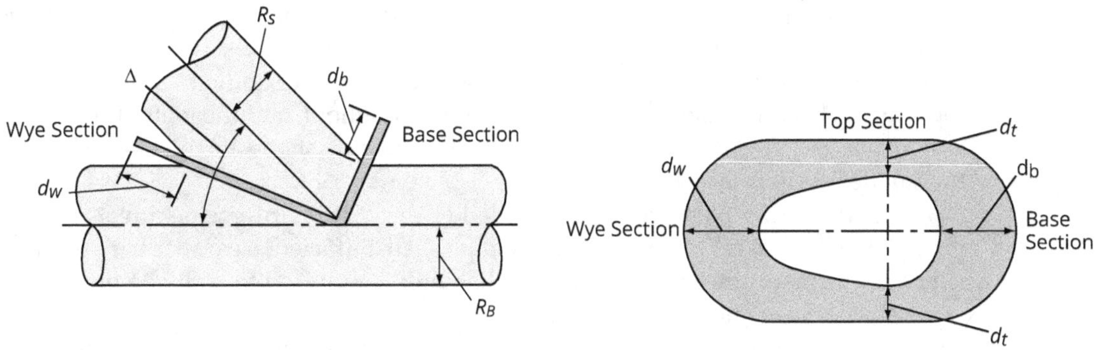

Source: Swanson et al. 1955.

Figure 7-17 Wye-branch plan and layout

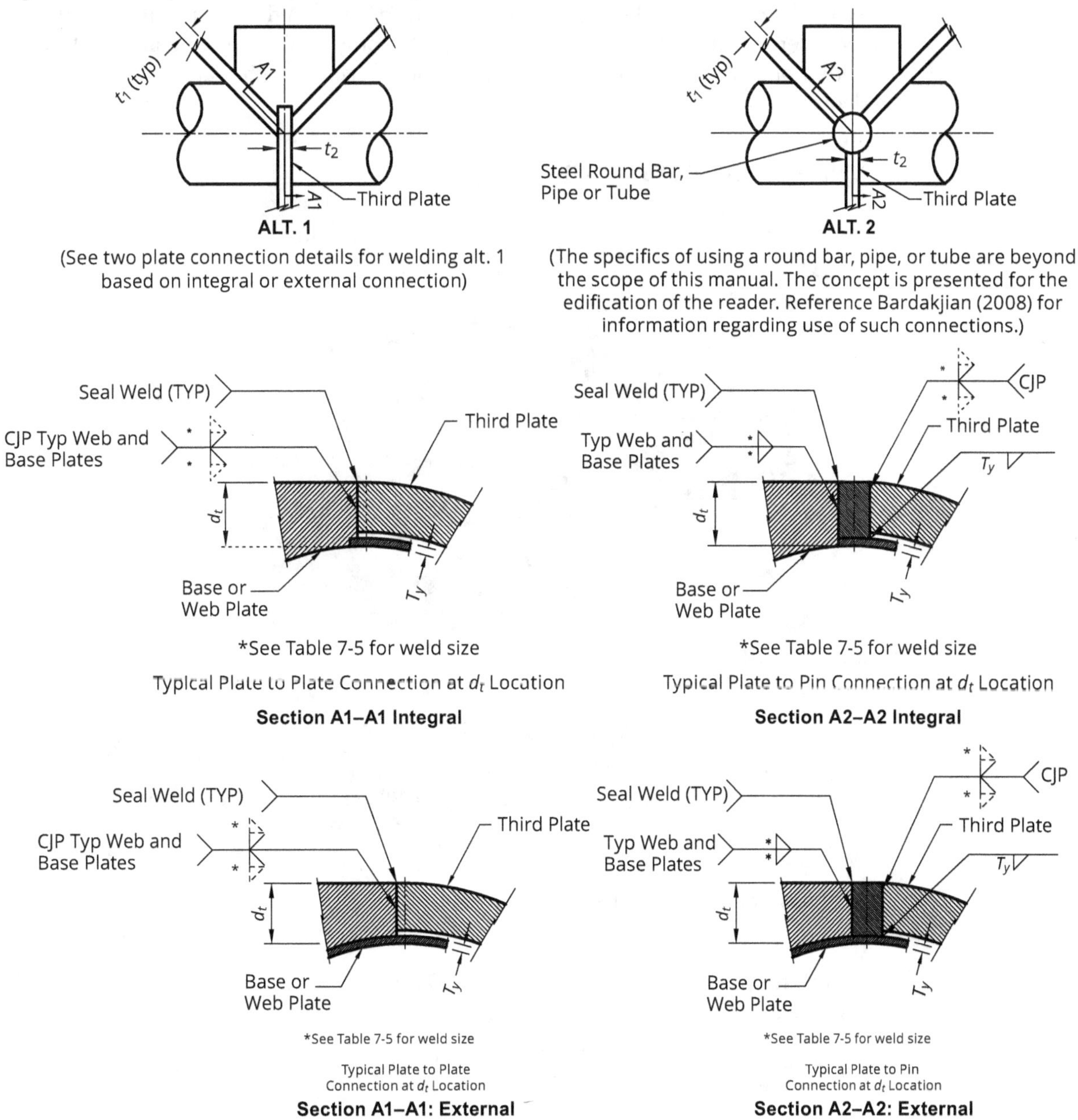

Figure 7-18 Plate configurations for third-plate design

If a third plate is desired as an addition to the two-plate design, its size should be dictated by the top depth d_t. Because the other two plates are flush with the inside surface of the pipe, however, the shell plate thickness plus clearance should be subtracted from the top depth. This dimension should be constant throughout, and the plate should be placed at right angles to the axis of the pipe, giving it a half-ring shape. Its thickness should equal the smaller of the main plates.

The third ring plate should be welded to the other reinforcement plates only at the top and bottom but remain free from the pipe shell. Sufficient clearance between the third ring and the main-line cylinder outside diameter is required to facilitate proper welding at junction of pipe and crotch plates.

Where a third plate has been accepted as advantageous, the connection configuration alternatives are as shown in Figure 7-18. The bar, pipe, or tube noted in alternate 2 of Figure 7-18 are for welding access and constructability only, and are not considered in the design. When a third plate is used, the clearance between the outside of the main-line pipe cylinder and the inside of the third ring need only be of sufficient size to allow for completion of any required welding between the adjacent crotch plates or other connective element.

Example 7-7: One-Plate Design

Figure 7-19 Illustration of one-plate design

D_o = 60 in. and $R_B = D_o/2 = 60/2 = 30$ in.

d_o = 42 in. and $R_S = d_o/2 = 42/2 = 21$ in.

$\Delta = 45°$ Crotch-plate material σ_Y = 30 ksi

Working pressure, 230 psi

Transient pressure, 300 psi

Test pressure, 325 psi

Check for proper pressure to calculate PDV. 325/1.5 = 217, which is < 230, so use 230 for PDV calculation.

$$\text{PDV} = \frac{Kpd_o^2}{D_o(\sin^2 \Delta)} = \frac{1.0(230)(42)^2}{60(\sin^2 45°)} = 13{,}524 \text{ therefore use crotch-plate reinforcment}$$

1.5 times the working pressure = (1.5)230 = 345 psi, which is larger than both the transient and test pressures. To simplify the use of the nomograph, let the design pressure = 350 psi.

Step 1a. With the larger pipe diameter 60 in. and the design pressure 350 psi, read the critical plate depth d from the Figure 7-12 nomograph ($t = 1$ in., $\Delta = 90°$):

$$d = 50 \text{ in.}$$

Step 1b. Since the plate material has 30-ksi yield strength, the 1-in. initial plate thickness cannot be adjusted.

Step 2. Using the deflection angle 45°, reference Figure 7-13 to find the factors on the *N*-factor curve that will convert the depth found in Step 1 to apply to a 45° wye branch ($t = 1$ in.):

$$d_w = N_w d = 2.4(50) = 120 \text{ in.}$$

$$d_b = N_b d = 1.2(50) = 60 \text{ in.}$$

Step 2b. With the ratio of the smaller pipe radius divided by the larger pipe radius $(R_S/R_B) = (21/30) = 0.70$ and the deflection angle ($\Delta = 45°$), use Figure 7-14 to find the *Q* factors that give the crotch depths for a single-plate pipe wye stiffener ($t = 1$ in.):

$$Q_w = 0.52$$

$$Q_b = 0.66$$

$$d'_w = 0.52(120) = 62.4 \text{ in.}$$

$$d'_b = 0.66(60) = 39.6 \text{ in.}$$

Step 3. Because the depth of d'_w is still greater than $30t$, Eq 7-7 should be used:
Try a thickness of 1½ in.:

$$d = d_1(t_1/t)^{(0.917 - \Delta/360)}$$

$$d = d_1(1/1.5)^{(0.917 - 45/360)} = d_1(2/3)^{0.792}$$

$$= d_1(0.725)$$

$$d'_w = 62.4(0.725) = 45 \text{ in.}$$

$$d'_b = 39.6(0.725) = 29 \text{ in.}$$

Step 4. Find the top depth d'_t from the curve for one-plate design in Figure 7-15:

For $d'_b = 29$ in., $d'_t = 18$ in.

Final results:

Thickness of reinforcing plate,	$t = 1.5$ in.
Depth of plate at acute crotch,	$d'_w = 45$ in.
Depth of plate at obtuse crotch,	$d'_b = 29$ in.
Depth of plate at top and bottom,	$d'_t = 18$ in.

Outside radius of plate at each crotch equals the top depth plus the inside radius of the small pipe = $d'_t + R_S = 18 + 21 = 39$ in.

Example 7-8: Two-Plate Design

Figure 7-20 Site illustration of two-plate design

$D_o = d_o = 72$ in. and $R_B = R_S = D_o/2 = 72/2 = 36$ in.

$\Delta = 53°$ Crotch-plate material $\sigma_Y = 42$ ksi

Working pressure, 140 psi
Transient pressure, 215 psi
Test pressure, 225 psi

Check for proper pressure to calculate PDV. 225/1.5 = 150, which is > than 140, so use 150 for PDV calculation.

$$\text{PDV} = \frac{Kpd_o^2}{D_o(\sin^2 \Delta)} = \frac{1.0(150)(72)^2}{72(\sin^2 53°)} = 16{,}933 \text{ therefore use crotch-plate reinforcment.}$$

1.5 times the working pressure (1.5)140 = 210 psi, which is less than both the transient and test pressures. Therefore, the design pressure will be the larger of the transient and test pressures, or 225 psi.

Step 1a. With a pipe diameter of 72 in. and the design pressure 225 psi, read the critical depth of plate from the Figure 7-12 nomograph ($t = 1$ in., $\Delta = 90°$):

$$d = 49 \text{ in.}$$

Step 1b. Revise required plate thickness due to use of material with 42-ksi yield strength.

Revised plate thickness, $t = (30/42)(1) = 0.714$ in.

Step 2a. Using the N-factor curves in Figure 7-13, find the two factors at $\Delta = 53°$:

$$d_w = 2.0(49) = 98 \text{ in.}$$
$$d_b = 1.1(49) = 53.9 \text{ in.}$$

Step 3. Because d_w is greater than 30 times the thickness of the plate, try $t = 1.75$ in. in the conversion equation:

$$d = d_1(t_1/t)^{(0.917 - \Delta/360)} = d_1(0.714/1.75)^{0.770}$$
$$= d_1(0.501)$$
$$d_w = 98(0.501) = 49.1 \text{ in.}$$
$$d_b = 53.9(0.501) = 27.0 \text{ in.}$$

Step 4. Read the top depth d_t from the two-plate design curve in Figure 7-15:

d_t = 14 in.

Final results:

Thickness of reinforcing plate,	t = 1.75 in.
Depth of plate at acute crotch,	d_w = 49.1 in.
Depth of plate at obtuse crotch,	d_b = 27.0 in.
Depth of plate at top and bottom,	d_t = 14 in.
Outside radius of plate at each crotch	= 50 in.

Example 7-9: Full-Size Cross Design

$D_o = d_o = 48$ in. and $R_B = R_S = D_o/2 = 48/2 = 24$ in.

$\Delta = 90°$ Crotch-plate material $\sigma_Y = 42$ ksi

Working pressure, 200 psi
Transient pressure, 250 psi
Test pressure, 275 psi

Check for proper pressure to calculate PDV. 275/1.5 = 183, which is < 200, so use 200 for PDV calculation.

$$\text{PDV} = \frac{Kpd_o^2}{D_o(\sin^2 \Delta)} = \frac{1.5(200)(48)^2}{48(\sin^2 90°)} = 14{,}400 \text{ therefore use crotch-plate reinforcment.}$$

1.5 times the working pressure (1.5)200 = 300 psi, which is greater than both the transient and test pressures. Therefore, the design pressure will be 1.5 (working pressure) = 300 psi.

Step 1a. With a pipe diameter of 48 in. and the design pressure 300 psi, read the critical depth of plate from the Figure 7-12 nomograph (t = 1 in., Δ = 90°):

d = 30 in.

Step 1b. Revise required plate thickness due to use of material with 42-ksi yield strength.

Revised plate thickness, t = (30/42)(1) = 0.714 in.

Step 2a. Using the N-factor curves in Figure 7-13, find the two factors at Δ = 90°:

d_w = 1.0(30) = 30 in.

d_b = 1.0(30) = 30 in.

Step 3. Because d_w is greater than 30 times the thickness of the plate, try t = 1.0 in. in the conversion equation:

$$d = d_1(t_1/t)^{(0.917 - \Delta/360)} = d_1(0.714/1.0)^{0.667}$$

$$= d_1(0.799)$$

d_w = 30(0.799) = 24.0 in.

d_b = 30(0.799) = 24.0 in.

Step 4. Read the top depth d_t from the two-plate design curve in Figure 7-15:

d_t = 10.0 in.

Final results:

Thickness of reinforcing plate,	t = 1 in.
Depth of plate at acute crotch,	d_w = 24.0 in.
Depth of plate at obtuse crotch,	d_b = 24.0 in.
Depth of plate at top and bottom,	d_t = 10.0 in.
Outside radius of plate at each crotch	= 34.0 in.

TRUE WYE DESIGN

For design of a true wye, as shown in Figure 7-21, the process is based on the application of principles of Swanson et al. (1955) and is completed in two separate stages. The first stage evaluates the plate dimensions at the acute intersection between the two branches of the wye, where the plate is designed for the thickness and d_w dimension based on an angle of 90°. The second stage evaluates the plate dimensions at the two obtuse intersections between the main-line and the branches, where the plate is designed for the thickness and d_b dimension based on an outlet with a centerline oriented 45° from the main-line pipe centerline. By symmetry, the second stage evaluation yields the dimensions of the obtuse angle plates on each side of the wye.

The design process will yield a plate thickness for the d_w plate that is thinner than that for the d_b plates. The designer has the option to use the d_b plate thickness for the d_w plate. Doing so allows the designer to reduce the depth of the d_w plate in accordance with the equation in design Step 3 of the nomograph design (as shown in Examples 7-7, 7-8, and 7-9). The economics of such a change should be evaluated based on the differences between the two thicknesses and resulting d_w dimensions. The convergence of the three plates is detailed in Figure 7-22. The bar, pipe, or tube noted in alternate 2 of Figure 7-22 are for welding access and constructability only, and are not considered in the design.

Example 7-10: True-Wye (Δ = 90°) Design

D_o = 96 in. and $R_B = R_S = D_o/2 = 96/2 = 48$ in.

Δ = 90° Crotch-plate material σ_Y = 36 ksi

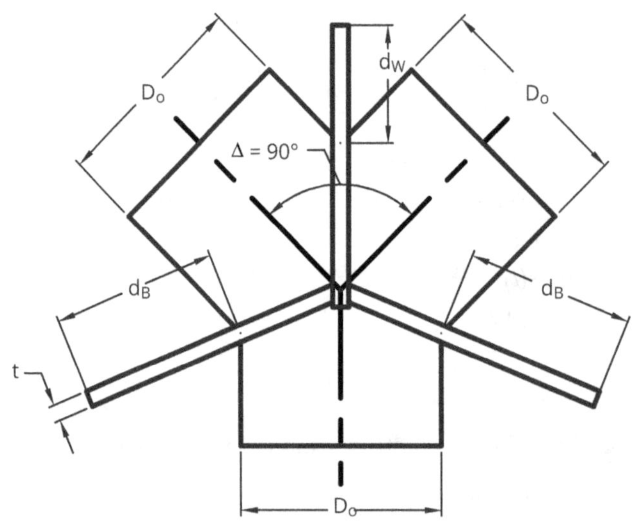

Figure 7-21 True-wye plan

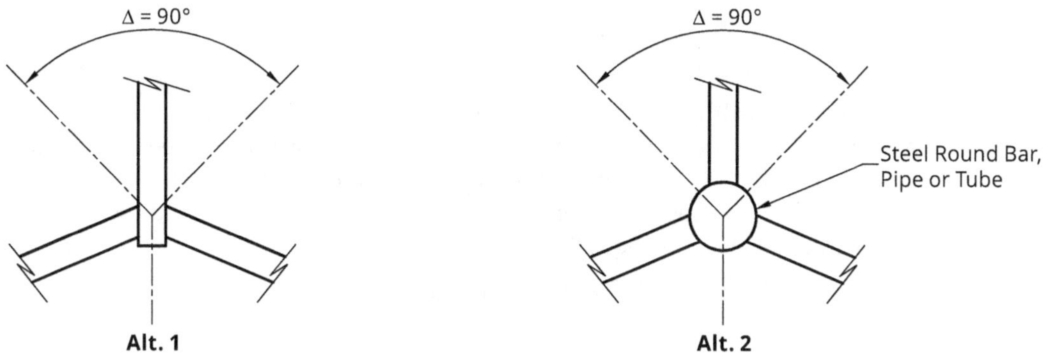

(See two plate connection details for welding Alt. 1 based on integral or external connection.

(The specifics of using a round bar, pipe or tube are beyond the scope of this manual. The concept is presented for the edification of the reader. Reference Bardakjian (2008) for information regarding use of such connections.)

Figure 7-22 Plate configurations for a true wye

Working pressure, 100 psi

Transient pressure, not defined in this example

Test pressure, not defined in this example

The design pressure equals 1.5 times the working pressure = (1.5)100 = 150 psi.

Stage 1: Design Plate Between the Two Branches of the Wye

Step 1a. With a pipe diameter of 96 in. and the design pressure 150 psi, read the critical depth of plate from the Figure 7-12 nomograph for t = 1 in.:

$$d = 64 \text{ in.}$$

Step 1b. Revise required plate thickness due to use of material with 36-ksi yield strength.

Revised plate thickness, $t = (30/36)(1) = 0.833$ in.

Step 2a. Using the N-factor curves in Figure 7-13, for $\Delta = 90°$, $N_w = 1.0$:

$$d_w = 1.0(64) = 64 \text{ in.}$$

Note that this stage involves only the plate between the two branches, which is associated with the d_w dimension only; the d_b dimension will be addressed in the second stage.

Step 3. Because d_w is greater than 30 times the 0.833-in. thickness of the plate, try t = 1.5 in. in the conversion equation:

$$d = d_1(t_1/t)^{(0.917 - \Delta/360)} = d_1(0.833/1.5)^{0.667}$$

$$= d_1(0.675)$$

$$d_w = 64(0.675) = 43.2 \text{ in.}$$

Stage 2: Design Plate Between the Mainline Cylinder and the Branches of the Wye

Step 1a. With a pipe diameter of 96 in. and the design pressure 150 psi, read the critical depth of plate from the nomograph for t = 1 in.:

$$d = 64 \text{ in.}$$

Step 1b. Revise required plate thickness due to use of material with 36-ksi yield strength.

Revised plate thickness, $t = (30/36)(1) = 0.833$ in.

Step 2a. Using the N-factor curves in Figure 7-13, for $\Delta = 45°$, $N_w = 2.4$ and $N_b = 1.2$:

$$d_w = 2.4(64) = 154 \text{ in.}$$
$$d_b = 1.2(64) = 77 \text{ in.}$$

Note that although this stage involves only the plate between the main-line cylinder and the branch of the wye, a representative d_w dimension must be evaluated to properly identify the thickness of the plate associated with the d_b dimension.

Step 3. Because d_w is greater than 30 times the 0.833-in. thickness of the plate, try t = 2.5 in. in the conversion equation:

$$d = d_1(t_1/t)^{(0.917 - \Delta/360)} = d_1(0.833/2.5)^{0.792}$$
$$= d_1(0.419)$$
$$d_w = 154(0.419) = 64.5 \text{ in.}$$
$$d_b = 77(0.419) = 32.3 \text{ in.}$$

Step 4. Read the top depth d_t from the two-plate 45° design curve in Figure 7-15 for d_b = 32.3 in.:

$$d_t = 16.5 \text{ in.}$$

Note that at the designer's option, the d_w plate thickness could be changed to 2.5 in. to match the d_b plates, and, per the equation in design Step 3, the revised d_w dimension would be:

$$d = d_1(t_1/t)^{(0.917 - \Delta/360)} = d_1(0.833/2.5)^{0.667}$$
$$= d_1(0.480)$$
$$d_w = 64(0.480) = 30.7 \text{ in.}$$

Final results:

Thickness of reinforcing plate between branches,	t = 1.5 in. (alternate t = 2.5 in.)
Depth of plate at acute crotch between branches,	d_w = 43.2 in. (alternate d_w = 30.7 in.)
Thickness of reinforcing plate between branches and main-line cylinder,	t = 2.5 in.
Depth of plate at obtuse crotch between branches and main-line cylinder,	d_b = 32.3 in.
Depth of plates at top and bottom,	d_t = 16.5 in.
Outside radius of plate at each crotch	= 64.5 in.

Handling and Shipping Lifting Holes

A common practice with crotch plates is to provide them with the excess plate area outside the design limit boundary as shown by the dashed lines in Figure 7-23. Given the geometric complexity associated with crotch-plate reinforced outlets, holes from which the fitting can be handled during manufacturing and installation or to secure the fitting during shipment are commonly incorporated into the plates in these excess areas and/or added at the top of the plate near the d_t section. The options shown are general in nature and not intended to preclude other viable configurations, as long as such configurations fall outside the design limit boundary. The specifics of the dimensions, shape, placement, and configuration of the holes are beyond the scope of this manual and should be addressed by a qualified designer. When the holes are supplied as shown in the top portion of Figure 7-23, a smooth transition between the border of the excess plate area containing the hole and the design limit border

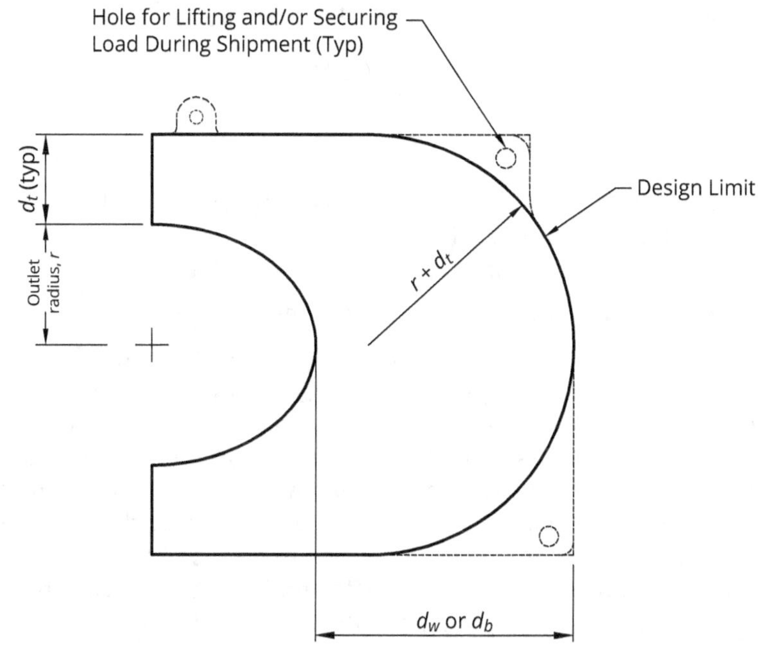

Figure 7-23 Common handling and shipping lifting hole configurations

is suggested to minimize corner stresses at these locations. Coating application or repair at areas where the holes are located must be addressed after the fitting is installed.

DESIGN OF ELLIPSOIDAL HEADS

This manual provides design guidelines for formed heads with 2:1 ellipsoidal shape. The geometric relationship of a 2:1 ellipsoidal shape is where the minor axis equals one-half of the major axis. An acceptable approximation of a 2:1 ellipsoidal head is where the knuckle radius is 0.17 times the diameter of the head and the spherical radius is 0.90 times the diameter of the head (ASME 2010 and see Figure 7-24). The length of straight flange (SF) on a head varies and is dependent on the capabilities of the manufacturer.

Design pressure for a head with pressure on the concave side shall be equal to the greater of p_w or $p_t/1.5$, where p_t is equal to the greater of transient pressure and test pressure. The allowable design stress shall not exceed 50 percent of the minimum specified yield strength of the head material at the design pressure for heads formed from one piece of plate (one-piece heads without construction seams), or for heads containing construction seams when such seams are welded and tested in accordance with ANSI/AWWA C200.

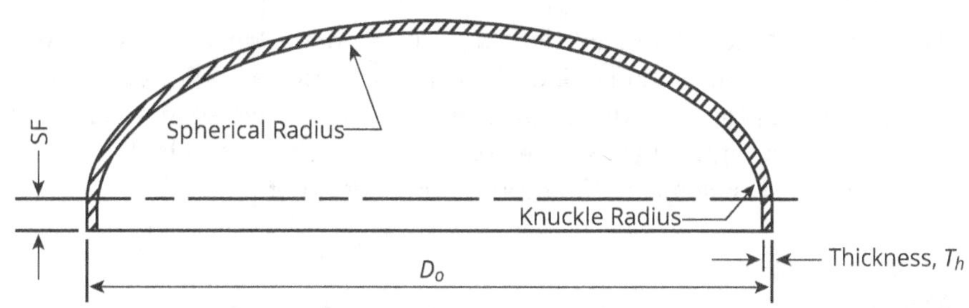

Figure 7-24 Approximation of an ellipsoidal head

The minimum thickness for a 2:1 ellipsoidal head with pressure on the concave side is

$$T_h = \frac{pD_o}{2s - 0.2p} \qquad \text{(Eq 7-8)}$$

Where:

T_h = minimum required thickness of head after forming, in.

p = design pressure, psi

D_o = outside diameter of head, in.

s = allowable design stress for head material, psi

For a head with pressure on the convex side, the head thickness shall be calculated using Eq 7-8 with a design pressure, $p = p_{cvx}$, equal to the greater of p_w or $p_t/1.3$, where p_t is equal to the greater of transient pressure and test pressure. The allowable design stress shall be equal to the lesser of 50 percent of the minimum specified yield strength of the head material and 18,000 psi. The resultant thickness shall then be multiplied by the corresponding value from Table 7-2. The multiplier values provided in Table 7-2 are valid for use with steels with minimum specified yield strength of 30,000 psi or higher. For pressures outside of the limits identified in Table 7-2 and for materials with specified minimum yield strength lower than 30,000 psi, refer to the ASME Code, Section VIII, Division 1, paragraph UG-33 (ASME 2010).

The design information included herein is not intended to preclude the use of other recognized codes, standards, or design procedures involving alternate configurations of heads, some of which include torispherical-, spherical-, or flat-type.

Table 7-2 Multiplier for pressure on the convex side of a head

p_{cvx} (psi)	Multiplier
$25 \le p_{cvx} \le 50$	4.9
$50 < p_{cvx} \le 75$	3.5
$75 < p_{cvx} \le 100$	2.9
$100 < p_{cvx} \le 125$	2.7
$125 < p_{cvx} \le 150$	2.6
$150 < p_{cvx} \le 200$	2.5
$200 < p_{cvx} \le 250$	2.3
$p_{cvx} > 250$	2.2

TESTING OF FITTINGS

ANSI/AWWA C200 (latest edition) describes nondestructive testing of weld seams for fittings and special sections. Special sections fabricated from previously hydrostatically tested straight pipe require testing of only those welded seams that were not previously tested in the straight pipe. Nondestructive testing methods include dye penetrant, magnetic particle, ultrasonic, radiographic (x-ray), or other methods.

JOINT HARNESSES

In areas where gasketed joint pipe and non-self-restraining couplings are subject to thrust resulting from internal pressure, an option to mitigate the thrust is to harness across the joint. Information for joint harness assemblies and tie rods to be used for given pipe diameters and select common design pressures are shown in Table 7-3. The design pressure noted in the table is intended to be the highest pressure condition to which the harness assembly will be subjected during service conditions, be that working, transient, test, or other pressure conditions. Harness design data applicable to sleeve couplings or other non–self-restrained gasketed-type joints are shown in Tables 7-3 through 7-5 and Figure 7-25. The joint harness system described here is applicable to single joints or multiple joints in series given that sufficiently long restraint rods are available to span the full series of joints. Further, the joint harness system described here is not intended to preclude the use of other restraint systems as applicable and agreed to by the purchaser.

The design process for harness assemblies utilizing both front and back rings is comprised of a two-part analysis. The first part of the analysis involves evaluating the load bearing capacity of the tie rods to determine the specific size and quantity required. Determining the size and number of harness rods is a trial-and-error process based on the allowable load of each specific size rod. The quantity of rods should be kept to even increments to maintain the assumed symmetry of the design process. The second part of the analysis involves evaluating the resulting stress developed in the front and back rings due to the rod loads. Front and back ring stresses are calculated based on uniform, equally spaced loads applied to the restraint rings using the applied analysis of Brockenbrough (1988).

The analyses for the assemblies in Table 7-3 are based on use of ASTM A36 steel (minimum yield strength, σ_Y, of 36 ksi) for the harness assembly rings. The selection of A36 steel was for convenience based on its common availability as plate material but is not intended to dissuade the designer from using other plate materials defined in ANSI/ AWWA C200. Given that the design pressures noted in Table 7-3 are maximum pressures to which the harness assemblies will be subjected, the allowable design stress for the front and back rings is 75 percent of the minimum yield strength of the rings' material. As ASTM A36 steel was the material chosen to design the rings in Table 7-3, the associated allowable stress is 27 ksi. The minimum cylinder thickness values, $T_{y\,min}$, presented in Table 7-3 for RR type designs as shown in Figure 7-25 are based on the larger of the following:

1. 0.135 in.
2. *D*/288
3. Calculated thickness based on an allowable stress of 17.5 ksi at a pressure equal to (design pressure)/1.5 for diameters ≤ 30 in. O.D., and 18 ksi at a pressure equal to (design pressure)/1.5 for diameters greater than 30-in. O.D.

The first two minimum thickness values shown above are not based on design principles. The first value represents a common minimum effective thickness for manufacturing double submerged arc-welded spiral pipe. The second value is a minimum practical thickness for handling if other factors do not govern. The minimum cylinder thickness values, $T_{y\,min}$, presented in Table 7-3 for P-type designs as shown in Figure 7-25 are as have been historically shown in this manual.

The diameters shown in Table 7-3 are nominal unless noted with "(OD)," which is the specific outside diameter of the steel cylinder for that size group. Where nominal diameters are shown, the calculated outside diameters will conservatively allow for the application of standard ANSI/AWWA C205 cement-mortar lining and two times the calculated minimum cylinder thicknesses shown in Table 7-3 without compromising the finished nominal diameter listed. The associated cylinder outside diameters, D_o, were calculated as follows:

- nominal diameter ≤ 20 in., nominal diameter plus 1 in.
- 20 in. < nominal diameter ≤ 36 in., nominal diameter plus 1.5 in.

Table 7-3 Tie rod schedule for harnessed joints

Pipe Diameter*	Design Pressure	Minimum Cylinder Thickness (T_{ymin}) Under Lug	Lug Type†	Tie Rod Diameter	Number Of Rods	Back Plate or Ring t_w	Front Plate or Ring t_w	Maximum Force
in.	*psi*	*in.*		*in.*		*in.*	*in.*	*lbf*
6⅝	50	0.188	P	⅝	2	0.188	0.188	1,724
(OD)	50	0.135	RR	⅝	2	0.188	0.188	1,724
	100	0.188	P	⅝	2	0.188	0.188	3,447
	100	0.135	RR	⅝	2	0.188	0.188	3,447
	150	0.193	P	⅝	2	0.188	0.188	5,171
	150	0.135	RR	⅝	2	0.188	0.188	5,171
	200	0.242	P	⅝	2	0.188	0.188	6,894
	200	0.135	RR	⅝	2	0.188	0.188	6,894
	250	0.282	P	⅝	2	0.188	0.188	8,618
	250	0.135	RR	⅝	2	0.188	0.188	8,618
	275	0.135	RR	⅝	2	0.188	0.188	9,480
	300	0.135	RR	⅝	2	0.188	0.188	10,341
8⅝	50	0.188	P	⅝	2	0.188	0.188	2,921
(OD)	50	0.135	RR	⅝	2	0.188	0.188	2,921
	100	0.194	P	⅝	2	0.188	0.188	5,843
	100	0.135	RR	⅝	2	0.188	0.188	5,843
	150	0.239	P	⅝	2	0.188	0.188	8,764
	150	0.135	RR	⅝	2	0.188	0.188	8,764
	200	0.291	P	⅝	2	0.188	0.188	11,685
	200	0.135	RR	⅝	2	0.188	0.188	11,685
	250	0.354	P	⅝	2	0.188	0.188	14,607
	250	0.135	RR	⅝	2	0.188	0.188	14,607
	275	0.135	RR	⅝	2	0.188	0.188	16,067
	300	0.135	RR	⅝	2	0.188	0.188	17,528
10¾	50	0.188	P	⅝	2	0.188	0.188	4,538
(OD)	50	0.135	RR	⅝	2	0.188	0.188	4,538
	100	0.242	P	⅝	2	0.188	0.188	9,076
	100	0.135	RR	⅝	2	0.188	0.188	9,076
	150	0.312	P	⅝	2	0.188	0.188	13,614
	150	0.135	RR	⅝	2	0.188	0.188	13,614

*Pipe diameters noted are nominal unless specifically noted as "OD." For nominal diameters, the outside diameter used in the calculation of the lug assembly is equal to the following: the nominal diameter plus 1 in. for pipe sizes ≤ 20 in.; the nominal diameter plus 1.5 in. for sizes > 20 in., but ≤36 in.; the nominal diameter plus 2 in. for sizes > 36 in., but ≤ 96 in.; and the nominal diameter plus 2.5 in. for sizes > 96 in. The minimum cylinder thickness is based on the larger of the following: 0.135 in.; D/288, and the calculated thickness based on a pressure equal to (design pressure)/1.5 and an allowable stress of 17.5 ksi and 18 ksi for diameters ≤ 30 in. OD. and > 30 in. OD, respectively.

†Lug types are defined as either individual plate lugs (P) or lug assemblies with both front and back rings (RR). Where P-type lugs are shown, RR-type lugs are acceptable.

Note: It is not recommended that harnessed flexible couplings be located immediately adjacent to pumps as this may cause undue stress on the pumps and pump base. If wrappers or pads are used, the minimum width or length shall not be less than the *A* or *X* dimensions in Figure 7-25, plus $1.56\sqrt{r_o T_y}$, or not less than *A* or *X* plus 2 in., whichever is greater.

Table continued next page

Table 7-3 Tie rod schedule for harnessed joints (continued)

Pipe Diameter*	Design Pressure	Minimum Cylinder Thickness ($T_{y\,min}$) Under Lug	Lug	Tie Rod Diameter	Number	Back Plate or Ring t_w	Front Plate or Ring t_w	Maximum Force
in.	*psi*	*in.*	Type†	*in.*	Of Rods	*in.*	*in.*	*lbf*
	200	0.386	P	¾	2	0.188	0.188	18,153
	200	0.135	RR	⅝	2	0.188	0.188	18,153
	250	0.466	P	¾	2	0.188	0.188	22,691
	250	0.135	RR	¾	2	0.188	0.188	22,691
	275	0.135	RR	¾	2	0.188	0.188	24,960
	300	0.135	RR	⅞	2	0.188	0.188	27,229
12¾	50	0.188	P	⅝	2	0.188	0.188	6,384
(OD)	50	0.135	RR	⅝	2	0.188	0.188	6,384
	100	0.286	P	⅝	2	0.188	0.188	12,768
	100	0.135	RR	⅝	2	0.188	0.188	12,768
	150	0.361	P	¾	2	0.188	0.188	19,151
	150	0.135	RR	¾	2	0.188	0.188	19,151
	200	0.447	P	⅞	2	0.188	0.188	25,535
	200	0.135	RR	¾	2	0.188	0.188	25,535
	250	0.540	P	⅞	2	0.188	0.188	31,919
	250	0.135	RR	⅞	2	0.188	0.188	31,919
	275	0.135	RR	⅞	2	0.188	0.188	35,111
	300	0.135	RR	1	2	0.188	0.188	38,303
14	50	0.135	RR	⅝	2	0.188	0.188	7,697
(OD)	100	0.135	RR	⅝	2	0.188	0.188	15,394
	150	0.135	RR	¾	2	0.188	0.188	23,091
	200	0.135	RR	⅞	2	0.188	0.188	30,788
	250	0.135	RR	1	2	0.188	0.188	38,485
	275	0.135	RR	1⅛	2	0.188	0.188	42,333
	300	0.135	RR	1⅛	2	0.188	0.188	46,181
16	50	0.135	RR	⅝	2	0.188	0.188	10,053
(OD)	100	0.135	RR	¾	2	0.188	0.188	20,106
	150	0.135	RR	⅞	2	0.188	0.188	30,159
	200	0.135	RR	1⅛	2	0.188	0.188	40,212
	250	0.135	RR	1¼	2	0.188	0.188	50,265
	275	0.135	RR	1¼	2	0.188	0.188	55,292

*Pipe diameters noted are nominal unless specifically noted as "OD." For nominal diameters, the outside diameter used in the calculation of the lug assembly is equal to the following: the nominal diameter plus 1 in. for pipe sizes ≤ 20 in.; the nominal diameter plus 1.5 in. for sizes > 20 in., but ≤36 in.; the nominal diameter plus 2 in. for sizes > 36 in., but ≤ 96 in.; and the nominal diameter plus 2.5 in. for sizes > 96 in. The minimum cylinder thickness is based on the larger of the following: 0.135 in.; D/288, and the calculated thickness based on a pressure equal to (design pressure)/1.5 and an allowable stress of 17.5 ksi and 18 ksi for diameters ≤ 30 in. OD. and > 30 in. OD, respectively.

†Lug types are defined as either individual plate lugs (P) or lug assemblies with both front and back rings (RR). Where P-type lugs are shown, RR-type lugs are acceptable.

Note: It is not recommended that harnessed flexible couplings be located immediately adjacent to pumps as this may cause undue stress on the pumps and pump base. If wrappers or pads are used, the minimum width or length shall not be less than the *A* or *X* dimensions in Figure 7-25, plus 1.56 $\sqrt{r_o T_y}$, or not less than *A* or *X* plus 2 in., whichever is greater.

Table continued next page

Table 7-3 Tie rod schedule for harnessed joints (continued)

Pipe Diameter*	Design Pressure	Minimum Cylinder Thickness (T_{ymin}) Under Lug	Lug Type†	Tie Rod Diameter	Number Of Rods	Back Plate or Ring t_w	Front Plate or Ring t_w	Maximum Force
in.	*psi*	*in.*		*in.*		*in.*	*in.*	*lbf*
	300	0.135	RR	1¼	2	0.188	0.188	60,319
16	50	0.135	RR	⅝	2	0.188	0.188	11,349
	100	0.135	RR	⅞	2	0.188	0.188	22,698
	150	0.135	RR	1⅛	2	0.188	0.188	34,047
	200	0.135	RR	1⅛	2	0.188	0.188	45,396
	250	0.135	RR	1¼	2	0.188	0.188	56,745
	275	0.135	RR	1¼	2	0.188	0.188	62,420
	300	0.135	RR	1¼	2	0.188	0.188	68,094
18	50	0.135	RR	⅝	2	0.188	0.188	12,723
(OD)	100	0.135	RR	1	2	0.188	0.188	25,447
	150	0.135	RR	1⅛	2	0.188	0.188	38,170
	200	0.135	RR	1¼	2	0.188	0.188	50,894
	250	0.135	RR	1¼	2	0.188	0.188	63,617
	275	0.135	RR	1⅜	2	0.188	0.188	69,979
	300	0.135	RR	1⅜	2	0.188	0.188	76,341
18	50	0.135	RR	⅝	2	0.188	0.188	14,176
	100	0.135	RR	1⅛	2	0.188	0.188	28,353
	150	0.135	RR	1⅛	2	0.188	0.188	42,529
	200	0.135	RR	1¼	2	0.188	0.188	56,706
	250	0.135	RR	1⅜	2	0.188	0.188	70,882
	275	0.135	RR	1½	2	0.188	0.188	77,970
	300	0.135	RR	1½	2	0.188	0.188	85,059
20	50	0.135	RR	⅞	2	0.188	0.188	15,708
(OD)	100	0.135	RR	1⅛	2	0.188	0.188	31,416
	150	0.135	RR	1¼	2	0.188	0.188	47,124
	200	0.135	RR	1⅜	2	0.188	0.188	62,832
	250	0.135	RR	1½	2	0.188	0.188	78,540
	275	0.135	RR	1½	2	0.188	0.188	86,394
	300	0.135	RR	1⅝	2	0.189	0.188	94,248
20	50	0.135	RR	⅞	2	0.188	0.188	17,318
	100	0.135	RR	1⅛	2	0.188	0.188	34,636
	150	0.135	RR	1⅜	2	0.188	0.188	51,954

*Pipe diameters noted are nominal unless specifically noted as "OD." For nominal diameters, the outside diameter used in the calculation of the lug assembly is equal to the following: the nominal diameter plus 1 in. for pipe sizes ≤ 20 in.; the nominal diameter plus 1.5 in. for sizes > 20 in., but ≤36 in.; the nominal diameter plus 2 in. for sizes > 36 in., but ≤ 96 in.; and the nominal diameter plus 2.5 in. for sizes > 96 in. The minimum cylinder thickness is based on the larger of the following: 0.135 in.; D/288, and the calculated thickness based on a pressure equal to (design pressure)/1.5 and an allowable stress of 17.5 ksi and 18 ksi for diameters ≤ 30 in. OD. and > 30 in. OD, respectively.

†Lug types are defined as either individual plate lugs (P) or lug assemblies with both front and back rings (RR). Where P-type lugs are shown, RR-type lugs are acceptable.

Note: It is not recommended that harnessed flexible couplings be located immediately adjacent to pumps as this may cause undue stress on the pumps and pump base. If wrappers or pads are used, the minimum width or length shall not be less than the *A* or *X* dimensions in Figure 7-25, plus 1.56 $\sqrt{r_o T_y}$, or not less than *A* or *X* plus 2 in., whichever is greater.

Table continued next page

Table 7-3 Tie rod schedule for harnessed joints (continued)

Pipe Diameter*	Design Pressure	Minimum Cylinder Thickness ($T_{y\,min}$) Under Lug	Lug Type†	Tie Rod Diameter	Number Of Rods	Back Plate or Ring t_w	Front Plate or Ring t_w	Maximum Force
in.	*psi*	*in.*		*in.*		*in.*	*in.*	*lbf*
	200	0.135	RR	1½	2	0.188	0.188	69,272
	250	0.135	RR	1½	2	0.188	0.188	86,590
	275	0.135	RR	1⅝	2	0.188	0.188	95,249
	300	0.135	RR	1¾	2	0.193	0.188	103,908
24	50	0.135	RR	⅝	4	0.188	0.188	22,619
(OD)	100	0.135	RR	¾	4	0.188	0.188	45,239
	150	0.135	RR	⅞	4	0.188	0.188	67,858
	200	0.135	RR	1	4	0.188	0.188	90,478
	250	0.135	RR	1⅛	4	0.188	0.188	113,097
	275	0.135	RR	1¼	4	0.188	0.188	124,407
	300	0.137	RR	1¼	4	0.188	0.188	135,717
24	50	0.135	RR	⅝	4	0.188	0.188	25,535
	100	0.135	RR	¾	4	0.188	0.188	51,071
	150	0.135	RR	1	4	0.188	0.188	76,606
	200	0.135	RR	1⅛	4	0.188	0.188	102,141
	250	0.135	RR	1⅛	4	0.188	0.188	127,676
	275	0.135	RR	1¼	4	0.188	0.188	140,444
	300	0.146	RR	1¼	4	0.188	0.188	153,212
26	50	0.135	RR	⅝	4	0.188	0.188	26,546
(OD)	100	0.135	RR	¾	4	0.188	0.188	53,093
	150	0.135	RR	1	4	0.188	0.188	79,639
	200	0.135	RR	1⅛	4	0.188	0.188	106,186
	250	0.135	RR	1¼	4	0.188	0.188	132,732
	275	0.136	RR	1¼	4	0.188	0.188	146,006
	300	0.149	RR	1¼	4	0.188	0.188	159,279
30	50	0.135	RR	⅝	4	0.188	0.188	35,343
(OD)	100	0.135	RR	⅞	4	0.188	0.188	70,686
	150	0.135	RR	1⅛	4	0.188	0.188	106,029
	200	0.135	RR	1¼	4	0.188	0.188	141,372
	250	0.143	RR	1⅜	4	0.188	0.188	176,715
	275	0.157	RR	1⅜	4	0.188	0.188	194,386

*Pipe diameters noted are nominal unless specifically noted as "OD." For nominal diameters, the outside diameter used in the calculation of the lug assembly is equal to the following: the nominal diameter plus 1 in. for pipe sizes ≤ 20 in.; the nominal diameter plus 1.5 in. for sizes > 20 in., but ≤36 in.; the nominal diameter plus 2 in. for sizes > 36 in., but ≤ 96 in.; and the nominal diameter plus 2.5 in. for sizes > 96 in. The minimum cylinder thickness is based on the larger of the following: 0.135 in.; D/288, and the calculated thickness based on a pressure equal to (design pressure)/1.5 and an allowable stress of 17.5 ksi and 18 ksi for diameters ≤ 30 in. OD. and > 30 in. OD, respectively.

†Lug types are defined as either individual plate lugs (P) or lug assemblies with both front and back rings (RR). Where P-type lugs are shown, RR-type lugs are acceptable.

Note: It is not recommended that harnessed flexible couplings be located immediately adjacent to pumps as this may cause undue stress on the pumps and pump base. If wrappers or pads are used, the minimum width or length shall not be less than the *A* or *X* dimensions in Figure 7-25, plus 1.56 $\sqrt{r_o T_y}$, or not less than *A* or *X* plus 2 in., whichever is greater.

Table continued next page

Table 7-3 Tie rod schedule for harnessed joints (continued)

Pipe Diameter*	Design Pressure	Minimum Cylinder Thickness ($T_{y\,min}$) Under Lug	Lug Type†	Tie Rod Diameter	Number Of Rods	Back Plate or Ring t_w	Front Plate or Ring t_w	Maximum Force
in.	*psi*	*in.*		*in.*		*in.*	*in.*	*lbf*
	300	0.171	RR	1½	4	0.188	0.188	212,058
30	50	0.135	RR	¾	4	0.188	0.188	38,966
	100	0.135	RR	1	4	0.188	0.188	77,931
	150	0.135	RR	1⅛	4	0.188	0.188	116,897
	200	0.135	RR	1¼	4	0.188	0.188	155,862
	250	0.146	RR	1⅜	4	0.188	0.188	194,828
	275	0.160	RR	1½	4	0.188	0.188	214,311
	300	0.175	RR	1½	4	0.188	0.188	233,793
36	50	0.135	RR	⅞	4	0.188	0.188	55,223
	100	0.135	RR	1¼	4	0.188	0.188	110,447
	150	0.135	RR	1⅜	4	0.188	0.188	165,670
	200	0.139	RR	1½	4	0.188	0.188	220,893
	250	0.174	RR	1⅝	4	0.188	0.188	276,117
	275	0.191	RR	1¾	4	0.188	0.188	303,728
	300	0.208	RR	1¾	4	0.188	0.188	331,340
42	50	0.153	RR	1⅛	4	0.188	0.188	76,027
	100	0.153	RR	1½	4	0.188	0.188	152,053
	150	0.153	RR	1⅝	4	0.188	0.188	228,080
	200	0.163	RR	1½	6	0.188	0.188	304,106
	250	0.204	RR	1⅝	6	0.188	0.188	380,133
	275	0.224	RR	1⅝	6	0.188	0.188	418,146
	300	0.244	RR	1¾	6	0.188	0.188	456,159
48	50	0.174	RR	1⅜	4	0.188	0.188	98,175
	100	0.174	RR	1⅝	4	0.188	0.188	196,350
	150	0.174	RR	1⅜	6	0.188	0.188	294,524
	200	0.185	RR	1⅝	6	0.188	0.188	392,699
	250	0.231	RR	1¾	6	0.188	0.188	490,874
	275	0.255	RR	1⅞	6	0.188	0.188	539,961
	300	0.278	RR	1¾	8	0.188	0.188	589,049
54	50	0.194	RR	1½	4	0.188	0.188	123,150
	100	0.194	RR	1⅜	6	0.188	0.188	246,301

*Pipe diameters noted are nominal unless specifically noted as "OD." For nominal diameters, the outside diameter used in the calculation of the lug assembly is equal to the following: the nominal diameter plus 1 in. for pipe sizes ≤ 20 in.; the nominal diameter plus 1.5 in. for sizes > 20 in., but ≤36 in.; the nominal diameter plus 2 in. for sizes > 36 in., but ≤ 96 in.; and the nominal diameter plus 2.5 in. for sizes > 96 in. The minimum cylinder thickness is based on the larger of the following: 0.135 in.; D/288, and the calculated thickness based on a pressure equal to (design pressure)/1.5 and an allowable stress of 17.5 ksi and 18 ksi for diameters ≤ 30 in. OD. and > 30 in. OD, respectively.

†Lug types are defined as either individual plate lugs (P) or lug assemblies with both front and back rings (RR). Where P-type lugs are shown, RR-type lugs are acceptable.

Note: It is not recommended that harnessed flexible couplings be located immediately adjacent to pumps as this may cause undue stress on the pumps and pump base. If wrappers or pads are used, the minimum width or length shall not be less than the *A* or *X* dimensions in Figure 7-25, plus 1.56 $\sqrt{r_o T_y}$, or not less than *A* or *X* plus 2 in., whichever is greater.

Table continued next page

Table 7-3 Tie rod schedule for harnessed joints (continued)

Pipe Diameter* in.	Design Pressure psi	Minimum Cylinder Thickness (T_{ymin}) Under Lug in.	Lug Type†	Tie Rod Diameter in.	Number Of Rods	Back Plate or Ring t_w in.	Front Plate or Ring t_w in.	Maximum Force lbf
	150	0.194	RR	1⅝	6	0.188	0.188	369,451
	200	0.207	RR	1¾	6	0.188	0.188	492,602
	250	0.259	RR	1¾	8	0.188	0.188	615,752
	275	0.285	RR	1⅞	8	0.188	0.188	677,327
60	50	0.215	RR	1⅛	6	0.188	0.188	150,954
	100	0.215	RR	1½	6	0.188	0.188	301,907
	150	0.215	RR	1¾	6	0.188	0.188	452,861
	200	0.230	RR	1¾	8	0.188	0.188	603,814
	250	0.287	RR	1¾	10	0.188	0.188	754,768
	275	0.316	RR	1¾	10	0.188	0.188	830,244
66	50	0.236	RR	1¼	6	0.188	0.188	181,584
	100	0.236	RR	1⅜	8	0.188	0.188	363,168
	150	0.236	RR	1⅝	8	0.188	0.188	544,752
	200	0.252	RR	1⅞	8	0.188	0.188	726,336
	250	0.315	RR	1⅞	10	0.188	0.188	907,920
	275	0.346	RR	2	10	0.188	0.188	998,712
72	50	0.257	RR	1⅜	6	0.188	0.188	215,042
	100	0.257	RR	1¾	6	0.188	0.188	430,084
	150	0.257	RR	1¾	8	0.188	0.188	645,126
	200	0.274	RR	1⅞	10	0.188	0.188	860,168
	250	0.343	RR	1⅞	12	0.188	0.188	1,075,210
	275	0.377	RR	2	12	0.188	0.188	1,182,731
78	50	0.278	RR	1½	6	0.188	0.188	251,327
	100	0.278	RR	1½	10	0.188	0.188	502,655
	150	0.278	RR	1¾	10	0.188	0.188	753,982
	200	0.296	RR	1⅞	12	0.188	0.188	1,005,310
	250	0.370	RR	2¼	10	0.188	0.188	1,256,637
	275	0.407	RR	2¼	10	0.188	0.188	1,382,301
84	50	0.299	RR	1⅝	6	0.188	0.188	290,440
	100	0.299	RR	1⅝	10	0.188	0.188	580,880
	150	0.299	RR	1⅞	10	0.188	0.188	871,321

*Pipe diameters noted are nominal unless specifically noted as "OD." For nominal diameters, the outside diameter used in the calculation of the lug assembly is equal to the following: the nominal diameter plus 1 in. for pipe sizes ≤ 20 in.; the nominal diameter plus 1.5 in. for sizes > 20 in., but ≤36 in.; the nominal diameter plus 2 in. for sizes > 36 in., but ≤ 96 in.; and the nominal diameter plus 2.5 in. for sizes > 96 in. The minimum cylinder thickness is based on the larger of the following: 0.135 in.; D/288, and the calculated thickness based on a pressure equal to (design pressure)/1.5 and an allowable stress of 17.5 ksi and 18 ksi for diameters ≤ 30 in. OD. and > 30 in. OD, respectively.

†Lug types are defined as either individual plate lugs (P) or lug assemblies with both front and back rings (RR). Where P-type lugs are shown, RR-type lugs are acceptable.

Note: It is not recommended that harnessed flexible couplings be located immediately adjacent to pumps as this may cause undue stress on the pumps and pump base. If wrappers or pads are used, the minimum width or length shall not be less than the *A* or *X* dimensions in Figure 7-25, plus 1.56 $\sqrt{r_o T_y}$, or not less than *A* or *X* plus 2 in., whichever is greater.

Table continued next page

Table 7-3 Tie rod schedule for harnessed joints (continued)

Pipe Diameter*	Design Pressure	Minimum Cylinder Thickness (T_{ymin}) Under Lug	Lug	Tie Rod Diameter	Number	Back Plate or Ring t_w	Front Plate or Ring t_w	Maximum Force
in.	*psi*	*in.*	Type†	*in.*	Of Rods	*in.*	*in.*	*lbf*
	200	0.319	RR	2	12	0.188	0.188	1,161,761
	250	0.398	RR	2¼	12	0.188	0.188	1,452,201
	275	0.438	RR	2¼	12	0.188	0.188	1,597,421
90	50	0.319	RR	1¾	6	0.188	0.188	332,381
	100	0.319	RR	1¾	8	0.188	0.188	664,761
	150	0.319	RR	1⅞	12	0.188	0.188	997,142
	200	0.341	RR	2¼	10	0.188	0.188	1,329,522
	250	0.426	RR	2¼	14	0.188	0.188	1,661,903
	275	0.469	RR	2¼	14	0.188	0.188	1,828,093
96	50	0.340	RR	1½	8	0.188	0.188	377,148
	100	0.340	RR	1⅞	8	0.188	0.188	754,296
	150	0.340	RR	2¼	8	0.188	0.188	1,131,445
	200	0.363	RR	2¼	12	0.188	0.188	1,508,593
	250	0.454	RR	2½	12	0.188	0.188	1,885,741
	275	0.499	RR	2½	12	0.188	0.188	2,074,315
102	50	0.363	RR	1½	8	0.188	0.188	428,837
	100	0.363	RR	1¾	12	0.188	0.188	857,674
	150	0.363	RR	2¼	10	0.188	0.188	1,286,512
	200	0.387	RR	2¼	14	0.188	0.188	1,715,349
	250	0.484	RR	2½	14	0.188	0.188	2,144,186
	275	0.532	RR	2½	14	0.250	0.250	2,358,605
108	50	0.384	RR	1¾	8	0.188	0.188	479,495
	100	0.384	RR	1¾	12	0.188	0.188	958,991
	150	0.384	RR	2¼	12	0.188	0.188	1,438,486
	200	0.409	RR	2½	12	0.188	0.188	1,917,982
	250	0.512	RR	2½	14	0.250	0.250	2,397,477
	275	0.563	RR	2½	16	0.250	0.250	2,637,225
114	50	0.405	RR	1¾	8	0.188	0.188	532,981
	100	0.405	RR	1⅞	12	0.188	0.188	1,065,962
	150	0.405	RR	2¼	12	0.188	0.188	1,598,943
	200	0.431	RR	2½	12	0.188	0.188	2,131,924

*Pipe diameters noted are nominal unless specifically noted as "OD." For nominal diameters, the outside diameter used in the calculation of the lug assembly is equal to the following: the nominal diameter plus 1 in. for pipe sizes ≤ 20 in.; the nominal diameter plus 1.5 in. for sizes > 20 in., but ≤36 in.; the nominal diameter plus 2 in. for sizes > 36 in., but ≤ 96 in.; and the nominal diameter plus 2.5 in. for sizes > 96 in. The minimum cylinder thickness is based on the larger of the following: 0.135 in.; D/288, and the calculated thickness based on a pressure equal to (design pressure)/1.5 and an allowable stress of 17.5 ksi and 18 ksi for diameters ≤ 30 in. OD. and > 30 in. OD, respectively.

†Lug types are defined as either individual plate lugs (P) or lug assemblies with both front and back rings (RR). Where P-type lugs are shown, RR-type lugs are acceptable.

Note: It is not recommended that harnessed flexible couplings be located immediately adjacent to pumps as this may cause undue stress on the pumps and pump base. If wrappers or pads are used, the minimum width or length shall not be less than the *A* or *X* dimensions in Figure 7-25, plus 1.56 $\sqrt{r_o T_y}$, or not less than *A* or *X* plus 2 in., whichever is greater.

Table continued next page

Table 7-3 Tie rod schedule for harnessed joints (continued)

Pipe Diameter*	Design Pressure	Minimum Cylinder Thickness ($T_{y\,min}$) Under Lug	Lug Type†	Tie Rod Diameter	Number Of Rods	Back Plate or Ring t_w	Front Plate or Ring t_w	Maximum Force
in.	*psi*	*in.*		*in.*		*in.*	*in.*	*lbf*
	250	0.539	RR	2¾	14	0.250	0.250	2,664,905
	275	0.593	RR	2¾	16	0.250	0.250	2,931,396
120	50	0.425	RR	1½	10	0.188	0.188	589,294
	100	0.425	RR	1⅞	14	0.188	0.188	1,178,588
	150	0.425	RR	2½	10	0.188	0.188	1,767,882
	200	0.454	RR	2¾	12	0.188	0.188	2,357,176
	250	0.567	RR	2¾	16	0.250	0.250	2,946,470
	275	0.624	RR	3	14	0.250	0.250	3,241,117
126	50	0.446	RR	1⅝	10	0.188	0.188	648,435
	100	0.446	RR	1⅞	14	0.188	0.188	1,296,869
	150	0.446	RR	2½	12	0.188	0.188	1,945,304
	200	0.476	RR	2½	16	0.188	0.188	2,593,738
	250	0.595	RR	3	14	0.250	0.250	3,242,173
	275	0.654	RR	3	16	0.250	0.250	3,566,390
132	50	0.467	RR	1¾	10	0.188	0.188	710,402
	100	0.467	RR	1⅞	16	0.188	0.188	1,420,805
	150	0.467	RR	2½	12	0.188	0.188	2,131,207
	200	0.498	RR	2¾	16	0.188	0.188	2,841,610
	250	0.623	RR	3	16	0.250	0.250	3,552,012
	275	0.685	RR	3	18	0.250	0.250	3,907,214
138	50	0.488	RR	1¾	10	0.188	0.188	775,198
	100	0.488	RR	2¼	12	0.188	0.188	1,550,396
	150	0.488	RR	2½	14	0.188	0.188	2,325,593
	200	0.520	RR	2¾	16	0.250	0.250	3,100,791
	250	0.650	RR	2¾	20	0.250	0.250	3,875,989
	275	0.716	RR	3	20	0.250	0.250	4,263,588
144	50	0.509	RR	1⅞	10	0.250	0.250	842,821
	100	0.509	RR	2¼	12	0.250	0.250	1,685,641
	150	0.509	RR	2½	16	0.250	0.250	2,528,462
	200	0.543	RR	2¾	18	0.250	0.250	3,371,282
	250	0.678	RR	3	18	0.250	0.250	4,214,103
	275	0.746	RR	3	20	0.250	0.250	4,635,513

*Pipe diameters noted are nominal unless specifically noted as "OD." For nominal diameters, the outside diameter used in the calculation of the lug assembly is equal to the following: the nominal diameter plus 1 in. for pipe sizes ≤ 20 in.; the nominal diameter plus 1.5 in. for sizes > 20 in., but ≤36 in.; the nominal diameter plus 2 in. for sizes > 36 in., but ≤ 96 in.; and the nominal diameter plus 2.5 in. for sizes > 96 in. The minimum cylinder thickness is based on the larger of the following: 0.135 in.; D/288, and the calculated thickness based on a pressure equal to (design pressure)/1.5 and an allowable stress of 17.5 ksi and 18 ksi for diameters ≤ 30 in. OD. and > 30 in. OD, respectively.

†Lug types are defined as either individual plate lugs (P) or lug assemblies with both front and back rings (RR). Where P-type lugs are shown, RR-type lugs are acceptable.

Note: It is not recommended that harnessed flexible couplings be located immediately adjacent to pumps as this may cause undue stress on the pumps and pump base. If wrappers or pads are used, the minimum width or length shall not be less than the *A* or *X* dimensions in Figure 7-25, plus 1.56 $\sqrt{r_o T_y}$, or not less than *A* or *X* plus 2 in., whichever is greater.

Table 7-4 Dimensions of joint harness tie rods and lugs for rubber-gasketed joints*†‡

Rod Diameter	T_s	Lug Type	*A*	*Y*	*W*	*X*	*HB*	*E*§	*HF*	Hole Diameter**
in.	*in.*		*in.*	*in.*	*in.*	*in.*	*in.*	*in.*	*in.*	*in.*
⅝	⅜	P	5	5	1⅜	5	3⅞	3	2	¾
¾	⅜	P	5	5	1½	5	4⅛	3⅛	2	⅞
⅞	½	P	5½	5	1⅝	5	4¼	3⅛	2	1
⅝	⅜	RR	5	Ring	1⅜	Ring	3⅞	3	2	¾
¾	⅜	RR	5	Ring	1½	Ring	4⅛	3⅛	2	⅞
⅞	½	RR	5½	Ring	1⅝	Ring	4¼	3⅛	2	1
1	½	RR	5¾	Ring	1¾	Ring	4⅝	3¼	2	1⅛
1⅛	½	RR	7	Ring	1⅞	Ring	5¼	3⅝	2½	1¼
1¼	⅝	RR	7½	Ring	2	Ring	5¼	3¾	2½	1⅜
1⅜	⅝	RR	8¾	Ring	2⅛	Ring	5¾	3¾	2½	1½
1½	¾	RR	10	Ring	2¼	Ring	6	3⅞	2½	1⅝
1⅝	¾	RR	10¾	Ring	2⅜	Ring	6¼	3⅞	2½	1¾
1¾	⅞	RR	12	Ring	2½	Ring	6½	4	2½	1⅞
1⅞	⅞	RR	13	Ring	2⅝	Ring	6⅝	4	2½	2
2	1	RR	14	Ring	2¾	Ring	7	4¼	2½	2¼
2¼	1	RR	15¾	Ring	3	Ring	7¾	4 9/16	2½	2½
2½	1¼	RR	17½	Ring	3¼	Ring	8¼	4¾	2½	2¾
2¾	1¾	RR	19¼	Ring	3½	Ring	8¾	4⅞	2½	3
3	2	RR	21	Ring	3¾	Ring	9¼	5	2½	3¼

* Use these dimensions with Figure 7-25 and Tables 7-3 and 7-5.

† See section on Joint Harnesses for design conditions covering maximum allowable pressure and placement spacing of the rods around the circumference of the pipe. The designs represented in Tables 7-3 are to resist longitudinal thrust only. Considerations for additional vertical, horizontal, or eccentric loadings are beyond the scope of this application.

‡ All fillet welds shall meet the minimum requirements of the American Institute of Steel Constructions specifications, with dimensions as noted in Table 7-5.

§ Dimension *E* in the above table has been adequate to provide clearance between the tie rod and the OD of the assembled coupling where the OD of the coupling is 4-in. to 5-in. larger than the OD of the pipe, as normally found in standard couplings through 72-in. diameter. For sleeve-type couplings designed for higher pressure and for diameters over 72 in., the *E* dimension should be checked by the designer for adequate clearance of the tie rod over the OD of the assembled coupling to be provided by the manufacturer.

** For harness rods 2-in. diameter and larger, the harness lug hole diameter is set at ¼-in. larger than the rod to allow for additional flexibility during assembly.

Table 7-5 Minimum fillet weld size for harness lug assembly and anchor ring attachment

Thickness, *t*, of thinner material being joined	Minimum fillet weld size, t_w or t_g
in.	*in.*
$t \leq$ ½ in.	3/16 in.
½ in. $< t \leq$ ¾ in.	¼ in.
$t >$ ¾ in.	5/16 in.

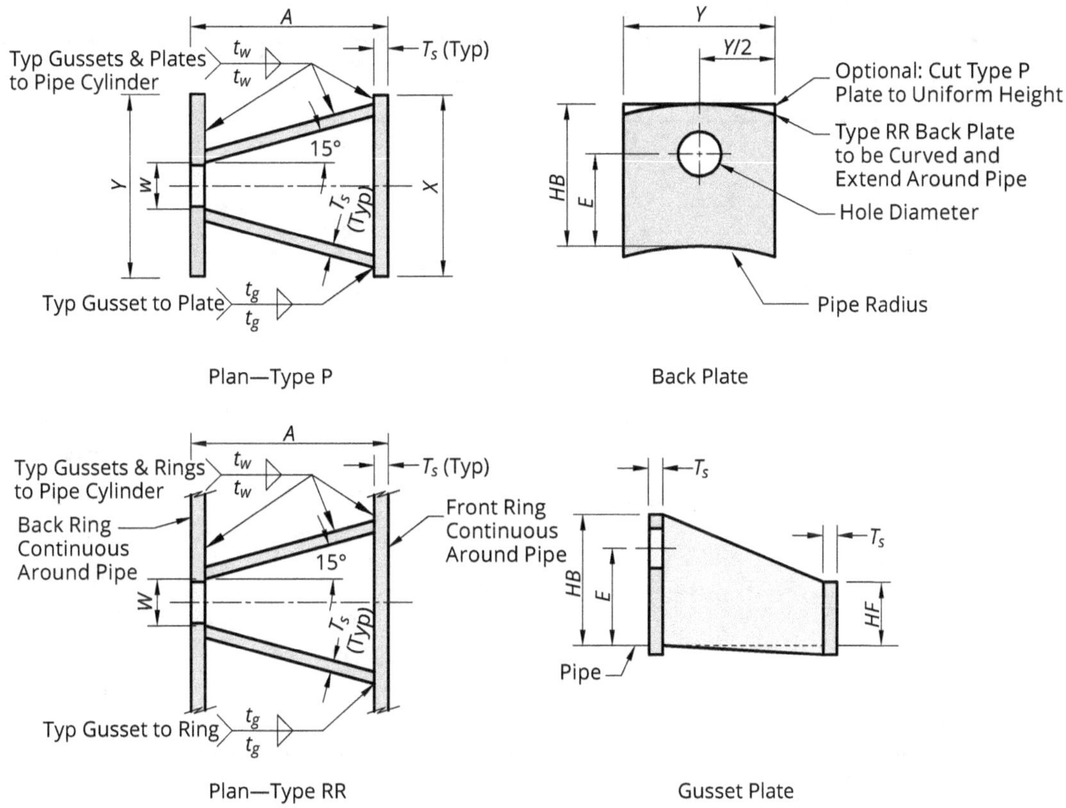

Notes:
1. See Tables 7-3 and 7-4 for dimensions.
2. See Joint Harnesses section for design conditions.
3. For harness lug type-RR, the gusset plates between the back ring and the front ring may be perpendicular to the front and back rings with a minimum clear distance between each pair of gusset plates dimension *W*.

Figure 7-25 Harness lug (top) and ring (bottom) detail

- 36 in. < nominal diameter ≤ 96 in., nominal diameter plus 2 in.
- nominal diameter > 96 in., nominal diameter plus 2.5 in.

The use of a smaller cylinder outside diameter for a given design represented in Table 7-3 is acceptable and yields a more conservative overall design. The calculated D_o was used for all design pressures in a nominal diameter group.

When the chosen design pressure from Table 7-3 is less than 1.5 times the working pressure, the design must be checked at working pressure to verify that the harness ring stresses do not exceed 50 percent of σ_Y of the ring material. The "Minimum Cylinder Thickness ($T_{y\,min}$) Under Lug" defined in Table 7-3 is the minimum thickness allowed under either ring assemblies or individual lugs. If wrappers or pads are used, $T_{y\,min}$ shall be the thickness of the wrapper or pad. Where $T_{y\,min}$ dictates the use of a wrapper due to insufficient parent pipe thickness, substitution of a cylinder with a thickness of at least $T_{y\,min}$ is acceptable in lieu of attaching a wrapper to the parent pipe, provided the cylinder meets the requirements of ANSI/AWWA C200 and the design requirements noted above.

The design process for the 48-in., 300-psi design pressure RR-type harness assembly in Table 7-3 is fully presented in appendix B for the reader's reference. The information shown in Table 7-3 is for a limited number of design pressure, material strength, and cylinder thickness combinations. The designer has the option to evaluate designs for other combinations in accordance with the process shown in appendix B. In such cases the

Table 7-6 Maximum allowable load per tie rod

Diameter		Net Area per Tie Rod*	Maximum Load per Tie Rod†
in.	Number of Threads per in.	*in.*^{2}	*lb*
⅝	11	0.226	9,040
¾	10	0.334	13,378
⅞	9	0.462	18,469
1	8	0.606	24,230
1⅛	8	0.790	31,618
1¼	8	1.000	39,988
1⅜	8	1.234	49,340
1½	8	1.492	59,674
1⅝	8	1.775	70,989
1¾	8	2.082	83,286
1⅞	8	2.414	96,565
2	8	2.771	110,825
2¼	8	3.557	142,292
2½	8	4.442	177,685
2¾	8	5.425	196,336
3	8	6.506	235,464

* The net area for tie rods has been calculated based on rods 1-in. diameter and larger having eight UN threads per inch, and rods smaller than 1-in. diameter having standard UNC threads.

† The maximum load per tie rod is based on an allowable stress in the rod of 40 ksi for rods ≤ 2.5-in. diameter and 36.2 ksi for rods > 2.5-in. diameter.

designer can use any steel material and grade listed in ANSI/AWWA C200, provided such material is available in a form conducive to the manufacturing process.

The information in Table 7-3 is based on harness rod data as follows: harness rods conforming to ASTM A193, Grade B7 or equal (see Table 7-6 for maximum allowable rod loads); nuts conforming to ASTM A194, Grade 2H or equal; lug material conforming to ASTM A36, Standard Specification for Carbon Structural Steel, or equal; stud bolts ⅝-in. through ⅞-in. diameter having UNC threads; stud bolts 1-in. diameter and larger having eight UN threads per inch; and a maximum allowable rod stress is equal to the minimum specified yield strength of the tie rod material divided by a safety factor of 2.625, at the design pressure noted. For ASTM A193 Grade B7 material and a safety factor of 2.625, the maximum allowable design stress in the tie rod is 40 ksi for rods less than or equal to 2.5-in. diameter and 36.2 ksi for rods larger than 2.5-in. diameter. The rod tensile area is defined as

$$\text{rod tensile stress area} = 0.7854[D - (0.9743/N)]^2 \quad \text{(Eq 7-9)}$$

Where:

D = nominal tie rod diameter, in.

N = number of threads per in.

The designer is cautioned regarding use of stainless-steel or other alloy steel rods and nuts in lieu of the ASTM A193 and ASTM A194 alloy steel materials defined above without an evaluation of resultant safety factors. Other steels exhibit different yield and tensile strengths compared to the defined alloy steel material. Indiscriminate replacement of the alloy steel fasteners with other steel fasteners will affect the inherent safety factor noted previously for the tie rods.

Harness lugs should be spaced equally around the pipe. Historically, unequal harness lug spacing has been used by some designers in certain circumstances. The suggested limitations and guidelines for use of unequal harness lug spacing are discussed in appendix C. Regardless of the spacing of the harness lugs, the pipe joint for which the lugs provide restraint must be fully assembled and in the desired position, including any necessary angular deflection, prior to tightening the nuts on the harness rods. In assembling the harness, the nuts shall be tightened gradually and equally at diametrically opposite sides until snug to prevent misalignment and provide the best potential for all rods to carry an equivalent load in service. The threads of the rods shall protrude a minimum of ½ in. from the nuts.

The end force values shown in Table 7-3 are the maximum values the harness assemblies are designed to withstand. The design pressure must include an anticipated allowance for transient pressure. The field test pressure must never exceed the design pressure.

Harness Lug Type-RR Attachment and Gusset Connection Fillet Weld Sizes

The harness lug type-RR attachment fillet weld size is calculated based on the design pressure of the associated harness ring assembly, but subject to the minimum sizes noted in Table 7-5 based on the harness lugs' and gusset plates' thicknesses and the steel cylinder thickness. For any individual lug, the effective angular length of fillet weld is limited to 30° or 360°/N_L, whichever is less. N_L is the number of lugs in a given harness assembly on a single pipe end. Conservatively, the fillet welds connecting the gusset plates to the steel cylinder have not been considered in the weld design and are to be sized as defined in Table 7-5. The fillet welds connecting the gusset plates to the front and back plates and rings are to be sized as defined in Table 7-5.

The resultant shear load that must be resisted by each circumferential fillet weld is given by

$$f_r = (f_b^2 + f_v^2)^{1/2} \qquad \text{(Eq 7-10)}$$

With:

$$f_b = \frac{M_r \alpha}{2\pi D_o \left(A - \dfrac{T_s}{2} \right)} \qquad \text{(Eq 7-11)}$$

$$M_r = \frac{p \pi D_o^2 E}{4{,}000 N_L} \qquad \text{(Eq 7-12)}$$

$$f_v = \frac{p D_o \alpha}{16{,}000 N_L} \qquad \text{(Eq 7-13)}$$

Where:

f_r = resultant shear force to be resisted by each front and back lug attachment fillet weld, kip/in.

f_b = unit shear force in each back lug attachment fillet weld to resist harness assembly bending moment, kip/in.

f_v = unit shear force in each front and back lug attachment fillet weld to resist longitudinal load from harness assembly, kip/in.

M_r = unit bending moment at lug, kip·in.

A = face-to-face dimension of harness lug assembly, in.

T_s = harness lug thickness, in.

p = design pressure, psi

α = angular influence factor = max[12, N_L]

D_o = pipe steel cylinder outside diameter, in.

E = height from cylinder outside diameter to harness rod centerline, in.

N_L = number of harness lugs

The sizes of the fillet welds are given by

$$t_w = \frac{f_r}{\left[(0.3)(\sigma_w)\left(\frac{\sqrt{2}}{2}\right)\right]}$$

$$t_g = \text{size per Table 7-5}$$

Where:

t_w = fillet weld size to attach lugs to steel cylinder, in.

σ_w = minimum tensile strength of welding electrode = 70 ksi

t_g = fillet weld size to attach gusset plates to front and back plates or rings, in.

Example 7-11: Harness Lug Type-RR Weld Attachment Design

Given a pipe with a 49.750-in. steel cylinder outside diameter, 0.304-in. wall thickness, six 1½-in. attached type-RR harness lugs, and a design pressure of 150 psi, evaluate the minimum size front and back ring fillet welds necessary to fabricate and attach the harness ring assembly to the steel cylinder.

Evaluate the appropriate angular influence factor, α, for $N_L = 6$,

$$\alpha = \max[12, 6] = 12$$

Calculate M_r,

From Table 7-4, E = 3⅞ in. for a 1½-in. lug.

$$M_r = \frac{150\pi(49.75)^2(3.875)}{4{,}000(6)} = 188.3 \text{ kip·in.}$$

Calculate f_b,

From Table 7-4, A = 10 in. and T_S = 0.75 in. for a 1½-in. lug.

$$f_b = \frac{188.3(12)}{2\pi(49.750)(10 - 0.750/2)} = 0.75 \text{ kip/in.}$$

Calculate f_v,

$$f_v = \frac{150(49.75)12}{16{,}000(6)} = 0.93 \text{ kip/in.}$$

Calculate f_r,

Back ring

$$f_r = (0.75^2 + 0.93^2)^{½} = 1.19 \text{ kip/in.}$$

Front ring

$$f_r = f_v = 0.93 \text{ kip/in.}$$

Then calculate t_w,

Back ring

$$t_w = \frac{1.19}{\left[(0.3)(70)\left(\frac{\sqrt{2}}{2}\right)\right]} = \frac{1.19}{14.849} = 0.08 \text{ in.}$$

Front ring

$$t_w = \frac{0.93}{\left[(0.3)(70)\left(\frac{\sqrt{2}}{2}\right)\right]} = \frac{0.93}{14.849} = 0.06 \text{ in.}$$

The weld sizes need to be checked against the minimum values in Table 7-5. The material thickness of the harness rings and gusset plates is ¾ in. and the steel cylinder thickness is 0.304 in. Therefore, from Table 7-5, the minimum fillet weld size for attaching the rings to the steel cylinder is 3⁄16 in., which is greater than both calculated weld sizes. From Table 7-5, the minimum weld size for attaching the gusset plates to the cylinder is 3⁄16 in., and the minimum fillet weld size for connecting the gusset plates to the rings is ¼ in.

Harness Lug Type-P Attachment and Gusset Connection Fillet Weld Sizes

The harness lug attachment fillet weld size is calculated based on the design pressure of the associated harness lug, but subject to the minimum sizes noted in Table 7-5 based on the harness lug plates' and gusset plates' thicknesses and the steel cylinder thickness. The fillet welds connecting the gusset plates to the steel cylinder are not considered in the design and are to be sized as defined in Table 7-5. The fillet welds connecting the gusset plates to the front and back plates are to be sized as defined in Table 7-5.

The resultant shear load that must be resisted by each of the front and back plate fillet welds is given by

$$f_r = (f_b^2 + f_v^2)^{1/2} \tag{Eq 7-14}$$

With:

$$f_b = \frac{M_r}{2Y\left(A - \frac{T_s}{2}\right)} \tag{Eq 7-15}$$

$$M_r = \frac{p\pi D_o^2 E}{4{,}000 N_L} \tag{Eq 7-16}$$

$$f_v = \frac{p\pi D_o^2}{8{,}000(X + Y)N_L} \tag{Eq 7-17}$$

Where:

f_r = resultant shear force to be resisted by each front and back plate attachment fillet weld, kip/in.

f_b = unit shear force in back plate attachment fillet weld to resist harness lug bending moment, kip/in.

f_v = unit shear force in each front and back plate attachment fillet weld to resist longitudinal load from lug, kip/in.

M_r = unit bending moment at lug, kip·in.

A = face-to-face dimension of harness ring assembly, in.

T_s = harness lug plate thickness, in.
p = design pressure, psi
D_o = pipe steel cylinder outside diameter, in.
E = height from cylinder outside diameter to harness rod centerline, in.
N_L = number of harness lugs
X = width of harness lug front plate, in.
Y = width of harness lug back plate, in.

The sizes of the fillet welds are given by

$$t_w = \frac{f_r}{\left[(0.3)(\sigma_w)\left(\frac{\sqrt{2}}{2}\right)\right]}$$

$$t_g = \text{size per Table 7-5}$$

Where:

t_w = fillet weld size to attach front, back, and gusset plates to steel cylinder, in.
σ_w = minimum tensile strength of welding electrode = 70 ksi
t_g = fillet weld size to attach gusset plates to front and back rings, in.

Example 7-12: Harness Lug Type-P Weld Attachment Design

Given a pipe with a 12.750-in. steel cylinder outside diameter, 0.375-in. wall thickness, two ¾-in. attached P-type lugs, and a design pressure of 150 psi, evaluate the minimum size fillet welds necessary to fabricate and attach the harness ring assembly to the steel cylinder.

Calculate M_r,
From Table 7-4, $E = 3\frac{1}{8}$ in. and $X = Y = 5$ in. for a ¾-in. lug.

$$M_r = \frac{150\pi(12.75)^2(3.125)}{4{,}000(2)} = 29.9 \text{ kip·in.}$$

Calculate f_b,
From Table 7-4, $A = 5$ in. and $T_S = 0.375$ in. for a ¾-in. lug.

$$f_b = \frac{29.9}{2(5)(5 - 0.375/2)} = 0.62 \text{ kip/in.}$$

Calculate f_v,

$$f_v = \frac{150\pi(12.75)^2}{8{,}000(5 + 5)(2)} = 0.48 \text{ kip/in.}$$

Calculate f_r,
Back plate

$$f_r = (0.62^2 + 0.48^2)^{1/2} = 0.78 \text{ kip/in.}$$

Front plate

$$f_r = f_v = 0.48 \text{ kip/in.}$$

Then calculate t_w,
Back plate

$$t_w = \frac{0.78}{\left[(0.3)(70)\left(\frac{\sqrt{2}}{2}\right)\right]} = \frac{0.78}{14.849} = 0.05 \text{ in.}$$

Front plate

$$t_w = \frac{0.48}{\left[(0.3)(70)\left(\frac{\sqrt{2}}{2}\right)\right]} = \frac{0.48}{14.849} = 0.03 \text{ in.}$$

The weld size needs to be checked against the minimum values in Table 7-5. The material thickness of the harness rings and gusset plates is ⅜ in. and the steel cylinder thickness is ⅜ in. Therefore, from Table 7-5, the minimum fillet weld size for attaching the rings to the steel cylinder is $^3\!/_{16}$ in., which is greater than the calculated weld size. From Table 7-5, the minimum weld size for attaching the gusset plates to the cylinder is $^3\!/_{16}$ in., and the minimum fillet weld size for connecting the gusset plates to the front and back plates is $^3\!/_{16}$ in.

ANCHOR RINGS

An anchor ring for use in a concrete anchor block or concrete wall is illustrated in Figure 7-26.

A ring shall be designed to accept dead-end thrust resulting from internal design pressure and other longitudinal loads as applicable. The information presented in Tables 7-7A, 7-7B, 7-7C, and 7-7D is based on longitudinal force due only to full dead-end thrust from internal pressure. The average bearing stress of the ring against the concrete encasement must not exceed 0.45 times the minimum specified 28-day compressive strength of the concrete. Where the pipe exits the structure wall, it may be necessary to increase the thickness of the steel cylinder or add wrapper plate reinforcement to maintain stresses within the acceptable limits defined below. The increased thickness or wrapper plate reinforcement must extend beyond the structure wall to limit longitudinal bending stresses in the steel cylinder. The design for anchor rings is adapted from the design analysis presented in ASCE Manuals and Reports on Engineering Practice (MOP) No. 79 (ASCE 2012), with allowable load and stress limits as noted below. The specific procedure is defined as follows based on

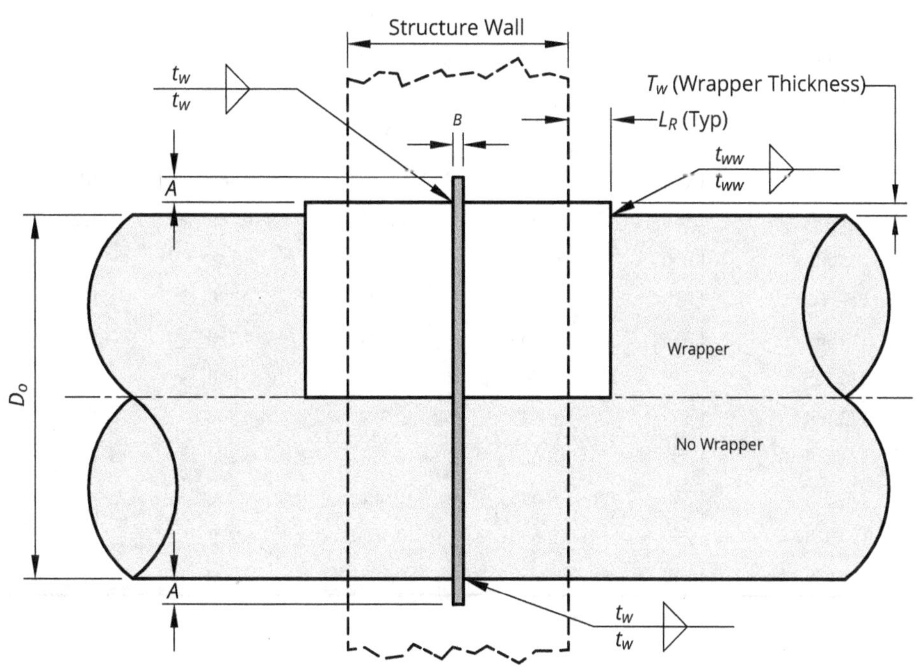

Figure 7-26 Anchor ring

Table 7-7A Dimensional information for anchor rings (100-psi maximum)

Nominal Diameter *in.*	D_o *in.*	Ring Height *A* *in.*	Ring Thickness *B* *in.*	T_{ymin} *in.*	Minimum Weld t_w *in.*	Extension of Shell Beyond Encasement, L_R *in.*	Permissible Load on Ring *lbf*
				100 psi			
6	6.625	0.500	0.188	0.135	0.188	2.0	3,447
8	8.625	0.500	0.188	0.135	0.188	2.0	5,843
10	10.750	0.500	0.188	0.135	0.188	2.0	9,076
12	12.750	0.500	0.188	0.135	0.188	2.5	12,768
14	14.000	0.500	0.188	0.135	0.188	2.5	15,394
14	14.938	0.500	0.188	0.135	0.188	2.5	17,525
16	16.000	0.500	0.188	0.135	0.188	2.5	20,106
16	16.938	0.500	0.188	0.135	0.188	2.5	22,531
18	18.000	0.500	0.188	0.135	0.188	3.0	25,447
18	19.000	0.500	0.188	0.135	0.188	3.0	28,353
20	20.000	0.500	0.188	0.135	0.188	3.0	31,416
20	21.000	0.500	0.188	0.135	0.188	3.0	34,636
20	22.000	0.500	0.188	0.136	0.188	3.0	38,013
24	24.000	0.500	0.188	0.144	0.188	3.5	45,239
24	25.063	0.500	0.188	0.148	0.188	3.5	49,333
26	26.000	0.500	0.188	0.151	0.188	3.5	53,093
30	31.125	0.500	0.188	0.166	0.188	4.0	76,087
36	37.125	0.500	0.188	0.185	0.188	4.5	108,249
42	43.500	0.750	0.250	0.238	0.188	5.5	148,617
48	49.563	0.750	0.313	0.258	0.188	6.0	192,928
54	55.563	0.750	0.313	0.277	0.188	6.5	242,467
60	61.688	1.000	0.375	0.329	0.188	7.5	298,871
66	67.750	1.000	0.375	0.349	0.188	8.5	360,503
72	73.750	1.000	0.375	0.368	0.188	9.0	427,183
78	79.875	1.000	0.438	0.388	0.188	9.5	501,085
84	85.875	1.063	0.438	0.415	0.188	10.0	579,193
90	91.938	1.188	0.500	0.451	0.188	11.0	663,858
96	98.000	1.250	0.500	0.479	0.188	11.5	754,296
102	104.063	1.313	0.500	0.506	0.188	12.0	850,508
108	110.063	1.375	0.563	0.534	0.250	13.0	951,412
114	116.125	1.438	0.563	0.561	0.250	13.5	1,059,111
120	122.188	1.500	0.625	0.589	0.250	14.0	1,172,583
126	128.250	1.563	0.625	0.616	0.250	15.0	1,291,828
132	134.313	1.750	0.688	0.661	0.250	16.0	1,416,846
138	140.375	1.750	0.688	0.680	0.250	16.5	1,547,638
144	146.438	1.813	0.750	0.708	0.250	17.0	1,684,203

Table 7-7B Dimensional information anchor rings (150-psi maximum)

Nominal Diameter *in.*	D_o *in.*	Ring Height *A* *in.*	Ring Thickness *B* *in.*	T_{ymin} *in.*	Minimum Weld t_w *in.*	Extension of Shell Beyond Encasement, L_R *in.*	Permissible Load on Ring *lbf*
			150 psi				
6	6.625	0.500	0.188	0.135	0.188	2.0	5,171
8	8.625	0.500	0.188	0.135	0.188	2.0	8,764
10	10.750	0.500	0.188	0.135	0.188	2.0	13,614
12	12.750	0.500	0.188	0.135	0.188	2.5	19,151
14	14.000	0.500	0.188	0.135	0.188	2.5	23,091
14	14.938	0.500	0.188	0.138	0.188	2.5	26,287
16	16.000	0.500	0.188	0.144	0.188	3.0	30,159
16	16.938	0.500	0.188	0.149	0.188	3.0	33,797
18	18.000	0.500	0.188	0.154	0.188	3.0	38,170
18	19.000	0.500	0.188	0.160	0.188	3.0	42,529
20	20.000	0.500	0.188	0.165	0.188	3.0	47,124
20	21.000	0.500	0.188	0.170	0.188	3.5	51,954
20	22.000	0.500	0.188	0.175	0.188	3.5	57,020
24	24.000	0.500	0.188	0.185	0.188	3.5	67,858
24	25.188	0.500	0.188	0.191	0.188	4.0	74,740
26	26.000	0.500	0.188	0.195	0.188	4.0	79,639
30	31.250	0.750	0.313	0.249	0.188	5.0	115,049
36	37.313	0.750	0.313	0.279	0.188	5.5	164,017
42	43.688	1.000	0.375	0.342	0.188	6.5	224,851
48	49.750	1.000	0.375	0.371	0.188	7.5	291,586
54	55.875	1.063	0.438	0.409	0.188	8.0	367,804
60	61.938	1.188	0.500	0.454	0.188	9.0	451,948
66	68.000	1.250	0.500	0.491	0.188	10.0	544,752
72	74.125	1.375	0.563	0.537	0.250	10.5	647,307
78	80.188	1.500	0.625	0.583	0.250	11.5	757,530
84	86.313	1.625	0.625	0.629	0.250	12.5	877,665
90	92.375	1.750	0.688	0.674	0.250	13.5	1,005,287
96	98.438	1.813	0.750	0.711	0.250	14.0	1,141,569
102	104.563	1.938	0.750	0.757	0.250	15.0	1,288,051
108	110.625	2.000	0.813	0.794	0.313	15.5	1,441,743
114	116.688	2.125	0.875	0.840	0.313	16.5	1,604,094
120	122.813	2.250	0.875	0.886	0.313	17.5	1,776,913
126	128.875	2.375	0.938	0.932	0.313	18.5	1,956,674
132	135.000	2.500	1.000	0.978	0.313	19.0	2,147,082
138	141.125	2.563	1.000	1.063	0.317*	20.5	2,346,330
144	147.125	2.688	1.063	1.063	0.328*	21.0	2,550,082

*Values are based on design and not minimums noted in Table 7-5.

Table 7-7C Dimensional information for anchor rings (200-psi maximum)

Nominal Diameter *in.*	D_o *in.*	Ring Height *A* *in.*	Ring Thickness *B* *in.*	T_{ymin} *in.*	Minimum Weld t_w *in.*	Extension of Shell Beyond Encasement, L_R *in.*	Permissible Load on Ring *lbf*
			200 psi				
6	6.625	0.500	0.188	0.135	0.188	2.0	6,894
8	8.625	0.500	0.188	0.135	0.188	2.0	11,685
10	10.750	0.500	0.188	0.135	0.188	2.0	18,153
12	12.750	0.500	0.188	0.149	0.188	2.5	25,535
14	14.000	0.500	0.188	0.158	0.188	2.5	30,788
14	15.000	0.500	0.188	0.165	0.188	3.0	35,343
16	16.000	0.500	0.188	0.172	0.188	3.0	40,212
16	17.000	0.500	0.188	0.179	0.188	3.0	45,396
18	18.000	0.500	0.188	0.185	0.188	3.5	50,894
18	19.063	0.500	0.188	0.192	0.188	3.5	57,079
20	20.000	0.500	0.250	0.198	0.188	3.5	62,832
20	21.125	0.750	0.250	0.238	0.188	4.0	70,099
20	22.000	0.750	0.250	0.244	0.188	4.0	76,027
24	24.000	0.750	0.313	0.258	0.188	4.5	90,478
24	25.313	0.750	0.313	0.266	0.188	4.5	100,644
26	26.000	0.750	0.313	0.271	0.188	4.5	106,186
30	31.438	1.000	0.375	0.333	0.188	5.5	155,244
36	37.500	1.000	0.375	0.372	0.188	6.5	220,893
42	43.875	1.125	0.438	0.430	0.188	7.5	302,381
48	50.000	1.250	0.500	0.485	0.188	8.5	392,699
54	56.125	1.375	0.563	0.541	0.250	9.5	494,803
60	62.250	1.563	0.625	0.605	0.250	10.5	608,693
66	68.375	1.688	0.688	0.660	0.250	11.5	734,369
72	74.438	1.813	0.750	0.715	0.250	12.5	870,369
78	80.563	2.000	0.813	0.779	0.313	13.5	1,019,497
84	86.688	2.125	0.875	0.835	0.313	14.5	1,180,410
90	92.813	2.313	0.938	0.899	0.313	15.5	1,353,109
96	98.938	2.500	1.000	0.963	0.313	16.5	1,537,594
102	105.125	2.563	1.000	1.063	0.315*	17.5	1,735,929
108	111.250	2.688	1.063	1.125	0.331*	18.5	1,944,106
114	117.250	2.813	1.125	1.125	0.346*	19.0	2,159,462
120	123.375	3.000	1.188	1.188	0.367*	20.0	2,390,970
126	129.500	3.125	1.250	1.250	0.383*	21.0	2,634,265
132	135.625	3.250	1.313	1.313	0.399*	22.0	2,889,345
138	141.750	3.438	1.375	1.375	0.419*	23.5	3,156,211
144	148.000	3.563	1.438	1.500	0.435*	25.0	3,440,672

*Values are based on design and not minimums noted in Table 7-5.

Table 7-7D Dimensional information for anchor rings (250-psi maximum)

Nominal Diameter *in.*	D_o *in.*	Ring Height *A* *in.*	Ring Thickness *B* *in.*	T_{ymin} *in.*	Minimum Weld t_w *in.*	Extension of Shell Beyond Encasement, L_R *in.*	Permissible Load on Ring *lbf*
			250 psi				
6	6.625	0.500	0.188	0.135	0.188	2.0	8,618
8	8.625	0.500	0.188	0.135	0.188	2.0	14,607
10	10.750	0.500	0.188	0.154	0.188	2.5	22,691
12	12.750	0.500	0.188	0.171	0.188	2.5	31,919
14	14.000	0.500	0.188	0.182	0.188	3.0	38,485
14	15.063	0.500	0.188	0.191	0.188	3.0	44,548
16	16.000	0.500	0.250	0.198	0.188	3.0	50,265
16	17.125	0.750	0.250	0.240	0.188	3.5	57,583
18	18.000	0.750	0.250	0.247	0.188	3.5	63,617
18	19.188	0.750	0.313	0.257	0.188	4.0	72,288
20	20.000	0.750	0.313	0.264	0.188	4.0	78,540
20	21.188	0.750	0.313	0.274	0.188	4.0	88,143
20	22.000	0.750	0.313	0.280	0.188	4.5	95,033
24	24.000	0.750	0.313	0.297	0.188	4.5	113,097
24	25.438	1.000	0.375	0.342	0.188	5.0	127,051
26	26.000	1.000	0.375	0.347	0.188	5.0	132,732
30	31.563	1.000	0.438	0.385	0.188	6.0	195,602
36	37.688	1.188	0.500	0.458	0.188	7.0	278,885
42	44.063	1.375	0.563	0.534	0.250	8.0	381,213
48	50.250	1.563	0.625	0.608	0.250	9.5	495,795
54	56.375	1.750	0.688	0.682	0.250	10.5	624,026
60	62.563	1.938	0.750	0.756	0.250	11.5	768,525
66	68.688	2.125	0.875	0.830	0.313	12.5	926,372
72	74.813	2.313	0.938	0.903	0.313	14.0	1,098,951
78	81.000	2.500	1.000	0.978	0.313	15.0	1,288,249
84	87.125	2.688	1.063	1.063	0.324*	16.0	1,490,443
90	93.375	2.875	1.125	1.188	0.350*	17.5	1,711,950
96	99.500	3.063	1.188	1.250	0.374*	18.5	1,943,910
102	105.625	3.188	1.250	1.313	0.394*	19.5	2,190,601
108	111.750	3.313	1.313	1.375	0.414*	20.5	2,452,025
114	118.000	3.500	1.438	1.500	0.429*	22.0	2,733,971
120	124.000	3.688	1.500	1.500	0.453*	22.5	3,019,071
126	130.250	3.875	1.563	1.625	0.479*	24.0	3,331,082
132	136.250	4.063	1.625	1.625	0.503*	25.0	3,645,045
138	142.500	4.250	1.688	1.750	0.528*	26.5	3,987,123
144	148.750	4.438	1.750	1.875	0.553*	28.0	4,344,540

*Values are based on design and not minimums noted in Table 7-5.

the design pressure being equal to the maximum pressure to which the pipe will be subjected at the location of the ring:

1. The ring transfers loads to the concrete encasement with a bearing pressure that varies linearly from zero at the free end of the ring to a maximum at the connection to the steel pipe cylinder. This design limits the average bearing pressure on the concrete to 0.45 times the minimum specified 28-day compressive strength of the concrete. The minimum 28-day compressive strength of the encasement concrete is assumed to be 4,500 psi.
2. The ring resists the full dead-end thrust force as calculated based on the design pressure.
3. The maximum bending stress in the ring is limited to 75 percent of the yield strength of the ring material, σ_Y, at design pressure. When the design pressure is not more than 1.5 times the working pressure, the design must be checked at working pressure to verify that the bending stress does not exceed 50 percent of σ_Y of the ring material. (Note: Arbitrarily increasing a thrust ring height beyond that shown in Tables 7-7A, 7-7B, 7-7C, and 7-7D in order for the ring to serve the dual purpose of a seep ring is not recommended. This action will increase the bending stress in the ring beyond the design limits used to generate the table values. Should such dual-purpose service be desired, the reader is directed to design the ring as defined below based on the desired seep ring height.)
4. Maximum principal and equivalent stresses in the steel pipe cylinder at the connection of the ring are limited to 75 percent of the lesser of the specified minimum yield strength of the steel pipe and wrapper reinforcing materials, σ_Y. When the design pressure is not more than 1.5 times the working pressure, the design must be checked at working pressure to verify that the stresses do not exceed 50 percent of σ_Y.

   The stresses in the steel pipe cylinder at the face of the concrete encasement include secondary bending stresses. Therefore, the maximum equivalent stress in the steel pipe cylinder at the face of the concrete encasement is limited to 90 percent of the lesser of the specified minimum yield strength of the steel pipe and wrapper reinforcing materials, σ_Y. When the design pressure is not more than 1.5 times the working pressure, the design must be checked at working pressure to verify that the stress does not exceed 67 percent of σ_Y.

   Equivalent stresses are calculated using the Hencky-von Mises theory.
5. Due to the embedment concrete at anchor rings, the pipe is fully restrained from longitudinal and circumferential growth due to internal pressure. Therefore, Poisson's stress due to internal pressure must be considered in the analysis.
6. The weld stress for attachment of the anchor ring and wrapper reinforcement is limited to 30 percent of the minimum tensile strength of the welding wire. The size of the fillet weld for attachment of the anchor ring and the wrapper reinforcement is equal to the greater of the calculated value and the size as required by Table 7-5.
7. The depth of anchor ring embedment in the concrete encasement must be sufficient to resist punching shear forces created by the transference of load from the ring to the concrete. Analysis for resistance of the concrete encasement to punching shear is beyond the scope of this manual,

but should be performed by qualified personnel. Information relevant to this design can be found in *MOP No. 79* (ASCE 2012).

Tables 7-7A, 7-7B, 7-7C, and 7-7D provide anchor ring dimensions for pipe sizes from 6 in. to 144 in., for four different maximum pressures. The values in the tables result from the following criteria: (1) The pipe material has a specified minimum yield strength of 35 ksi for diameters ≤ 24-in. nominal, and 36 ksi for diameters greater than 24-in. nominal; (2) the anchor ring material has a minimum yield strength, σ_Y, of 36 ksi; (3) E70XX welding electrodes are used to attach the anchor ring to the steel cylinder; (4) pipe outside diameters, other than for standard sizes, allow for the application of ANSI/AWWA C205 cement-mortar lining and maintain the nominal finished inside diameter; (5) minimum practical ring height of 0.50 in.; (6) minimum practical ring thickness of 0.188 in., with values increasing based on standard available plate thicknesses; (7) minimum practical wall thickness of 0.135 in.; and (8) minimum fillet weld size equal to the greater of that required by design and the minimum sizes noted in Table 7-5.

ANCHOR RING DESIGN

The following presents the design process for the ring in Table 7-7D for 60-in. diameter pipe. The procedure assumes longitudinal thrust due to internal pressure is the only applied load. Other longitudinal loads would need to be included in the analysis as appropriate.

Example 7-13: Anchor Ring Design

For the design process, assume a steel cylinder with a 61.750-in. outside diameter and a wall thickness of 0.257 in., fabricated from material with a specified minimum yield strength of 36 ksi. Calculate the anchor ring size, minimum cylinder thickness, and ring attachment minimum fillet weld size. The specified minimum yield strength of the anchor ring and any required reinforcement steel is 36 ksi. The encasement concrete has a 28-day minimum compressive strength of 4,500 psi. The working pressure is 150 psi, and the test pressure is 250 psi. The fillet weld required for attachment of the anchor ring and any required reinforcement assumes an E70XX grade electrode is used.

Step 1: Minimum Anchor Ring Height

The design pressure equals the maximum internal pressure in the pipe, which is defined as 250 psi. The minimum ring height required to yield the desired average bearing stress, σ_a, of the anchor ring on the concrete encasement is given by

$$D'_r = D_o \sqrt{(p/\sigma_a) + 1}$$

Where:

σ_a = average bearing stress of anchor ring on the concrete, $\leq 0.45f'_c$, psi

p = design pressure for analysis, psi

D_o = steel cylinder outside diameter, in.

$D'r$ = anchor ring minimum outside diameter, in.

f'_c = concrete minimum specified 28-day compressive strength, psi

Based on a design pressure of 250 psi, the anchor ring minimum outside diameter is:

$$D'_r = 61.75 \sqrt{\frac{250}{0.45(4{,}500)} + 1} = 65.45 \text{ in.} \quad \text{For simplicity, let } D_r = 65.50 \text{ in.}$$

Therefore, the anchor ring height $A = (D_r - D_o)/2 = (65.50 - 61.75)/2 = 1.875$ in.

Step 2: Minimum Anchor Ring Thickness

The unit bending moment in the anchor ring is a maximum at the connection to the steel cylinder. For the assumed triangular bearing, the bending moment in the ring at the connection to the steel cylinder is given by

$$M_r = \left(\frac{\pi p D_o^2}{4}\right)\left(\frac{1}{\pi D_o}\right)\left(\frac{A}{3}\right)\left(\frac{1}{1{,}000}\right) = \frac{p D_o A}{12(1{,}000)}$$

Where:

M_r = unit bending moment in anchor ring, in.-kip/in.

A = anchor ring height, in.

Verify that the working pressure is less than (design pressure)/1.5.

250/1.5 ≈ 167 > 150 psi, so the working pressure need not be evaluated in the design.

Therefore,

$$M_r = \frac{250(61.75)(1.875)}{12(1{,}000)} = 2.412 \text{ in.-kip/in.}$$

The minimum thickness of the anchor ring, B', is given by

$$B' = \sqrt{\frac{6M_r}{\sigma_r}}$$

Where:

B' = anchor ring minimum thickness, in.

σ_r = bending stress in the ring at the ring/pipe connection, ksi

The bending stress in the anchor ring shall not exceed 75 percent of the specified minimum yield strength of the ring material at design pressure.

Therefore,

$$B' = \sqrt{\frac{6(2.412)}{0.75(36)}} = 0.732 \text{ in.}$$

Let B = 0.750 in.

Step 3: Bending Moment in the Steel Cylinder

The bending moment, M_r, must be resisted by internal bending moments in the steel cylinder. To maintain static equilibrium, half of the bending moment must be resisted by the steel cylinder on each side of the ring. Therefore, the unit longitudinal bending moment in the steel cylinder on each side of the anchor ring is equal to

$$M_1 = \frac{M_r}{2}, \text{ or } \frac{p D_o A}{24(1{,}000)}$$

Where:

M_1 = unit bending moment in the steel cylinder on each side of the anchor ring, in.-kip/in.

Therefore,

$$M_1 = \frac{2.412}{2} = 1.206 \text{ in.-kip/in.}$$

Step 4: Longitudinal Stress in the Steel Cylinder/Wrapper

In a biaxial stress condition, the conservative result is achieved when one component stress is positive and the other is negative. Given a primary tensile hoop stress, the longitudinal stress in the biaxial condition must therefore be evaluated as a negative. Note that in this analysis the longitudinal stress evaluation includes a Poisson's ratio component of hoop stress, though, that is positive in every case. Therefore, the longitudinal stress in the steel cylinder, which is completely restrained from movement by the surrounding concrete encasement, is given by

$$\sigma_1 = -\frac{pD_o}{4T_y(1{,}000)} + \nu_s \frac{pD_o}{2T_y(1{,}000)} - \frac{6M_1}{T_y^2} \leq 0.75\,\sigma_Y$$

Where:

σ_1 = longitudinal stress in the steel cylinder at the anchor ring, ksi

T_y = wall thickness of steel cylinder,* in.

ν_s = Poisson's ratio for steel = 0.3

Therefore,

$$\sigma_1 = \left(-\frac{250(61.75)}{4(0.257)1{,}000} + 0.3\frac{250(61.75)}{2(0.257)1{,}000} - \frac{6(1.206)}{(0.257)^2}\right) = -116 \text{ ksi} \gg -27 \text{ ksi}$$

The longitudinal stress exceeds the allowable stress so wrapper reinforcement or increased cylinder thickness is required. Evaluate increasing the cylinder thickness to 0.756 in., with a revised pipe outside diameter of 62.563 in. Recalculating Steps 1 through 4 yields a revised M_1 = 1.263 in.-kip/in.

$$\sigma_1 = \left(-\frac{250(62.563)}{4(0.756)1{,}000} + 0.3\frac{250(62.563)}{2(0.756)1{,}000} - \frac{6(1.263)}{(0.756)^2}\right) = -15.328 \text{ ksi} < -27 \text{ ksi}$$

A 0.756-in.-thick cylinder is adequate based on longitudinal stress.

CAUTION: *When the designer chooses to use wrapper reinforcement, it is not acceptable to simply add a sufficient thickness wrapper to the steel cylinder to achieve the thickness calculated by evaluating increasing the cylinder thickness alone. If wrapper reinforcement is desired, the calculations must be performed specific to that application to yield the correct wrapper reinforcement thickness. Generally, the combined thickness of the steel cylinder and the wrapper will be in excess of the value achieved by evaluating only thickening the cylinder. For instance, in this example problem, to achieve the required 0.756-in. cylinder thickness, the designer cannot simply add a 0.50-in. wrapper to the 0.257-in. cylinder thickness. The wrapper thickness is additive to the mainline cylinder thickness for hoop and longitudinal stress calculations only, but is used by itself for the bending stress portion of the calculations. Therefore, for this example, the designer would have to provide a 0.756-in. thick cylinder or a 0.756-in. thick wrapper over the 0.257-in. cylinder.*

Step 5: Circumferential Stress in the Steel Cylinder

The circumferential stress in the steel cylinder, which is completely restrained from movement by the surrounding concrete encasement, is given by

$$\sigma_2 = \frac{pD_o}{2T_y} + v_s\left(\frac{pD_o}{4T_y} + \frac{6M_1}{T_y^2}\right)$$

Where:

σ_2 = circumferential stress in the steel cylinder,* ksi

* When wrapper reinforcement is used, the combined thickness of the steel cylinder and wrapper shall be used for T_y in the first and second terms, but only the wrapper thickness shall be used for T_y in the third term.

† See note in Step 4.

Therefore,

$$\sigma_2 = \frac{250(62.563)}{2(0.756)1{,}000} + 0.3\left(\frac{250(62.563)}{4(0.756)1{,}000} + \frac{6(1.263)}{(0.756)^2}\right) = 15.874 \text{ ksi}$$

Step 6: Equivalent Stress in the Steel Cylinder at the Ring Attachment

The equivalent stress at this location shall not exceed $0.75\sigma_Y$, and is calculated by the Hencky-von Mises theory as follows:

$$\sigma_{eq} = \sqrt{\sigma_1{}^2 + \sigma_2{}^2 - \sigma_1\sigma_2} \le 0.75\sigma_Y$$

Where:

σ_{eq} = equivalent stress, ksi

Therefore,

$$\sigma_{eq} = \sqrt{(-15.328)^2 + 15.874^2 - (-15.328)(15.874)} = 27.0 \text{ ksi} \le 27 \text{ ksi O.K.}$$

Step 7: Equivalent Stress in the Steel Cylinder at the Encasement Face

Confinement of the steel cylinder by the concrete encasement will result in a longitudinal secondary bending stress in the steel cylinder at the location where the cylinder leaves the encasement. The secondary bending stress in the steel cylinder at the encasement face is given by

$$\sigma_b = 1.82\left(\frac{pD_o}{2T_y}\right)\left(\frac{1}{1{,}000}\right)$$

Where:

σ_b = secondary bending stress in steel cylinder,† ksi

Therefore:

$$\sigma_b = 1.82\left(\frac{250(62.563)}{2(0.756)}\right)\left(\frac{1}{1{,}000}\right) = 18.83 \text{ ksi}$$

The equivalent stress at this location shall not exceed $0.9\sigma_Y = 0.9(36) = 32.4$ ksi, and is calculated by the Hencky-von Mises theory as follows:

$$\sigma_{eq} = \sqrt{\sigma_h{}^2 + (\sigma_b + \sigma_L)^2 - \sigma_h(\sigma_b + \sigma_L)} \le 0.9\sigma_Y$$

Where:

σ_{eq} = equivalent stress, ksi

σ_h = hoop stress in cylinder, ksi = 250(62.563)/[2(0.756)]/1,000 = 10.34 ksi

σ_b = compressive secondary bending stress in steel cylinder, ksi (assuming compressive stress yields a conservative analysis)

σ_L = compressive longitudinal stress in steel cylinder = $\sigma_h/2$, ksi (assuming compressive stress yields a conservative analysis)

Therefore,

$$\sigma_{eq} = \sqrt{(10.34)^2 + (-18.83 - 5.17)^2 - 10.34(-18.83 - 5.17)} = 30.5 \text{ ksi} \le 32.4 \text{ ksi O.K.}$$

Note: When wrapper reinforcement is used, the thickness used to calculate σ_h, σ_b, and σ_L is the sum of the cylinder thickness and wrapper thickness.

Step 8: Extension of Reinforcement Beyond the Concrete Encasement

When use of wrapper reinforcement or increased thickness reinforcement of the steel cylinder is required by the above analyses, such increased thickness must extend beyond the limit of the concrete encasement. The distance that the reinforcement or increased thickness must extend beyond the concrete encasement is given by

$$L_R = 2.33\left(\frac{D_o T_y}{2}\right)^{1/2}$$

Where:

L_R = length that reinforcement or increased thickness must extend beyond concrete encasement, in.

Therefore,

$$L_R = 2.33\left(\frac{62.563(0.756)}{2}\right)^{1/2} = 11.33 \text{ in.} \quad \text{Round up to } L_R = 11.50 \text{ in.}$$

Step 9: Anchor Ring Fillet Weld Attachment Size

Anchor rings shall be attached by fillet welding both sides of the ring to the steel cylinder or wrapper. The resultant shear load that must be resisted by each weld is given by

$$f_r = (f_b^2 + f_v^2)^{1/2}$$

With:

$$f_b = \frac{M_r}{\left(B + \frac{t_w}{2}\right)} \text{ and } f_v = \frac{pD_o}{8(1{,}000)}$$

Where:

f_r = resultant shear force to be resisted by each fillet weld, kip/in.

f_b = unit shear force in fillet weld to resist anchor ring bending, kip/in.

f_v = unit shear force in fillet weld to resist direct shear from anchor ring, kip/in.

t_w = fillet weld size, in.

B = anchor ring actual thickness, in. = 0.75 in.

From above, M_r = 2.412 in.-kip/in.

$$f_b = \frac{2.412}{\left(0.750 + \frac{t_w}{2}\right)} \text{ and } f_v = \frac{250(62.563)}{8(1{,}000)} = 1.955 \text{ kip/in.}$$

Therefore,

$$f_r = \left(\left(\frac{2.412}{0.750 + \frac{t_w}{2}}\right)^2 + (1.955)^2\right)^{1/2}$$

The size of the fillet weld is given by

$$t_w = \frac{f_r}{\left[(0.3)(\sigma_w)\left(\frac{\sqrt{2}}{2}\right)\right]}$$

Where:

t_w = fillet weld size, in.

σ_w = minimum tensile strength of welding electrode = 70 ksi

Since f_r is a function of t_w, solving directly for t_w yields a quadratic equation that is difficult to solve. In lieu of a direct solution, a simpler iterative process is used to achieve the required fillet weld size. The process begins by assuming the attachment fillet weld size is equal to the steel cylinder thickness. Using either the initial or reinforced cylinder thickness as the first assumption will yield the same resultant weld size at the end of the iterative process.

First, calculate f_r.

$$f_r = \left(\left(\frac{2.412}{0.750 + \frac{0.756}{2}}\right)^2 + (1.955)^2\right)^{1/2} = 2.897 \text{ kip/in.}$$

Then, calculate t_w.

$$t_w = \frac{2.897}{\left[(0.3)(70)\left(\frac{\sqrt{2}}{2}\right)\right]} = \frac{2.897}{14.849} = 0.195 \text{ in.}$$

Use the first solution to again solve the two equations and repeat until achieving adequate closure on a value.

$$f_r = \left(\left(\frac{2.412}{0.750 + \frac{0.195}{2}}\right)^2 + (1.955)^2\right)^{1/2} = 3.453 \text{ kip/in.}$$

$$t_w = \frac{3.453}{14.849} = 0.233 \text{ in.}$$

The weld sizes of the first two iterations are reasonably close, but better closure will be achieved with a third iteration.

$$f_r = \left(\left(\frac{2.412}{0.750 + \frac{0.233}{2}}\right)^2 + (1.955)^2\right)^{1/2} = 3.402 \text{ kip/in.}$$

$$t_w = \frac{3.402}{14.849} = 0.229 \text{ in.}$$

The third iteration achieves adequate closure on the value of 0.229 in., but this weld size needs to be checked against the minimum values in Table 7-5. The material thickness of the anchor ring is ¾ in. and the steel cylinder thickness is at least 0.756 in. Therefore, from Table 7-5, the minimum fillet weld size for attaching the ring to the steel cylinder is t_w = ¼ in.

Step10: Wrapper Reinforcement Fillet Weld Attachment Size

Had reinforcement been chosen in lieu of increased cylinder thickness, the wrapper must be attached by fillet welding both sides of the reinforcement to the steel cylinder. The resultant shear load that must be resisted by each weld is f_v, and the fillet weld size is determined as follows:

$$t_{ww} = \frac{f_v}{\left[(0.3)(\sigma_w)\left(\frac{\sqrt{2}}{2}\right)\right]}$$

Where:

t_{ww} = wrapper reinforcement fillet weld size, in.

Therefore,

$$t_{ww} = \frac{f_v}{\left[(0.3)(\sigma_w)\left(\frac{\sqrt{2}}{2}\right)\right]} = \frac{1.955}{\left[0.3(70)\left(\frac{\sqrt{2}}{2}\right)\right]} = 0.132 \text{ in.}$$

This weld size needs to be checked against the minimum values in Table 7-5. The material thickness of the steel cylinder is 0.257 in. The material thickness of the wrapper plate would be approximately 0.756 – 0.257 = 0.499 in., or ½ in. Therefore, from Table 7-5, the minimum fillet weld size for attaching the wrapper reinforcement to the steel cylinder is t_{ww} = 3⁄16 in.

The results of the example yield an anchor ring that is 0.750-in. thick by 1.875-in. tall. The steel cylinder thickness must be increased to a minimum of 0.756 in. for a length sufficient to extend past each end of the encasement by 11.5 in. The ring must be attached to the wrapper reinforcement by a ¼-in. double fillet weld. The wrapper reinforcement, if used, must be attached to the steel cylinder by a 3⁄16-in. double-fillet weld.

OUTLETS

Outlets from steel mains can be easily arranged in any desired location according to size, shape, or position. Outlets are welded to the main line with or without supplemental reinforcement depending on the results of the design analyses relative to the service conditions. All outlets should be checked to determine whether reinforcement is required. Attachment of outlets can be done in the shop during the fabrication of the pipe, at trenchside, or after the pipe is installed. Shop lining and coating of outlets and pipe are satisfactory and typically more economical than work done in the field.

If required for hydraulic efficiency, a reducer may be welded to the main pipe with the outlet welded to the reducer. In such cases, the design analysis for reinforcing the shell must be performed using the larger diameter of the reducer for the main pipe size.

The end of the outlet should be prepared to receive the valve or fitting to be attached. This may call for a flange, a grooved or shouldered end for a mechanical coupling, a plain end for a flexible coupling joint, a grooved spigot end for a bell-and-spigot joint, a threaded end, or other required end.

BLOWOFF CONNECTIONS

Outlets for draining a pipeline should be provided at low points in the profile and upstream of line valves located on a slope. Short dips, such as those occurring in many pipelines in city streets when a line must pass under a large drain or other structure, can often be dewatered by pumping, when necessary.

The exact location of blowoff outlets is frequently determined by opportunities to dispose of the water. Where a pipeline crosses a stream or drainage structure, there usually will be a low point in the line; but if the pipeline goes under the stream or drain, it cannot be completely drained into the channel. In such a situation, a blowoff connection should be located at the lowest point that will drain by gravity and provide easy means for pumping out the section below the blowoff.

Blowoffs are generally attached tangentially to the bottom of the main but can be attached radially, and must, of course, be provided with a shutoff valve. If the pipeline is aboveground, the valve should be attached directly to the outlet nozzle on the bottom of the pipeline. A pipe attached to the valve will be necessary to route the discharge to an appropriate location. The discharge pipe frequently requires installation of an elbow near

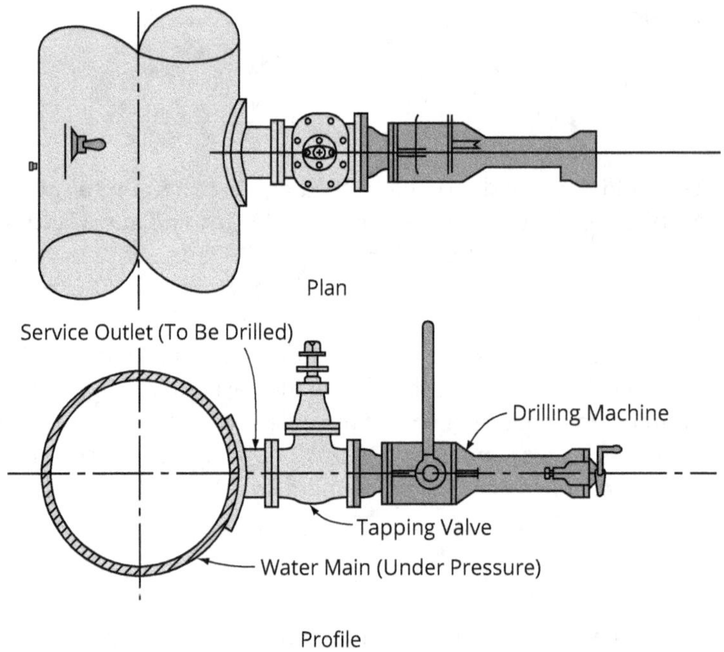

Procedure: (1) Weld outlet and reinforcing collar (if required) to main; (2) bolt on gate valve, adapter (if required), and drilling machine; (3) insert tool and drill hole in main; and (4) withdraw tool, close gate, and remove machine.

Figure 7-27 Tapping main under pressure

the blowoff valve, which must be securely blocked to avoid stresses on the attachment to the pipeline.

Usually the blowoff will be belowground. Because the operating nut of the valve must be accessible from the surface, the valve cannot be under the main but may be set with the stem vertical and just beyond the side of the pipeline.

MANHOLES

The most common type of manhole for access in waterworks is circular, having a short, flanged neck and a flat, bolted cover. Such manholes are commonly 24 to 36 in. in diameter.

Manholes should be placed in accessible locations. They provide access to the inside of the pipeline for many purposes besides inspection. In general, they will be most useful if located close to valves and sometimes close to the low points that might need to be pumped out for inspection or repair.

AIR-RELEASE VALVES AND AIR/VACUUM VALVES

Air-release and air/vacuum valves are installed to vent accumulated air from the waterline so that the pipe's flow capacity is not impaired or to admit air into the waterline to avoid the creation of a vacuum, respectively. AWWA Manual M51, Air-Release, Air/Vacuum, and Combination Air Valves, provides a full scope of information relative to air valves. M51 guides operators in the selection, installation, and maintenance of air valves in waterline applications, including information regarding valve types, valve location, valve orifice sizing, water hammer, operation, and safety. For additional information, see ANSI/AWWA C512, Air Release, Air/Vacuum, and Combination Air Valves for Water and Wastewater Service (latest edition).

MISCELLANEOUS CONNECTIONS AND OTHER APPURTENANCES

Special tapping machines for mains under pressure are available and have been used for many years. Figure 7-27 illustrates the method. The reinforcing pad can be eliminated unless required by design. The outlet is ordinarily a piece of extra-heavy standard-weight pipe with an AWWA standard plate flange attached. The tapping valve is special and allows proper clearance for the cutter on the drilling machine.

As an alternate to welding on an outlet, ANSI/AWWA C223, Fabricated Steel and Stainless-Steel Tapping Sleeves, provides information for outlets mechanically fastened to a main. ANSI/AWWA C223 defines the requirements and specific limitations of such connections.

LAYOUT OF PIPELINES

The logistics of surveying and laying out a pipeline are affected by both the size of the line and its location. More detail and care are necessary as the size of the line increases and as the line passes from rural to urban areas.

In general, a plan and a profile layout, as well as certain other details, are necessary for any water pipeline. The layout can be a drawing showing the installation location of pipes, a stick figure layout, a numerical laying schedule, a standard orthographic diagram, or other method clearly defining the required information. A layout should show clearly and completely the essential details for each pipe piece. In addition, the layout should show the necessary data for the proper assembly sequence and for spotting of pipe specials and sections. Regardless of the layout format chosen, the following information should be included:

1. Horizontal and vertical distances, either directly or by survey station and elevation (if slope distances are given on the layout, this fact should be stated).
2. Location (point of intersection) and degree of angles or bends, both horizontal and vertical. When horizontal and vertical bends occur at the same point of intersection, the resultant combined angle should be noted.
3. Curves shall have the following information provided or sufficient information provided so that all other information can be calculated: angle, direction, radius, beginning and ending stations, curve length, and tangent length. Station equations resulting from use of curves should be noted.
4. Points of intersection with pipe centerline for tees, wyes, crosses, or other branches, including direction—right- or left-hand, up or down—or angle of flow, as viewed from the inlet end.
5. Location and lengths of all valves, pumps, or other inserted fittings not supplied by the pipe manufacturer.
6. Location of adjacent or interfering installations or structures.
7. Location and length of all sections of casing including size and type of casing and position of the carrier within the casing.
8. Any special requirements affecting the manufacture of the pipe or the installation procedures.

Pipe may be identified by a consecutive-piece number system, by using another system in accordance with the common practice of the pipe manufacturer, or as established by mutual agreement between the purchaser and the manufacturer. A requirement for consecutive numbering and installation of straight pieces of uniformly cut length is uneconomical if the pieces are interchangeable in the line. Unique special sections should be marked to show their specific location in the layout. (Note: General marking requirements are provided in the relevant AWWA standards.)

Fabrication or "shop" drawings may accompany the layout. Shop drawings should include:

1. Dimensional details or descriptions of all specials, including other data required to supplement AWWA standards (see the "Information Regarding Use of This Standard" section of the relevant standard).
2. Details, dimensions, and class designation or other description of all flanges and mechanical field joints.

GOOD PRACTICE

The standard-dimension fittings described in ANSI/AWWA C208 (latest edition) should be used whenever possible. If drawings are not used in purchasing, the designation of fittings is always necessary. Design data should be used to determine if reinforcement is needed. When necessary, special welded steel pipe fittings can be fabricated to meet unusual requirements and severe service conditions. When special steel-pipe fittings are designated, they should be accompanied with drawings to show their exact configuration. The fitting or special configurations and associated design procedures presented in this manual are not sufficiently comprehensive to address all possible configurations as dictated by industry requirements. For the design of configurations beyond the scope of this manual, the designer is directed to other recognized codes, standards, manuals, or design methods based on the applicability of each to the specific configuration of interest.

REFERENCES

American Iron and Steel Institute. 1992. *Steel Plate Engineering Data—Volume 4, Buried Steel Penstocks*. Washington, DC: American Iron and Steel Institute.

American National Standards Institute/American Water Works Association (ANSI/AWWA) C200, Steel Water Pipe, 6 In. (160 mm) and Larger. Latest edition. Denver, CO: AWWA.

ANSI/AWWA C205, Cement-Mortar for Protective Lining and Coating for Steel Water Pipe—4 In. (100 mm) and Larger—Shop Applied. Latest edition. Denver, CO: AWWA.

ANSI/AWWA C207, Steel Pipe Flanges for Waterworks Service, Sizes 4 In. Through 144 In. (100 mm Through 3,600 mm). Latest edition. Denver, CO: AWWA.

ANSI/AWWA C208, Dimensions for Fabricated Steel Water Pipe Fittings. Latest edition. Denver, CO: AWWA.

ANSI/AWWA C223, Fabricated Steel and Stainless Steel Tapping Sleeves. Latest edition. Denver, CO: AWWA.

ANSI/AWWA C512, Air Release, Air/Vacuum, and Combination Air Valves for Water and Wastewater Service. Latest edition. Denver, CO: AWWA.

American Society for Testing and Materials (ASTM). 2012a. ASTM A36/A36M, Standard Specification for Carbon Structural Steel. Philadelphia: ASTM.

ASTM. 2012b. ASTM A193/A193M-12b, Standard Specification for Alloy-Steel and Stainless Steel Bolting Materials for High Temperature or High Pressure Service and Other Special Purpose Applications. Philadelphia: ASTM.

ASTM. 2013. ASTM A194/A194M, Standard Specification for Carbon and Alloy Steel Nuts for Bolts for High-Pressure or High-Temperature Service, or Both. Philadelphia: ASTM.

American Society of Civil Engineers (ASCE). 2012. *ASCE Manuals and Reports on Engineering Practice (MOP) No. 79, Steel Penstocks*, 2nd Ed. Reston, VA: ASCE.

American Society of Mechanical Engineers (ASME). 2003. ASME B1.1, Unified Inch Screw Threads (UN & UNR Thread Form). New York: ASME.

ASME. 2010. Boiler and Pressure Code, Section VIII, Rules for Construction of Pressure Vessels, Division 1. New York: ASME.

American Water Works Association (AWWA) Manual M51, *Air-Release, Air/Vacuum, and Combination Air Valves*. Latest edition. Denver, CO: AWWA.

American Welding Society (AWS). AWS D1.1/D1/1M: 2010, *Structural Welding Code—Steel*. Miami, FL: American Welding Society.

Arch, C. 1980. *Outlet Reinforcement*. Baldwin Park, CA: United Concrete Pipe.

Bardakjian, H., and M. Zarghamee. 2008. *Design of a 120 in.-Diameter Steel Bifurcation With a Small Acute Angle for a High-Pressure Penstock*. Reston, VA: ASCE.

Barnard, RE. 1948. Design Standards for Steel Water Pipe. *Jour. AWWA*, 40(1):24.

Brockenbrough, R.L. 1988. Personal Communication. Pittsburgh, PA: R.L. Brockenbrough & Assoc. Inc.

Goit, LB. 1949. Steel Pipeline Appurtenances. *Jour. AWWA*, 41(1):47.

Ruud, F.O. 1964. *Stress Analysis of Wye Branches*. Engineering Monograph 32. Denver, CO: US Bureau of Reclamation.

Swanson, H.S., et al. 1955. Design of Wye Branches for Steel Pipe. *Jour. AWWA*, 47(6):581.

Timoshenko, S. 1940. *Strength of Materials. Part II*. New York: Van Nostrand Co.

Young, W., and R. Budynas. 2002. *Roark's Formulas for Stress and Strain*. 7th Ed. New York: McGraw-Hill.

Chapter 8

Thrust Restraint for Buried Pipelines

THRUST FORCES

When a water transmission or distribution buried pipeline is under internal pressure, unbalanced thrust forces develop at changes of cross-sectional area (such as reducers), at changes of direction in the pipeline (such as bends, wyes, tees, etc.), and at pipeline terminations (such as bulkheads). Thrust forces of primary importance are (1) hydrostatic thrust due to internal pressure of the pipeline and (2) hydrodynamic thrust due to changing momentum of flowing water. Because most waterlines operate at relatively low velocities, the hydrodynamic thrust is insignificant and is usually ignored. For example, the hydrodynamic force created by water flowing at 8 ft/sec is less than the hydrostatic force created by 1 psi.

HYDROSTATIC THRUST

Hydrostatic thrust is a function of internal pressure, cross-sectional area of the pipe or outlet, and piping configuration. Typical examples of hydrostatic thrust are shown in Figure 8-1. The magnitude of thrust forces for tees (Figure 8-1c), wyes (Figure 8-1d), and bulkhead or dead end (Figure 8-1b) is equal to the product of the internal pressure and the cross-sectional area of the pipe or outlet, or

$$T = p\,A \text{ or } = p\,A_o \quad \text{(Eq 8-1)}$$

Where:

T = the thrust force, lb

p = maximum internal pressure including any anticipated transient pressure or static test pressure if greater than operating pressure, psi

A = cross-sectional area of the pipe, in.^{2}

A_o = cross-sectional area of the tee or wye outlet, in.^{2}

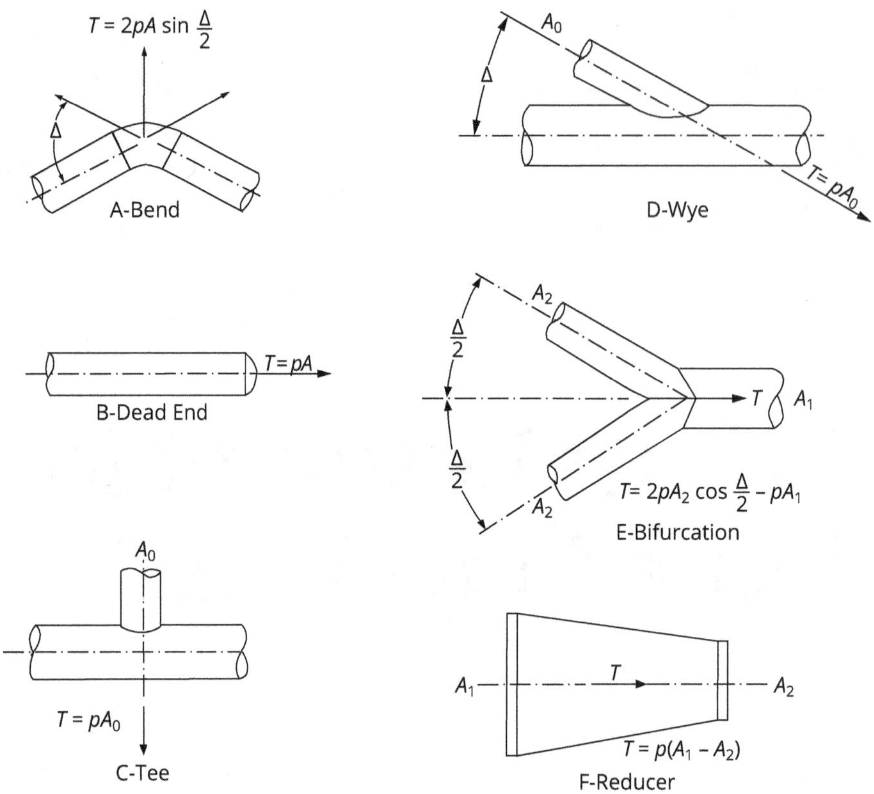

Figure 8-1 Hydrostatic thrust *T* applied by fluid pressure to typical fittings

At bends (Figure 8-1a), thrust is also a function of deflection angle, Δ, and the resultant thrust force T is

$$T = 2\,pA \sin(\Delta/2) \qquad \text{(Eq 8-2)}$$

Where:

Δ = the deflection angle of the bend

At bifurcations (Figure 8-1e), the resultant thrust force T is

$$T = 2\,pA_2 \cos(\Delta/2) - pA_1 \qquad \text{(Eq 8-3)}$$

Where:

Δ = the deflection angle of the wye

A_2 = cross-sectional area of the wye branch, in.^{2}

A_1 = cross-sectional area of the main piping, in.^{2}

At reducers (Figure 8-1f), the resultant thrust T is

$$T = p(A_1 - A_2) \qquad \text{(Eq 8-4)}$$

Where:

A_1 = cross-sectional area of the larger-diameter reducer end, in.^{2}

A_2 = cross-sectional area of the smaller-diameter reducer end, in.^{2}

THRUST RESISTANCE

Methods to restrain the thrust forces may be provided by an external reaction from a concrete thrust block or by the development of axial frictional resistance between the pipe and the soil through restrained or harnessed joints. The concepts of both methods are different, thus both methods should not be combined. Exceptions to this restriction include reinforced concrete thrust collar shear keys or steel thrust collars in chamber walls.

In a fully restrained system, no additional analysis is required if the joints are designed for transmission of longitudinal forces from one pipe unit to the next.

THRUST BLOCKS

Concrete thrust blocks are usually classified as bearing type as shown in Figure 8-2 or the lower thrust block of Figure 8-3 or gravity type as shown in the upper thrust block in Figure 8-3. The bearing-type thrust blocks increase the ability of fittings to resist movement by increasing the lateral bearing area for horizontal thrust or vertical bearing area for downward vertical thrust. The gravity type increases the weight of the fitting and pipe assembly to resist or provide counterweight for the thrust. Bearing-type thrust blocks can be designed based on the safe bearing capacity of the soil or the passive soil pressure behind the thrust block, which is beyond the scope of this manual.

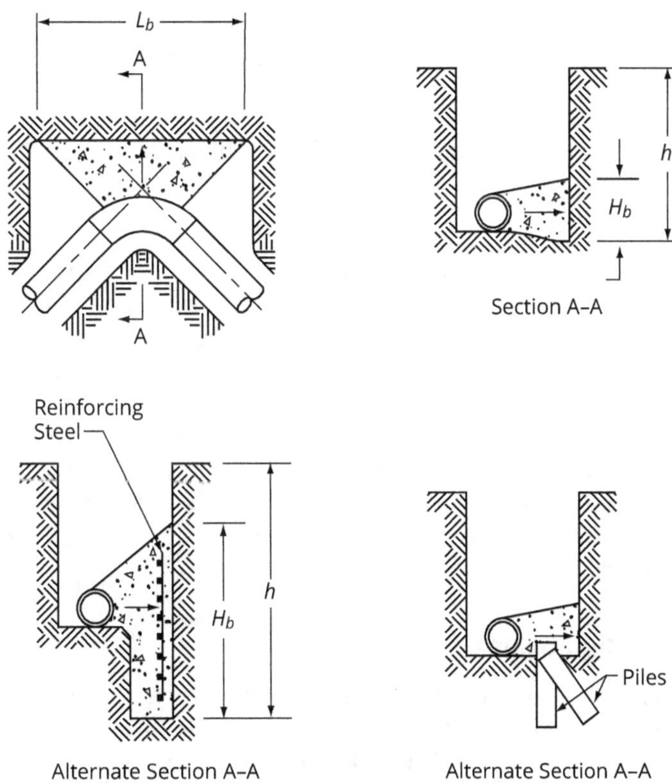

Source: AWWA M45.

Figure 8-2 Typical thrust blocking of a horizontal bend

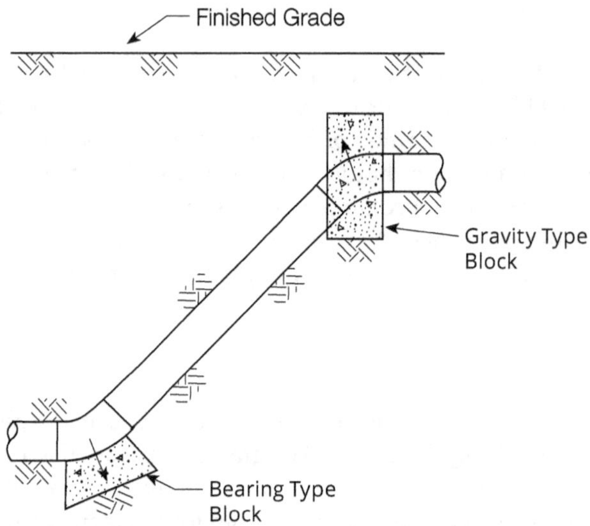

Figure 8-3 Typical thrust blocking of vertical bends with bearing-type and gravity-type blocks

Calculation of Bearing-Type Size

Thrust block size can be calculated based on the lateral or vertical bearing capacity of the soil:

$$\text{Area of Block} = L_b \times H_b = T/\sigma \qquad \text{(Eq 8-5)}$$

Where:

$L_b \times H_b$ = area of bearing surface of thrust block, ft^2

T = thrust force, lbf

σ = soil bearing capacity, psf

If it is impractical to design the block for the thrust force to pass through the geometric center of the soil-bearing area, then the design should be evaluated for stability.

After calculating the thrust block size based on the safe lateral or vertical bearing capacity of soil, the shear resistance of the passive soil wedge behind the thrust block should be checked because it may govern the design. For a thrust block having its height, H_b, greater than one-half the distance from the ground surface to the base of the block, h, the design of the block is generally governed by shear resistance of the soil wedge behind the thrust block. Determining the value of the safe lateral bearing and shear resistance of the soil is beyond the scope of this manual. Consulting a qualified geotechnical engineer is recommended.

Typical Configurations

Determining the safe bearing value, σ, is the key to sizing a thrust block. Values can vary from less than 1,000 lbf/ft^2 for very soft soils to several tons per ft^2 for solid rock. Knowledge of local soil conditions is necessary for proper sizing of thrust blocks. Figure 8-2 shows several details for distributing thrust at a horizontal bend. Section A-A is the more common detail, but the other methods shown in the alternate section A-A may be necessary in weaker soils. Figure 8-3 shows typical thrust blocking of vertical bends. Design of the block for a bottom bend is the same for a horizontal bend, but the block for a top bend must be sized to adequately resist the vertical component of thrust with dead weight of the block, bend, water in the pipe, and overburden.

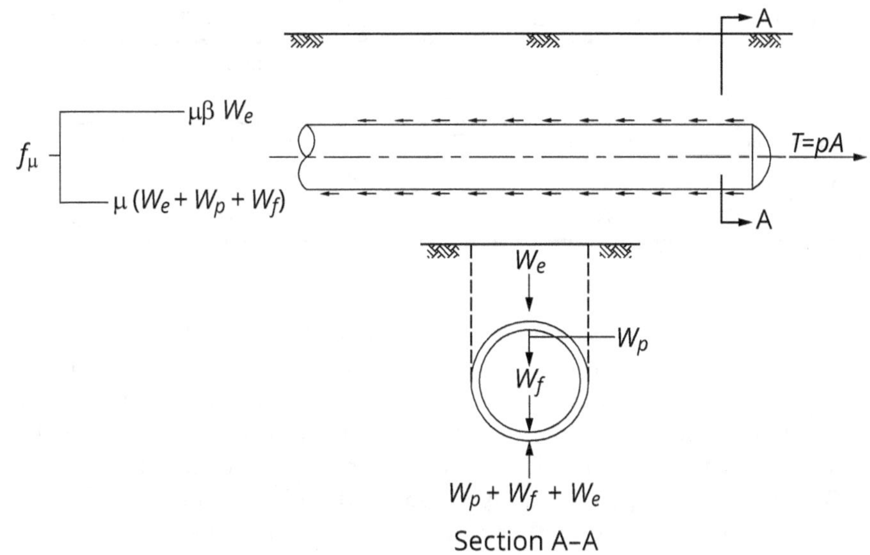

Figure 8-4 Horizontal frictional forces that resist horizontal thrust $T = pA$

THRUST RESTRAINT WITH WELDED OR HARNESSED JOINTS FOR *pA* HORIZONTAL THRUST

Thrust force at bulkheads or dead ends, tees, and valves is equal to the internal pressure times the area of the pipe, *pA*. A restraint system with welded or harnessed joints may be used to resist the thrust force through the development of friction between the pipe and the soil surrounding it. A restraint system with the horizontal thrust force *pA* is resisted by the frictional resistance force acting along the longitudinal axis of the pipe and surrounding soil. The frictional resistance is assumed to be distributed along the restrained length of the pipeline. Figure 8-4 depicts the horizontal frictional forces that resist horizontal thrust *pA*.

The frictional resistance per linear foot, f_μ, is expressed by

$$f_\mu = \mu\,[(1+\beta)\,W_e + W_p + W_f] \qquad \text{(Eq 8-6)}$$

Where:

μ = coefficient of friction
W_p = weight of pipe, lb/lin ft
W_f = weight of fluid in pipe, lb/lin ft*
β = shallow cover factor
W_e = soil prism weight above the pipe, lb/lin ft = $\gamma\, D_o\, H$

Where:

γ = unit weight of backfill, lb/ft^3*
D_o = pipe outside diameter, ft
H = depth of cover, ft

The coefficient of friction, μ, depends on the type, compaction, and moisture content of the backfill soil and the coating type (roughness). Field tests conducted in 1988 (Bardakjian 1991) on 6-in.-diameter bare steel pipe installed in compacted granular soil

* In conditions where the pipe is fully submerged, the weight of the water in the pipe should not be considered. The unit weight of the backfill in a submerged condition should not be reduced for buoyant conditions. Further, the reduction in pipe weight in a submerged condition has an insignificant effect on the calculation of frictional resistance.

Table 8-1 Unit weight of soil, lb/ft³, based on type of soil and relative compaction

Type of Soil*	Standard AASHTO relative compaction		
	85%	90%	95%
Fine grained soils with less than 25% sand content (CL, ML, CL-ML)	100 lb/ft³	105 lb/ft³	110 lb/ft³
Coarse grained soils with fines (SM, SC)	110 lb/ft³	115 lb/ft³	120 lb/ft³
Coarse-grained soils with little or no fines (SP, SM, GP, GW)	130 lb/ft³	135 lb/ft³	140 lb/ft³

*Soil type symbols are from the Unified Classification System.

(sand) resulted in a calculated friction coefficient value of 0.34 based on equating the minimum longitudinal pull force to twice the soil prism load. In 1981, US Steel Corporation conducted laboratory tests on 2-in.-diameter steel tubing in sand boxes to determine the frictional characteristics of bare and tape-coated pipe assuming the friction acts all around the perimeter of the pipe. The calculated friction coefficient for the bare steel pipe was 0.342 and for the tape-coated pipe, measured with the spiral direction, was 0.319 for one tape type and 0.431 for another tape type. In 1965 field tests (Price 1965) were conducted on 16-in.-diameter mortar-coated pipe installed in dry and saturated granular and clayey backfill soils. The calculated coefficient of friction, μ, varied between 0.52 for saturated clay to 1.32 for dry tamped sand. Tests were completed in 2012 (Alam et al. 2013) on 14-in. pipes coated with cement mortar, tape, and polyurethane at various simulated depths and in three distinct soil types. From this testing, it was found that polyurethane and tape dielectric coatings generate similar calculated friction values. The use of friction coefficients of 0.32 and 0.50 is recommended for steel pipe with dielectric coatings and steel pipe with mortar coatings, respectively.

The unit weight of soil is dependent on the type of backfill soil and degree of compaction. Guidelines for unit weight of soils and compaction developed from limited testing for the types of soil and compaction levels listed in Table 6-1 are given in Table 8-1. Actual unit weight of soil may vary. Consulting a qualified geotechnical engineer is recommended.

When the pipe has low cover, the soil on the top of the pipe may move with the pipe. Soil resistance against movement of the pipe is provided, in part, along the sides of the soil block directly above the pipe, rather than at the pipe-to-soil interface along the top surface of pipe. Hence, the shallow cover factor, β, which cannot exceed 1, may be expressed by the following (Zarghamee et al. 2004):

$$\beta = K_o \tan \varphi \,(12\, H/D_o + 0.50)^2 / [\mu\, (12\, H/D_o + 0.107)] \leq 1 \qquad \text{(Eq 8-7)}$$

Where:

K_o = 1 − sin φ = coefficient of lateral soil pressure

φ = angle of internal friction, in degrees (varies between 20° and 45° depending on the soil characteristics)

H = depth of cover, ft

D_o = outside diameter of pipe, in.

Restrained length of horizontal pA force L can be expressed by

$$L = pA/f_\mu \qquad \text{(Eq 8- 8)}$$

GASKETED JOINTS WITH SMALL DEFLECTIONS

Pipe with deflected or mitered joints is used to lay around long-radius curves. Pipe is laid on chords as shown in Figure 8-5. The pipeline centerline radius for deflected pipe is

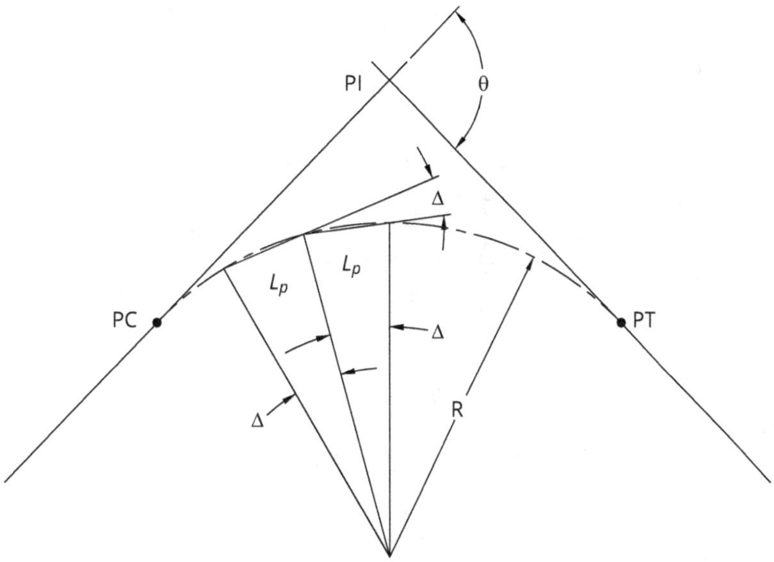

PC = point of curve
PI = point of intersection
PT = point of tangent
θ = total deflection angle of curve
Δ = angular deflection per pipe section
L_p = calculated centerline pipe laying length
R = pipeline certerline radius

Figure 8-5 Pipe alignment through a curve

$$R = L_p / [2 \sin \Delta/2] \qquad \text{(Eq 8-9)}$$

Where:

Δ = angular deflection per pipe section

L_p = calculated centerline pipe laying length

Thrust restraint is normally not required at rubber-gasket joints of mitered pipe (less than 5°) or standard pipe installed with small angular deflections since the thrust is usually low.

Small Horizontal Deflections With Joints Free to Rotate

Thrust at deflected joints on long-radius horizontal curves is resisted by friction on the pipe as shown in Figure 8-6. The total friction, F, developed is equal to the thrust and acts in the opposite direction. Additional restraint is not required when

$$T \leq \mu L_p [(1+\beta) W_e + W_p + W_f] \qquad \text{(Eq 8-10)}$$

Where:

$T = 2 pA \sin \Delta/2$, lb

L_p = length of pipe section, ft

Design Example for Long-Radius Horizontal Curves

Given

D = 48 in., D_o = 49.5 in.

Field-test pressure, p_t = 225 psi

Length of standard pipe section = 40 ft

Curve radius, R = 457 ft

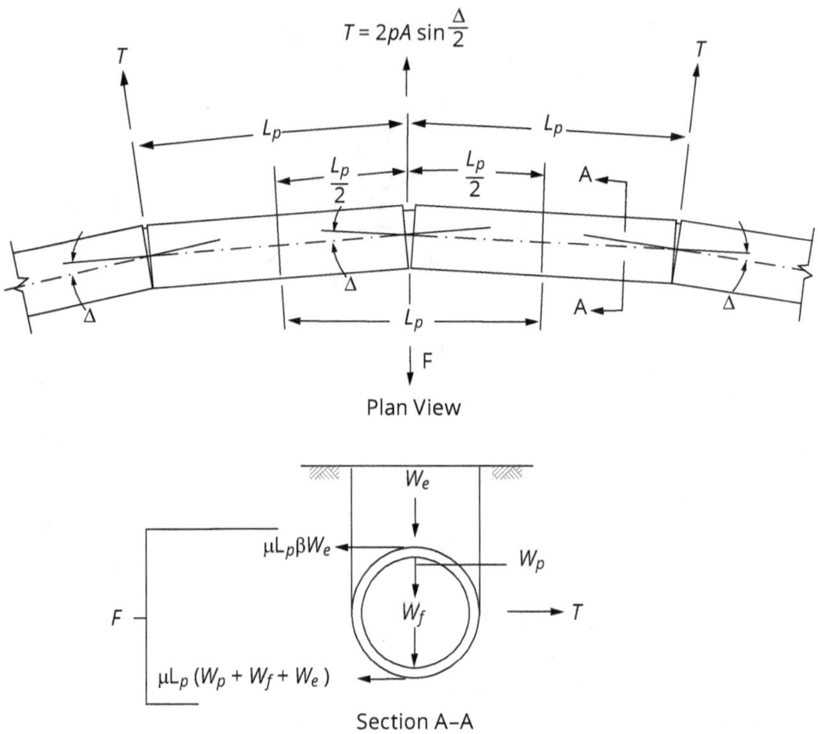

Figure 8-6 Restraint of thrust at deflected gasketed joints on long-radius horizontal curves

Deflection angle, Δ = 5° (from Eq 8-9)
Lay length of miter pipe, L_p = 40 – [49.5/2 (12)] (tan 5°) = 39.82 ft
Depth of cover, H = 6 ft
Unit weight of backfill soil = 115 lb/ft^3
Internal friction angle of backfill soil, φ = 20°
Pipe weight, W_p = 400 lb/ft
Water weight, W_f = 784 lb/ft
W_e = 6 (49.5/12) (115) = 2,846 lb/ft
Friction coefficient, μ = 0.30

Design Check

Reference	
Equation 8-7	β = (1 – sin 20) (tan 20) 12(6)/49.5 + 0.5]^2/[0.30 (12 (6)/49.5+ 0.107)] = 1.95 > 1, therefore = 1
Equation 8-10	T = 2 (225) (49.5)^2/4 (sin 5/2) = 37,758 lb
	$\mu\, L_p\, [W_p + W_f + (1+ \beta)\, W_e]$ = 0.3 (39.82) [400 + 784 + 2 (2846)] = 82,141 lb
	Since 82,141 > 37,758, therefore there is no need to restrain the joints.

THRUST RESTRAINT WITH WELDED OR HARNESSED JOINTS FOR HORIZONTAL BENDS

The two methods that have been used historically to determine the restraint length at each leg of a horizontal bend are (1) based on thrust force of pA sin Δ/2, which assumes that the system moves laterally in the opposite direction of the resultant force (see Figure 8-7), and (2) based on thrust force of pA (1 – cosΔ), which assumes that the system moves axially

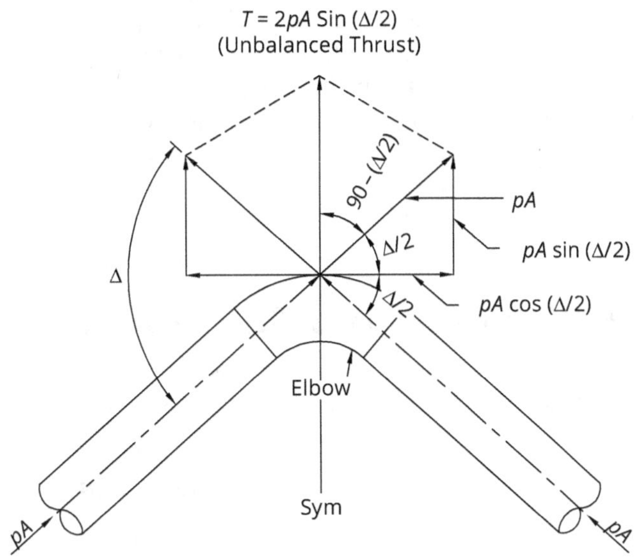

Figure 8-7 Unbalanced thrust at horizontal bends, $T = 2pA \sin \Delta/2$

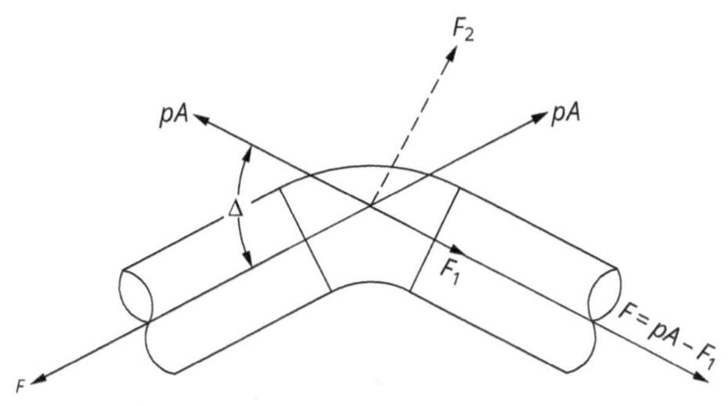

NOTE: For clarity only the forces on one leg are shown. Forces on both legs are identical.

Figure 8-8 Unbalanced axial thrust, $F = pA\,(1 - \cos \Delta)$ plus unbalanced thrust normal to axial thrust, $F_2 = pA \sin \Delta$

only and that the unbalanced force of $pA \sin \Delta$ load is resisted by the passive soil resistance (see Figure 8-8). Both methods ignore the effect of the bending movement against the soil.

It is generally recognized that a buried pipe has to move through the soil to develop frictional resistance forces. It also has to move against the soil in order to develop lateral (passive) resistance forces, which in combination with the frictional resistance forces resist the unbalanced thrust. Axial and transverse pipe movements cause additional pipe stresses (axial, shear, and bending) on the pipe at or near the unbalanced forces.

A case study (Bardakjian 2011) was conducted to calculate the required restraint length, bend displacements, and resulting shear and bending stresses in buried continuous steel pipelines in normal soil conditions. That study supported the historical performance of steel water pipe that the combined bending and axial stresses do not control the design of the continuous steel pipeline. The case study also showed that the use of $pA\,(1 - \cos \Delta)$ procedure compared to the study procedure produced longer restrained lengths for bend angles greater than 60° and shorter restrained length for bend angles less than 60° based on frictional resistance alone. However, due to the steel pipe being furnished in relatively long sections, the number of welded joints for shallow bend angles in most cases will be the same as required by the $pA\ (1 - \cos \Delta)$ procedure.

Therefore, the restraint length for each leg of the bend, L, is

$$L = \frac{pA(1 - \cos \Delta)}{\mu[(1 + \beta)\, W_e + W_p + W_f]} \qquad \text{(Eq 8-11)}$$

SMALL VERTICAL DEFLECTIONS WITH JOINTS FREE TO ROTATE

Uplift thrust at deflected joints on long-radius vertical curves is resisted by the combined dead weight $W_p + W_f + W_e$, as shown in Figure 8-9. Additional restraint is not required when

$$T \leq L_p\,(W_p + W_f + W_e)\cos\,(\alpha - \Delta/2) \qquad \text{(Eq 8-12)}$$

Where:

α = slope angle, in degrees

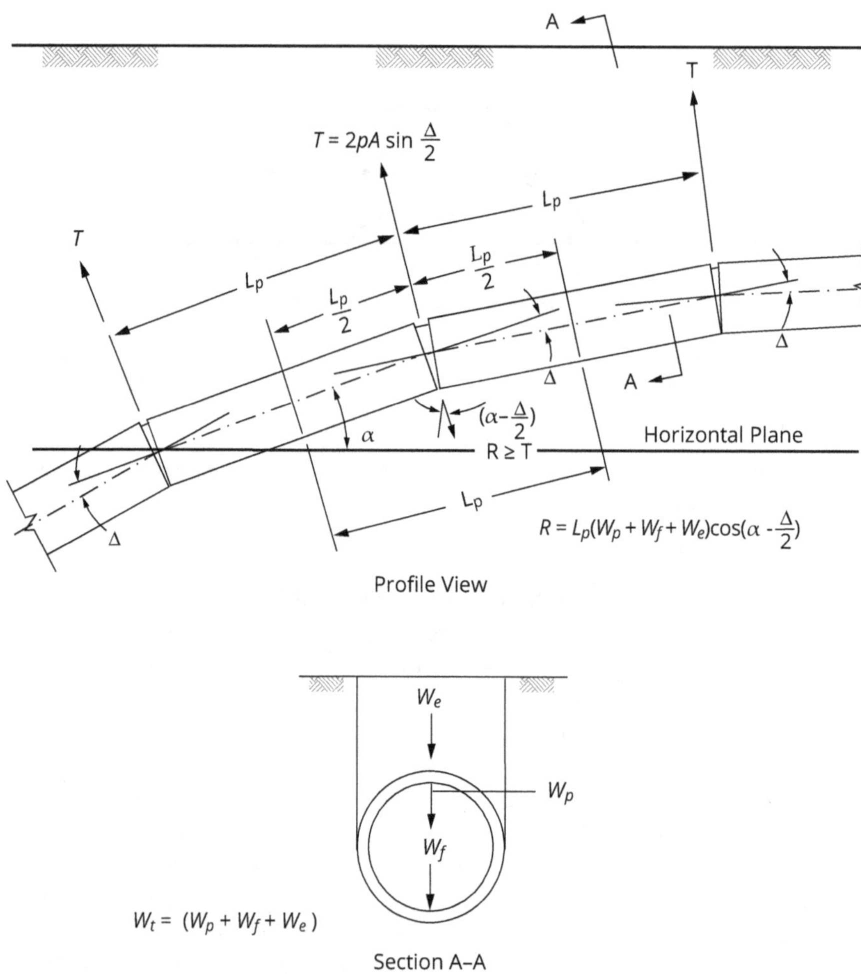

Figure 8-9 Restraint of uplift thrust at deflected joints on long-radius vertical curves

Downward thrust at deflected joints on long-radius vertical curves is resisted by bearing on the bottom of the pipe. There is seldom need to investigate thrust in this direction for properly bedded pipe.

THRUST RESTRAINT WITH WELDED OR HARNESSED JOINTS FOR VERTICAL BENDS

The design procedure for vertical bends with downward thrust (pushing down into the soil) is similar to the design procedure for horizontal bends. For vertical bends subjected to uplift thrust (pushing up and out of the soil), the thrust must be resisted by the dead weight of the pipe, fluid, and soil.

The total uplift thrust at the vertical bend is 2 pA sin $\Delta/2$ and the dead weight resistance per foot of pipe = $(W_p + W_f + W_e)$ cos $(\alpha - \Delta/2)$.

Therefore, the restraint length, L, for each leg of a vertical uplift bend is

$$L = pA\,(\sin \Delta/2)\,/\,[(W_p + W_f + W_e)\cos(\alpha - \Delta/2)] \qquad \text{(Eq 8-13)}$$

Types of Restrained Joints

Generally, there are two types of restrained joints: (1) welded and (2) harnessed. Both joint types as well as other restraining options are discussed in greater detail in chapter 6.

Restraint for Steep Slopes

When necessary to install a pipeline on a steep slope, it may be desirable to use anchor blocks, harnessed joints, or welded joints to keep the pipe from separating because of downhill sliding. Although the pipe may be capable of resisting downhill movement because of its own frictional resistance with the soil, the backfilling operation can sometimes provide enough additional downhill force to open a joint.

Special Restraint Conditions

Many special conditions require thrust restraint including (1) overlapping restraint length as a result of close proximity of fittings; (2) connections to pipe in casings; and (3) connections to structures.

Overlapping Restraint Length. In many configurations, fittings may be close to one another so that adjacent calculated restrained lengths overlap. The restraint should be calculated independently for each fitting, and the resultant lengths should not be added together.

Casing Pipe. Pipe installed in a casing without a grouted annular space cannot be considered as effectively restrained. Pipe installed in a casing with a fully grouted annular space may be considered effectively restrained, depending on the characteristics of the grout as placed.

Connections to Structures. Special provisions need to be employed to resist any residual thrust at structures and such provisions are beyond the scope of this manual.

REFERENCES

Alam, S., E.N. Allouche, C. Bartlett, A. Sherpa, and B. Keil. 2013. Experimental Evaluation of Soil-Pipe Friction Coefficients for Coated Steel Pipes. In *Proc. of ASCE 2013 Pipelines*, Ft. Worth, Texas. Reston, VA: ASCE.

Bardakjian, H. 1991. Draft Report, ACPPA Thrust Restraint Research. South Gate, CA.

Bardakjian, H. Corresponding Author for Subcommittee on Continuous Pipelines, ASCE Task Committee on Thrust Restraint Design of Buried Pipelines. 2011. Thrust Restraint Design and Analysis for Continuous Pipelines—Case Study of Steel Pipe Using M9, M11, and Fiberglass Analyses. In *Proc. ASCE 2011 Pipelines,* Seattle, Washington, July 2011. Reston, VA: ASCE.

Price, R.E. 1965. *Determination of Longitudinal Force Necessary to Move Pipe in Backfilled Trench.* Wharton, NJ: Wharton Research Center, International Pipe and Ceramics Corporation.

US Steel Corporation. 1981. *Soil Friction Tests on Tape-Coated Steel Pipe.*

Zarghamee, M.S., D.W. Eggers, R.P Ojdrovic, and D.P Valentine. 2004. Thrust Restraint Design of Concrete Pressure Pipe. *Journal of Structural Engineering,* ASCE, January, pp. 95–107.

Chapter 9

Pipe on Supports

This chapter is intended to address buried pipelines with sections that are installed aboveground. Due to topography within the pipeline right-of-way, these aboveground installations may be economical or environmentally advantageous. Where design of the aboveground portion of the pipeline is required to address pipe on a slope, conditions for temperature change or thrust, design of supports or penstocks, design procedures in manuals like *ASCE MOP No. 79* (ASCE 2012) should be consulted.

Pipe is supported in various ways, depending on size, circumstances, and economics. Pipe acting as a selfsupporting bridge may rest on suitably padded concrete saddles (Figures 9-1 and 9-2) or may be supported by ring girders or flange rings welded to the pipe (Figures 9-3 through 9-5). The kind of support selected may be determined by installation conditions or by economics. With saddle design, the cost of the pipeline can normally be reduced from the cost of ring girder construction, while providing greater flexibility with installation.

Small pipe within structures may be held by adjustable hangers or brackets or attached to building members. When subjected to temperature changes causing considerable longitudinal movement, steel pipe is frequently set on concave rollers. Data on adjustable hangers and rollers have been published (Roark 1954).

SADDLE SUPPORTS

There has been very little uniformity in the design or spacing of saddle supports. Spans have gradually increased because experience has proven that such increases were safe and practical. In general, the ordinary theory of flexure applies when a circular pipe is supported at intervals, is held circular at and between the supports, and is completely filled. If the pipe is only partially filled and the cross section at points between supports becomes out-of-round, the maximum fiber stress is considerably greater than indicated by the ordinary flexure formula, being highest for the half-filled condition (Schorer 1933).

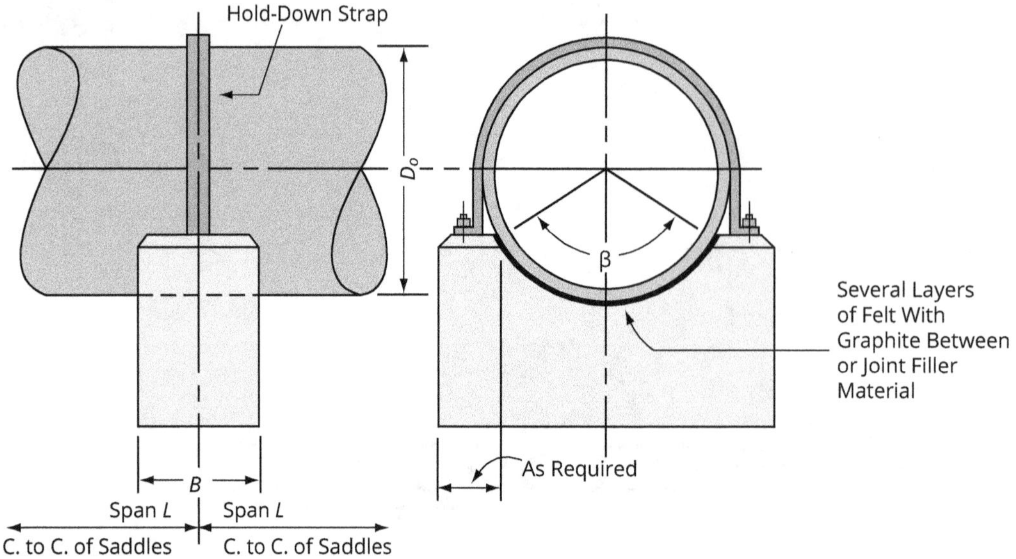

Pipe acting as a self-supporting bridge may rest on suitably padded concrete saddles.

Source: Barnard 1948.

Figure 9-1 Details of concrete saddle

Figure 9-2 Saddle supports for 78-in. pipe

Figure 9-3 Ring girders provide support for 54-in.-diameter pipe

The rings are supporting a 54-in.-diameter pipe laid on a slope.

Figure 9-4 Expansion joints between stiffener rings

This block anchors a 66-in.-diameter pipe against longitudinal movement.

Figure 9-5 Anchor block

In the case of a pipe carrying internal pressure where the ends are fully restrained, the Poisson ratio effect of the hoop stress, which produces axial tension, must be added to the flexural stress to obtain the total beam stress. A maximum deflection or sag of $1/360$ of the span length between supports has traditionally been used. However, structural steel guidelines only specify that deflection not impair the serviceability of the structure, and values as high as 1/240 of the span length can be used in certain circumstances (AISC 2010 and ASCE 2012).

Saddle supports may cause relatively high local stresses both longitudinally and circumferentially in unstiffened, comparatively thinwall pipe at the tips and edges of the supports. The highest local stresses are circumferential bending stresses at the saddle tips. Stresses vary with the load, the diameter–wall thickness ratio, and the angle of contact with the pipe. In practice, the contact angle varies from 90° to 120°. Contact angles greater than 120° are typically not used due to potentially increased difficulty during installation, coupled with the limited stress reduction benefits of using saddles over 120°. Supports are typically designed using a radius greater than that of the cradled pipe in order to account for pipe out-of-roundness and diametral tolerances. For equal load, the stresses are less for a large contact angle than for a small one, and interestingly, their intensity is practically independent of the width of the saddle (dimension B, Figure 9-1). The width of the saddle may, therefore, be that which is most desirable for good pier design.

Because saddle supports cause critical points of stress in the metal adjacent to the saddle edges, it is frequently more economical to increase the wall thickness of the pipe when it is overstressed than to provide stiffening rings. This is especially true where pipe sizes are 36 in. in diameter and smaller. Even a small increase in wall thickness has a great stiffening effect. The whole length of the span may be thickened, or only a short length at the saddle support need be thickened. The minimum length of reinforcement can be calculated as

$$L_r = B + 2\left(\frac{\sqrt{r_o t}}{1.285}\right) \quad \text{(Eq 9-1)}$$

Where:

L_r = total length of saddle reinforcement, in.

r_o = pipe radius, in.

t = pipe wall thickness, in.

B = saddle width, in.

When pipe lengths resting on saddles are joined by flanges or mechanical couplings, the strength and position of the joints must be carefully evaluated to safely address any bending and shear forces while remaining tight. Ordinarily it is advisable to place joints at, or as near as practicable to, the point of zero bending moment in the span or spans. The bending and shear forces are minimized by placing the coupling between two supports with a typical maximum spacing of one pipe diameter. Manufacturers of mechanical joints should be consulted regarding the use of their joints on self-supporting pipe spans.

Secure anchorages may be required at intervals in multiplespan installations to limit vertical, axial, or lateral movement to acceptable levels.

Research (Stokes 1965) has shown that, for pipelines supported by saddles, secondary stresses at the supports are large enough to create critical conditions only near the saddle tips. The highest stress is the circumferential bending stress, which tends to decrease as the internal pressure increases. Therefore, the critical condition is usually with the pipe full but at zero pressure. This stress can be calculated from

$$\sigma_{cs} = k\frac{P}{t^2}\ln\left(\frac{r_o}{t}\right) \quad \text{(Eq 9-2)}$$

Where:

σ_{cs} = local bending stress at saddle, psi

k = 0.02 – 0.00012 (β-90), contact angle factor

β = contact angle, degrees (see Figure 9-1)

P = total saddle reaction, lb

r_o = pipe radius, in.

t = pipe wall thickness or wrapper thickness, in.

If a longitudinal stress exists near the saddle tips, such as a thermal stress or the beam bending stress at that depth on the pipe, designate its calculated value as σ_{ls} (see Eq 9-5). Then calculate the equivalent stress, σ_e:

$$\sigma_e = \sqrt{\sigma_{cs}^2 + \sigma_{ls}^2 - \sigma_{cs}\sigma_{ls}} \quad \text{(Eq 9-3)}$$

This stress (σ_e) must not exceed the yield point. It is not necessary to apply a safety factor because tests have shown that because this is a very localized condition, the resulting design will have a practical safety factor of two.

The bending stress when the pipe is under pressure is calculated by multiplying σ_{cs} by a reduction factor (RF) calculated from

$$RF = \frac{\tanh(\beta)}{\beta} \quad \text{(Eq 9-4)}$$

Where:

$\beta = 1.1(r_o/t)(\sigma_h/E)^{1/2}$

σ_h = hoop stress, psi

E_S = modulus of elasticity, psi (30,000,000 for steel)

tanh = hyperbolic tangent

The hoop stress equals the sum of the membrane stress caused by pressure (usually tension) and the membrane stress at the tip of the cradle caused by the supported load (usually compression). It must be added to the reduced bending stress to get the total circumferential stress. It is usually not necessary to make this calculation because the zero pressure condition controls the design. The constant of 1.1 in the reduction factor was experimentally calibrated for a 150° saddle and is considered reasonable for a 120° saddle.

As with all support systems, the maximum beam bending stress for the pipe span must be calculated and limited to a suitable allowable stress. It is usually not necessary to add the beam bending stress at the bottom of the pipe at the support (e.g., at an intermediate support in a continuous span arrangement) to a secondary saddle stress, as was sometimes done in past procedures, because Stokes has shown that these stresses are much smaller than those given in Eq 9-2. As mentioned previously, if the pipe is under pressure and the ends are restrained, the Poisson's ratio effect of the hoop stress ($0.30\sigma_h$) must be added to the beam flexural stress. The total longitudinal tension stress (σ_t) is calculated as

$$\sigma_t = \sigma_y + 0.30\sigma_h \quad \text{(Eq 9-5)}$$

Example of Combined Stresses:

49.750-in. outside diameter (24.875-in. radius) by 0.313-in. wall pipe

42-ksi yield steel (σ_Y = 42,000 psi)

40,000-lb total vertical reaction on 120° saddle (calculated using shear force and bending moment diagrams for the given support scenario)

7,800-psi longitudinal stress in compression assuming a +40°F thermal cycle (see chapter 6)

$$\sigma_T = 40°\text{F} \times 195 \text{ psi/°F} = 7{,}800 \text{ psi}$$

1,600-psi longitudinal bending stress at saddle tips (compressive at outside); $D/4$ below center of pipe for 120° saddle angle (calculated using shear force and bending moments for the given support scenario)

$$k = 0.02 - 0.00012 \times (120 - 90) = 0.0164$$

$$\sigma_{cs} = 0.0164 \times \frac{40{,}000}{(0.313)^2} \times \ln\left(\frac{24.875}{0.313}\right) = \pm\, 29{,}300 \text{ psi}$$

$$\sigma_{ls} = \sigma_T + \sigma_b \quad \text{(Eq 9-6)}$$

$$\sigma_{ls} = -7{,}800 \text{ psi} + (-1{,}600 \text{ psi}) = -9{,}400 \text{ psi}$$

$$\sigma_e = \left[(29{,}300)^2 + (-9{,}400)^2 - (-9{,}400 \times 29{,}300)\right]^{1/2} = 35{,}000 \text{ psi}$$

$$35{,}000 \text{ psi} < 42{,}000 \text{ psi}$$

Beam stresses must still be checked by Eq 9-5.

The flexural stress σ_y should be calculated in the usual manner: $\sigma_y = M_l r_o / I_S$. In single spans, this stress is maximum at the center between supports and may be quite small over the support if flexible joints are used at the pipe ends. In multiplespan cases, the flexural stress in rigidly joined pipe will be that determined by continuous beam theory.

PIPE DEFLECTION AS BEAM

In the design of free spans of pipe, the theoretical deflection should be determined in order to judge flexibility or ascertain that the deflection does not exceed an acceptable upper limit. Freely supported pipe sometimes must be laid so that it will drain fully and not pool water between supports. The allowable deflection or sag between supports must be found to determine the necessary grade.

In any given case, the deflection is influenced by the conditions of installation. The pipe may be a single span or may be continuous over several supports. The ends may act as though free or fixed. In addition to its own weight and the weight of the water, the pipe may carry the weight of insulation or other uniform load. Concentrated loads, such as valves, other appurtenances, or fittings, may be present between supports.

The maximum theoretical deflection for a simple span can be determined using

$$y = 22.5\frac{WL^3}{E_S I_s} \qquad \text{(Eq 9-7)}$$

Where:

y = maximum deflection at center of span, in.

W = total load on span, lbf

L = length of span, ft

E_S = modulus of elasticity, psi; 30,000,000 psi for steel pipe

I_s = moment of inertia of pipe, in.^{4} = $\pi(D_o{}^4 - D_I{}^4)/64$

Except for some changes in unit designation, this is the standard textbook formula for uniformly distributed load and free ends. It can be used for concentrated loads at the center of the span, and it can be applied to other end conditions by applying a correction factor described later in this chapter.

METHODS OF CALCULATION

The following methods of calculating deflection are based on the formulas commonly found in textbooks for the cases given. Maximum deflection in a given case can be calculated by first assuming that the load is uniformly distributed and the ends are free. This is case 1 below. Later this result can be modified if the load is concentrated or the ends are fixed (cases 2, 3, and 4 below). The deflection for case 1 may be calculated using Eq 9-4. Note that in cases 1 and 2 the load W is the total uniformly distributed load on the span, but in cases 3 and 4 it is the load concentrated at the center of the span.

The four most commonly encountered conditions, with their corresponding deflection factors, are

Case 1: If the load W is uniformly distributed and the ends are free, the deflection is calculated using Eq 9-5.

Case 2: If the load W is uniformly distributed but the ends are fixed, the deflection is 0.2 times that for case 1.

Case 3: If the load W is concentrated at the center and the ends are free, the deflection is 1.6 times that for case 1.

Case 4: If the load W is concentrated at the center and the ends are fixed, the deflection is 0.4 times that for case 1.

The deflections caused by different loads are additive. Therefore, if a uniformly loaded pipe span contains a concentrated load, the calculated deflection for the latter is added to that for the uniform load, and the total sag in the pipe is the sum of the two deflections.

GRADIENT OF SUPPORTED PIPELINES TO PREVENT POCKETING

If intermittently supported pipelines are to drain freely, they must contain no sag pockets. To eliminate pockets, each downstream support level must be lower than its upstream neighbor by an amount that depends on the sag of the pipe between them. A practical average gradient of support elevations to meet this requirement may be found by using the following formula (Wilson and Newmark 1933):

$$G = \frac{4y}{L} \qquad \text{(Eq 9-8)}$$

Where:

G = gradient, ft/ft

L = span, ft

y = midspan deflection from pipe dead load without weight of water, ft

The elevation of one end should be higher than the other by an amount equal to four times the deflection calculated at midspan of the pipe.

Example: If the deflection of an insulated, 20-in. OD, 0.375-in. wall thickness pipe is 0.033 ft in a simple, freeended 50-ft span, what should be the grade of a series of 50-ft spans to allow drainage?

Solution:

$$G = 4(0.033)/50 = 0.0027 \text{ ft/ft}$$

It has been suggested (Roark 1954) that in the interest of satisfactory operation, the calculated theoretical deflection should be doubled when determining the slope of the pipeline gradient. If this were done in the preceding example, the grade used would be 0.0054 ft/ft.

SPAN LENGTHS AND STRESSES

The span length to be used in any particular situation is frequently dependent on economics. Longer spans result in fewer piers and typically lower cost, but they may require substantially heavier support rings or ring girders and greater pipe wall thicknesses over the supports and at midspan, which could materially offset any savings on the decreased number of piers. These factors, together with required distances between anchor points (changes in direction or slope), dictated by field conditions will influence the determination of span length. All of these factors are considered in making preliminary layouts, which will lead to the selection of a final layout.

Span lengths of pipe joined by sleeve-type or split-sleeve–type couplings may be limited by the allowable axial movement based on the anticipated temperature change.

Stresses considered between supports are

1. Longitudinal stresses caused by beam bending.
2. Longitudinal stresses caused by longitudinal movement under temperature changes and internal pressure.
3. Circumferential (hoop) stress as a result of internal pressure.
4. Equivalent stress based on the Hencky-von Mises theory of failure.

Stresses considered at supports are

1. Circumferential stresses in supporting ring girder as a result of bending and direct stresses and tensile stress due to internal pressure.
2. Longitudinal stresses in the shell at support caused by beam bending, and stresses in the shell as a result of longitudinal movement of the shell under temperature changes and internal pressure.
3. Bending stresses imposed by the rigid ring girder.
4. Equivalent stress based on the Hencky-von Mises theory of failure.

For a pipeline laid out as a continuous beam between an anchor and an expansion joint, several combinations of span lengths should be studied to determine the optimum span lengths for existing conditions. Span lengths between supports and length of cantilevered sections adjacent to the expansion joint should be proportioned so that the longitudinal bending moment at the supports is equal to or approaches the moment for a fixed-end beam, $M = \pm\, WL^2/12$. More importantly, the slope or deflection of the free end of the cantilevered section should be equal to that of the free end of the adjoining cantilevered section ensuring minimal shear at the joint so that the connecting expansion joint will operate freely. The moments, reactions at the supports, and bending stresses are readily computed for any point along the continuous beam. Combined with these longitudinal stresses are the stresses due to longitudinal forces imposed on the shell in overcoming the forces of friction at the supports and expansion joints. The friction force is usually considered to be 500 lbf per circumferential ft. The stresses from these frictional forces are small but are combined with the longitudinal bending stresses when considering the combination of longitudinal and circumferential stresses. The circumferential or pr/t stress in the shell between supports is computed and combined with the longitudinal stress in accordance with the Hencky-von Mises theory.

$$\sigma_e^2 = \sigma_x^2 - \sigma_x\sigma_y + \sigma_y^2 \qquad \text{(Eq 9-9)}$$

Where σ_e equals equivalent stress and σ_x and σ_y are principal stresses. The equivalent stress is not permitted to exceed 50 percent of the minimum specified yield strength at working pressure and 75 percent of the minimum specified yield strength at transient/test pressure. This analysis may frequently result in a thick plate wrapper being used at supports.

A pipeline installed aboveground is supported either on concrete saddles or on piers, and in the latter case, ring girders or support rings can be provided to transfer the beam reactions through rocker assemblies, roller assemblies, or bearing plates to the concrete piers.

DESIGN EXAMPLE

The following is an example design of a pipe on supports. Note that in the interest of simplicity, the pipe is assumed to not be subjected to unusual loading (seismic loads, snow loads, etc.), although the basic procedure presented herein can be easily modified to accommodate such cases.

Example: Determine the required wall thickness for a 48-in. nominal diameter steel pipe with cement-mortar lining, flexible coating, 150-psi working pressure with a full 14.7-psi vacuum pressure, with a 50-ft span in a 120° saddle support, ΔT = 70°F, and saddle width, B = 12 in.

From chapter 4 example:

Outside diameter = 49.75 in.

Wall thickness for pressure (working and transient) = 0.187 in.

Wall thickness for handling = 0.200 in.

σ_Y = specified minimum yield = 40,000 psi

σ_U = specified minimum tensile = 60,000 psi

p_{cr} = 11.2 psi

Calculate wall thickness for full vacuum:

Step 1: Vacuum Design—A wall thickness of 0.200 in. provides an allowable external pressure 13.2 psi for a true circle and 11.2 psi with 1 percent out-of-roundness. For a full vacuum, try 0.250-in. wall thickness (chapter 4 example).

Using a wall thickness of 0.250 in. and 0.500-in. cement lining in Eq 4-5 for a round pipe,

p_c = 17.3 psi

t_a = 0.250 in. (steel) + 0.067 in. (cement lining equivalent) = 0.317 in.

$m = r_o/t_a = 78.47$

$(p_{cr})^2 - (\sigma_Y/m + (1 + 6m\Delta x)p_c)p_{cr} + \sigma_Y p_c/m = 0$ (Eq 4-7)

$(p_{cr})^2 - (40{,}000/78.47 + ((1 + 6 \times 78.47 \times 0.01)17.3)p_{cr} + (40{,}000 \times 17.3/78.47) = 0$

$(p_{cr})^2 - (608.5)(p_{cr}) + 8818.7 = 0$

$p_{cr} = [608.5 \pm (608.5^2 - 4(8818.7))^{½}]/2$

$p_{cr} = (608.5 \pm 578.8)/2$ = 14.9 psi or 594 psi;

by inspection 14.9 is correct

This value has a safety factor of 1.01 against a perfect vacuum of 14.7 psi.

Step 2: Check Beam Bending and Combined Stresses at Midspan—Midspan stress due to bending is

$$\sigma_y = M_l r_o / I_s$$

Where:

M_l = longitudinal bending moment at midspan = $w(L)^2/9$

w = weight of water, pipe, and any additional loads; for this example, the weight of water and pipe = 84.75 lb/lin in.

I_s = moment of inertia of the steel shell = $\pi \times$ (outside radius^4 – inside radius^4) / 4 = 11,908 in.^{4}

L = length of pipe between supports = 50 ft = 600 in.

r_o = outside radius of pipe = 24.875 in.

M_l = 84.75 lb/in × (600 in.)^2/9 = 3,390,000 lb in.

σ_y = 3,390,000 lb in.×24.875 in./11,908 in.^{4} = 7,082 psi

Hoop stress with 0.250-in. wall thickness is

$\sigma_h = \sigma_x = pD_o/2t$ = 150 psi × 49.750 in./2(0.250 in.) = 14,925 psi

Combined stress with hoop stress and bending stress is checked per Hencky-von Mises

$\sigma_e^2 = \sigma_x^2 - \sigma_x\sigma_y + \sigma_y^2$

$\sigma_e = ((14{,}925)^2 - (14{,}925 \times 7{,}082) + (7{,}082)^2)^{½}$ = 12,930 psi < 50 percent of yield or 20,000 psi

Check combined stresses due to temperature change and Poisson's stress

Thermal stress from Eq 6-2 $\Delta\sigma_T = 195 \times \Delta T$

$\Delta\sigma_T$ = 195 × 70 = 13,650 psi

Poisson's stress = 0.3 × Hoop stress = 0.3 × 14,925 psi = 4,480 psi

Combined longitudinal forces are then

σ_y = bending stress + thermal stress + Poisson's stress = 7,082 psi + 13,650 psi + 4,480 psi = 25,210 psi < 90 percent of minimum specified tensile stress or 54,000 psi (90 percent of 60,000 psi tensile steel)

Thermal stress is considered a secondary stress (see chapter 6 for explanation). A conservative limitation of secondary stress acting alone is 90 percent of the minimum specified tensile strength of the steel cylinder material (Luka and Ruchti 2008).

Step 3: Check Stress at Saddle Tips.

$$\sigma_{cs} = (kP/t^2) \times \ln (r_o/t)$$

Where:

$k = 0.02 - 0.00012(\beta - 90)$

β = contact angle in degrees = 120°

P = total saddle reaction = $1.143wL$ = 1.143 × 84.75 lb/lin in. × 600 in. = 58,100 lb

σ_{cs} = 0.0164 × 58,100 lb /(0.250 in.)^2 × ln (24.875 in. / 0.250 in.) = 70,100 psi

This stress is more than the yield strength and therefore a shorter length or a thicker shell at the support is necessary. Try a 0.500-in.-thick wrapper at the support. Total thickness is then 0.250 in. + 0.500 in. = 0.750 in.

σ_{cs} = 0.0164 × 58,100 lb/(0.750 in.)^2 × ln (24.875 in. / 0.750 in.) = 6,120 psi, which is less than 40,000 minimum specified yield

Step 4: Check Combined Stress at Supports—Check longitudinal stress caused by longitudinal movement under temperature changes and internal pressure.

Stress due to bending at the support with 0.750-in. wall thickness is

$$\sigma_y = M_l r_o / I_s$$

Where:

$M_l = w(L)^2/9$

w = weight of water, pipe, and any additional loads. For this example, the weight of water and pipe = 99 lb/lin in.

I_s = moment of inertia of the steel shell = π × (outside radius^4 – inside radius^4)/4 = 34,658 in^4

M_l = 99 lb/in × (600 in.)^2/9 = 3,960,000 lb in.

σ_y = 3,960,000 × 24.875/34,658 = 2,842 psi

Hoop stress with 0.750 in. wall is

$\sigma_h = \sigma_x = pD_o/2t$ = 150 psi×49.750/2(0.750) = 4,975 psi

Temperature stress from Eq 6-2 $\Delta\sigma_T = 195 \times \Delta T$

$\Delta\sigma_T$ = 195 × 70 = 13,650 psi

Poisson's stress = 0.3 hoop stress

Therefore Poisson's stress is = 0.3 × 4,975 psi = 1,490 psi

Combined longitudinal forces are then

σ_y = bending stress + thermal stress + Poisson's stress = 2,842 psi + 13,650 psi + 1,490 psi = 17,980 psi

Combined stress with hoop stress and longitudinal stress is checked per Hencky-von Mises

$\sigma_e^2 = \sigma_x^2 - \sigma_x\sigma_y + \sigma_y^2$

σ_e = ((4,975)^2 – (4,975 × 17,980) + (17,980)^2)$^{1/2}$ = 16,080 psi < 50 percent of yield or 20,000 psi

Step 5: Calculate Length of Wrapper.

The wrapper must be thickened for a distance of $1.56(r_o t)^{1/2}$

Thicker cylinder length = $1.56(24.875 \times 0.750)^{1/2} = 6.75$ in.

RING GIRDERS

For conditions where thickening of the pipe wall is not possible or is not practical, ring girders can be a reasonable solution. While ring girders can be used to address a variety of design issues, they are most commonly used to increase the length of spans between supports.

A satisfactory and rational design for ring girder construction was presented by Herman Schorer (1933) and fully described in Roark (1954) and AISI (1983). Higher maximum allowable design stresses can be used when the ring girder analysis is based on the more comprehensive formulas and coefficients published in the Boulder Canyon Project, Final Reports, USBR Bulletin No. 5, Part 5, *Penstock Analysis and Stiffener Ring Design* (US Bureau of Reclamation 1944).

Concrete Piers

Concrete piers should be designed for the vertical reactions at the support, for longitudinal forces resulting from frictional resistance as a result of longitudinal strain (Poisson's ratio) and temperature movements, and for lateral forces caused by wind and earthquake forces. The resultant of all forces under the most unfavorable conditions should intersect the base within the middle third to ensure that the footing is in bearing (compression) throughout. The pier must be stable against sliding. The vertical component of the resultant of all forces should not be less than the horizontal component of all forces divided by the coefficient of sliding friction at the base of the pier. The friction coefficient may vary from 0.35 to 0.65, depending on the underlying material. The base of the pier should be placed below the frost line. Steel reinforcement of concrete piers is usually limited to that required for temperature and shrinkage crack control.

Concrete Anchors

Pipelines supported aboveground and having expansion joints, unrestrained split-sleeve–type couplings, or sleeve-type couplings require anchors at all points of changes in slope or alignment. Where expansion joints are used, a spacing of more than 500 ft between anchors and expansion joints is not normally desirable because of the accumulation of longitudinal forces and the desirability of more fixed points during installation. Buried pipelines with welded joints normally do not require anchors at points of changes in slope or alignment. Buried pipelines with sleeve-type couplings, unrestrained split-sleeve–type couplings, or other gasketed field joints require anchors similar to those required for an aboveground installation.

RING-GIRDER CONSTRUCTION FOR LOWPRESSURE PIPE

General designs for two types of longspan pipe of the flow line variety are shown in Figure 9-6.

Type 1 Pipe

Usually recommended for crossing canals and other low places where a single length of pipe for spans up to 60 ft can be used, type 1 pipe may be made and shipped from the

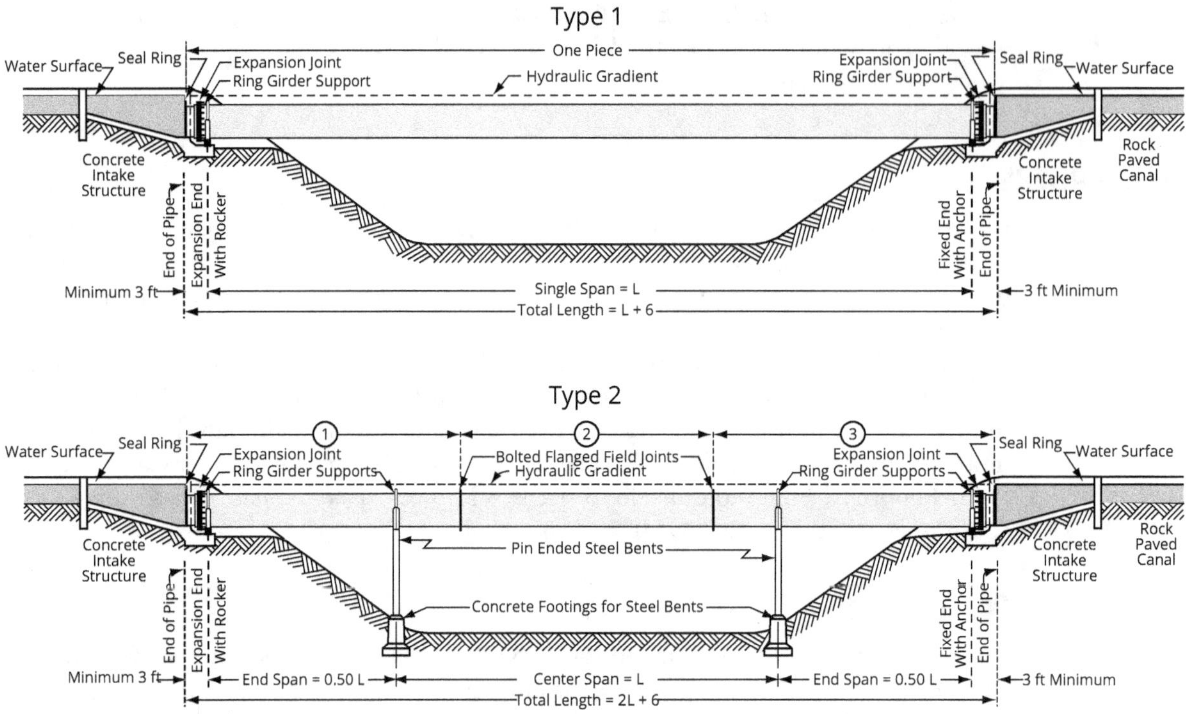

Figure 9-6 Long-span steel pipe for low pressures

factory in one length or in two lengths; in the latter case, a welded joint must be made in the field at the time of installation.

Type 2 Pipe

Used for crossing highways, canals, or rivers, where the length of the crossing makes it necessary to install two intermediate supporting columns, type 2 pipe is designed in three lengths with flanges welded to the ends of each length at points of contraflexure, together with expansion joints for both intake and outlet. This type is normally used for crossings 60 ft to 132 ft, with end spans half the length of the center span.

INSTALLATION OF RING GIRDER SPANS

In addition to proper design, longspan ring girder–supported steel pipelines require careful field erection, particularly in regard to alignment and camber; avoidance of movement caused by temperature differences on opposite sides of the pipe; and correct welding procedure. The following suggestions will be helpful, and more information has been published.

Pipes such as these (see Figure 9-7) that may be exposed to low temperatures can affect the ability of the steel to resist brittle fracture. See section on "Effects of Cold Working on Strength and Ductility," in chapter 1. Steel should be properly selected, detailed, and welded to mitigate this effect.

Concrete Footings

Before assembling the pipe, concrete footings (but not the intake or outlet boxes) should be poured. If the pipe is supported on rollers, a pocket is left at the top of the footings as a base for the roller bed plates. If steel bents are used, anchor bolts are set in concrete footings to anchor the lower end of the pinended steel bents or the base plates. The concrete

Figure 9-7 Ring girders on 111-in. pipe

footings should be sized to allow for grouting these supporting members to their proper height.

Expansion Joints

Expansion joints are installed in longspan steel pipe to allow for expansion or contraction caused by temperature changes. These joints are placed near the concrete headwalls, and expansion joint limit rods and packing bolts should be left entirely loose until the concrete has been allowed to set for at least 2 weeks. If expansion joints are tightened before concrete is poured, the pipe may pull loose from the green concrete. After concrete has set thoroughly, expansion joints are tightened and all danger of damage from pipe movement is eliminated.

To protect the expansion joint during shipment, the manufacturer may have to tack-weld steel ties to the inside of the pipe, tying the two pieces of pipe together across the joint. When this is done, the steel ties must be removed from the pipe as soon as it is set in place and before the concrete is poured.

Assembling Pipe

Pipe being assembled should be supported by a temporary framework between piers. All bolts except expansion joint bolts should be tightened. When the pipe is in place, concrete intake and outlet boxes should be poured. Bed plates for the rollers or pinended steel bents can then be grouted in place to the proper height. Temporary supports and blocking should be removed before the pipe is filled with water, otherwise the structure will be subjected to undue stress.

REFERENCES

American Institute of Steel Construction (AISC). 2010. *Specification for Structural Steel Buildings*. AISC 360-10. Chicago: American Institute of Steel Construction.

American Iron and Steel Institute (AISI). 1983. Welded Steel Pipe. In *Steel Plate Engineering Data,* Vol. 3. Washington, DC: AISI.

American Society of Civil Engineers (ASCE). 2012. *ASCE Manual of Practice No. 79. Steel Penstocks*. Reston, VA: ASCE.

Barnard, R.E. 1948. Design Standards for Steel Water Pipe. *Jour. AWWA,* 40(1):24.

Luka, J.L., and G. Ruchti. 2008. Axial Joint Design for Welded Buried Steel Pipe. In *Proc. ASCE Pipeline Conference, Atlanta, GA: Pipelines 2008*. Reston, VA: American Society of Civil Engineers (ASCE).

Roark, R.J. 1954. *Formulas for Stress and Strain*. New York: McGraw-Hill Book Co.

Schorer, H. 1933. Design of Large Pipelines, *Trans. ASCE,* 98:101.

Stokes, R.D. 1965. *Stresses in Steel Pipelines at Saddle Support*. Civil Engineering Transactions, The Institution of Engineers, Australia, October 1965.

US Bureau of Reclamation. 1944. *Penstock Analysis and Stiffener Ring Design*. Bull. 5, Part V. Tech. Invest. Boulder Canyon Project, Final Design Reports. Denver, CO: US Bureau of Reclamation.

Wilson, W.M., and N.M. Newmark. 1933. *The Strength of Thin Cylindrical Shells as Columns*. Bull. 255. Urbana, IL: Univ. of Illinois, Engrg. Exp. Stn.

The following references are not cited in the text.

AISI. 1982. *Steel Penstocks and Tunnel Liners. Steel Plate Engineering Data,* Vol. 4. Washington, DC: AISI.

Bier, P.J. 1940. Siphon Self-Supporting in Long Spans. *Engineering News-Record,* 124:852.

Bier, P.J. 1949. *Welded Steel Penstocks—Design and Construction*. Engineering Monograph 3. Washington, DC: US Bureau of Reclamation.

Cates, W.H. 1950. Design Standards for Large-Diameter Steel Water Pipe, *Jour. AWWA,* 42:860.

Crocker, S., ed. 1954. *Piping Handbook,* 4th ed. New York: McGraw-Hill Book Co.

Foster, H.A. 1949a. Formulas Facilitate Design of Ring-Supported Pipes. *Civ. Engrg.,* 19:629.

Foster, H.A. 1949b. Formulas Indicate Earthquake Forces in Design of Ring Girder-Supported Pipes. *Civ. Engrg.,* 19:697.

Timoshenko, S. 1936. *Theory of Elastic Stability,* 1st ed. Engrg. Soc. Monographs. New York: McGraw-Hill Book Co.

Younger, J.E. 1935. *Structural Design of Metal Airplanes*. New York: McGraw-Hill Book Co.

Chapter 10

Principles of Corrosion and Corrosion Protection

Corrosion is the deterioration or degradation of a material's mechanical or physical properties that results from a chemical or electrochemical reaction with a material's environment. The process of corrosion can be complex and detailed explanations even more so. Detailed technical publications on the subject are available including *Corrosion Basics: An Introduction*, 2nd Edition (NACE 2006), and *Corrosion Prevention by Protective Coatings*, 2nd Edition (Munger and Vincent 1999). AWWA Manual M27, *External Corrosion Control for Infrastructure Sustainability*, is also a general reference.

To better understand various methods for protecting steel pipe from corrosion, it is necessary to understand the basic conditions required for corrosion to occur in a piping system. This chapter and chapter 11 explain these various corrosion protection methodologies and how they disrupt the continuity of the corrosion cell. Although many of these methods apply to all metallic pipes or castings, both chapters specifically address corrosion and corrosion protection of steel piping systems.

GENERAL CORROSION THEORY

Entropy is a thermodynamic principle that, in brief, states that all materials eventually change to the state that is most stable under prevailing conditions. Most structural metals, having been converted from an ore, tend to revert to an ore. This reversion is an electrochemical process—a chemical change accompanying the passage of an electrical current. Such a combination is termed an *electrochemical cell*.

The electrochemical cell shown in Figure 10-1 illustrates the current flow that takes place when corrosion of metals occurs. An electrochemical corrosion cell consists of four components: an anode, a cathode, an electrolyte, and a metal path between the anode and cathode. If any one of these four components is missing, corrosion cannot occur.

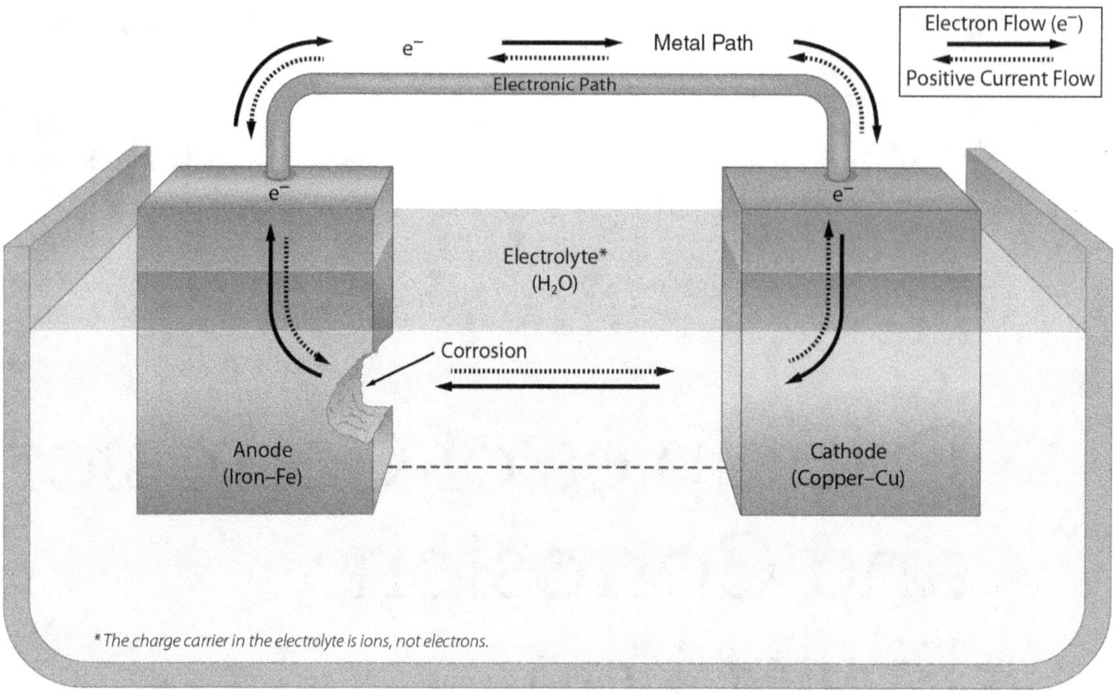

Source: Courtesy of NACE International.

Figure 10-1 Electrochemical corrosion cell

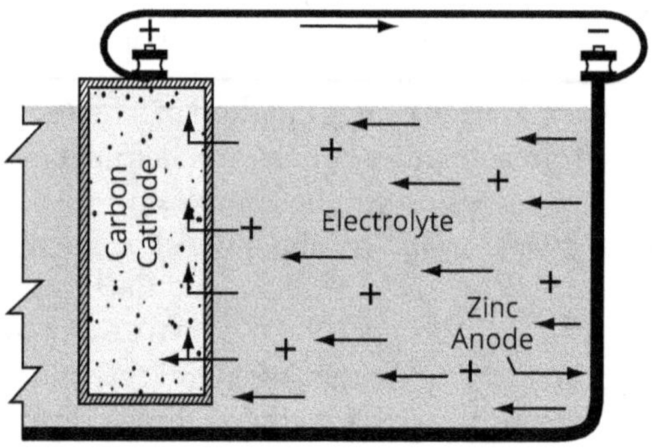

Figure 10-2 Electrochemical corrosion cell—alkaline flashlight battery

An alkaline battery is an example of a common corrosion cell (Figure 10-2). When the battery is connected in a metallic circuit, current flows from the zinc case (anode) into the electrolyte, changing metallic zinc to zinc ions while simultaneously releasing electrons that travel though the metallic path to the cathode. Conventional current flow is from the zinc (anode), through the electrolyte to the carbon rod (cathode), and back to the zinc anode through the metallic path. Zinc will be consumed (corroded) in proportion to the magnitude of corrosion current discharged, but the carbon rod is protected from corrosion.

Released electrons from the anodic reaction travel through the metallic path (electron current flow) to accumulate at the cathode. There they react with hydrogen ions in the electrolyte to form hydrogen gas in acidic solutions or hydroxyl ions in aerated and/or alkaline conditions. Cathodic reaction products are called *polarization* and impede or slow the anodic reaction rate, thereby reducing the magnitude of corrosion current and related metal losses. Polarization can be removed (depolarization) by dissolved oxygen, water flow, movement through the electrolyte, bacteria, and so on, but can also be maintained using cathodic protection (CP) to provide long-term corrosion protection to the cathode or steel pipe. If cathodic protection is removed or becomes nonfunctional, polarization will dissipate and corrosion activity on a pipe of structure may resume.

TYPICAL CORROSION CELLS

Several forms of corrosion occur on metals. Some of the more common corrosion cell types are

- General corrosion—a uniform type of corrosion that occurs over the entire surface of a metal due to the existence of very small corrosion cells on the metal surface. It commonly occurs on uncoated metallic pipe in atmospheric exposures.
- Localized or pitting corrosion—in localized attack, all or most of the metal loss occurs at discrete areas. Pitting and crevice corrosion are considered localized corrosion.
- Galvanic or dissimilar metal corrosion—a galvanic cell develops when two (or more) dissimilar metals are connected together in an electrolyte. Because of a potential difference between the metals, current will discharge from the more active or negative metal (anode) causing corrosion. The greater the potential difference between the metals, the more rapidly the anode corrodes.
- Dealloying—dissolution of one or more alloy components of a metal, such as graphitization corrosion associated with cast or ductile iron pipe.

Environmental factors can increase or decrease the rate of corrosion within a corrosion cell by increasing the chemical reaction rate at the anode or cathode. Environmental factors that will affect corrosion rates are

- Velocity effects—impingement, cavitation, or erosion corrosion is an attack accelerated by high-velocity flow effects. Cavitation, flow-assisted erosion corrosion, and fretting are considered velocity effect corrosion.
- Microbiologically influenced corrosion (MIC)—certain bacteria and other microbes can create corrosive environments internally or externally on pipelines.
- Concentration cells—various types of concentration cells can occur because of differences in oxygen, metal ions, pH, and other electrolyte components.
- Temperature change—corrosion rates will increase or decrease in proportion to temperature, with below 32°F having no corrosion and 212°F, the greatest corrosion rate at atmospheric pressure. A common rule is that corrosion rates can double or halve for each 50°F change in temperature relative of room temperature.

Dissimilar Metal (Galvanic) Corrosion Cell

Dissimilar metal (galvanic) corrosion occurs when two components of dissimilar metals are electrically connected and exposed to a common electrolyte. Dissimilar metal corrosion is also commonly referenced as *galvanic corrosion.* Figure 10-3 shows a dissimilar metal corrosion cell between stainless-steel (cathode) and carbon steel fasteners (anode).

Some common examples of galvanic corrosion can also be found in Figures 10-4 through 10-7. The galvanic series shown in Table 10-1 identifies metals that will be anodic and those that will be cathodic in a dissimilar metal corrosion cell. Anodic metals (top of list) will corrode if connected in a circuit to a metal listed beneath it in the galvanic series and are in a common electrolytic, such as water, soil, or concrete.

Figure 10-3 Dissimilar metal corrosion between stainless-steel base metal and carbon steel fasteners

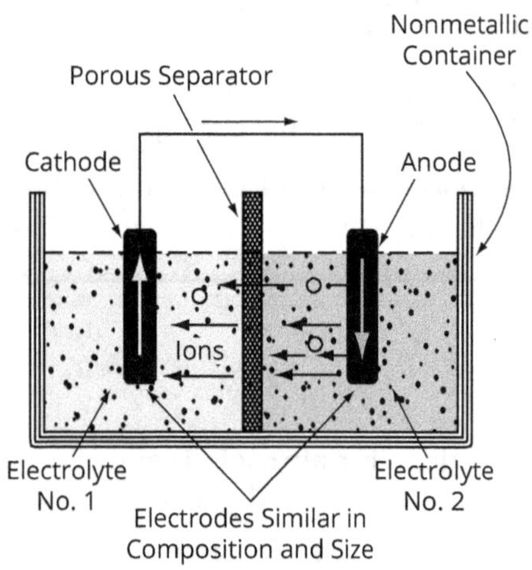

Figure 10-4 Corrosion cell—dissimilar electrolytes typical of car battery

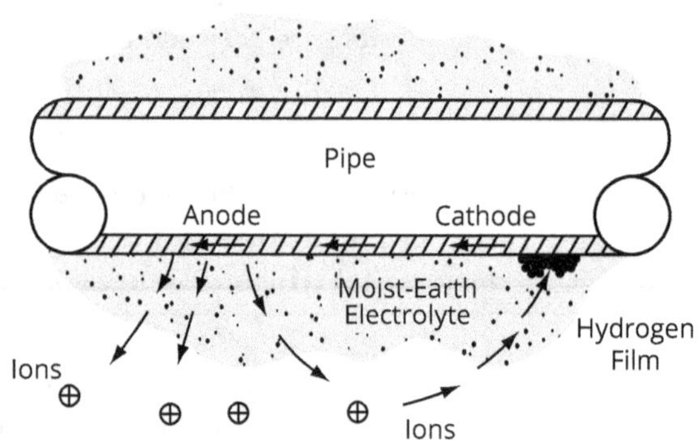

Moist earth is the electrolyte; two areas on the uncoated pipe are anode and cathode; pipe wall takes the place of wire in Figures 10-2 and 10-4. Pipe wall at anode will corrode like the zinc battery case; pipe wall at cathode site will not corrode but will tend to be protected as polarization develops and if not removed, will tend to build resistance to current flow and thereby eliminate or slow the corrosion of pipe wall at anode site. Coatings prevent cell formation.

Figure 10-5 Galvanic cell on embedded pipe without protective coating

The order of metals in Table 10-1 is known as the *galvanic series;* which is true for neutral electrolytes. Changes to the electrolyte, such as temperature or pH, may cause some metals to shift position or even reverse positions relative to other metals in the table. For example, zinc is listed above iron in the table and will corrode when connected to iron in freshwater at normal temperature. However, when the temperature of water is above 140°F, the metal order will reverse and iron now corrodes to protect zinc. Steel will shift its position to a more noble value in the galvanic series when embedded in concrete. Therefore, the table cannot be used to predict the performance of all metal combinations under all conditions but in normal water works operations the table is appropriate for use.

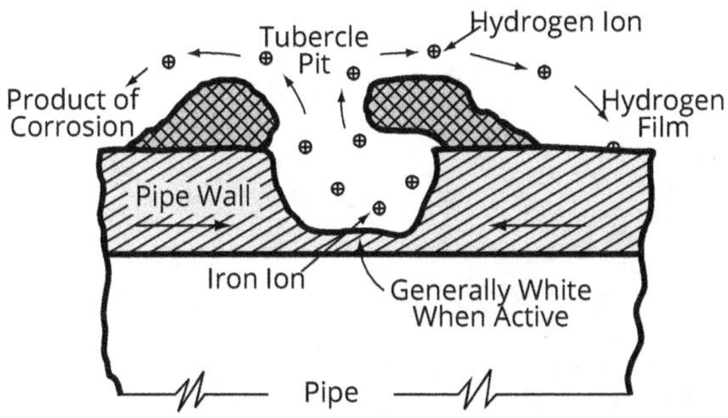

Detail of uncoated pipe wall at anode in Figure 10-5 is shown. As current leaves surface of anode, corrosion occurs as metal (ions) go into the electrolyte where they combine with other components in the electrolyte to form various compounds. These compounds can form various scales or rust in the case of iron or iron alloys, which may be protective to the remaining metal. These scales can provide some resistance to the movement of current from the pipe into the electrolyte but do not stop the corrosion process.

Figure 10-6 Galvanic cell—pitting action

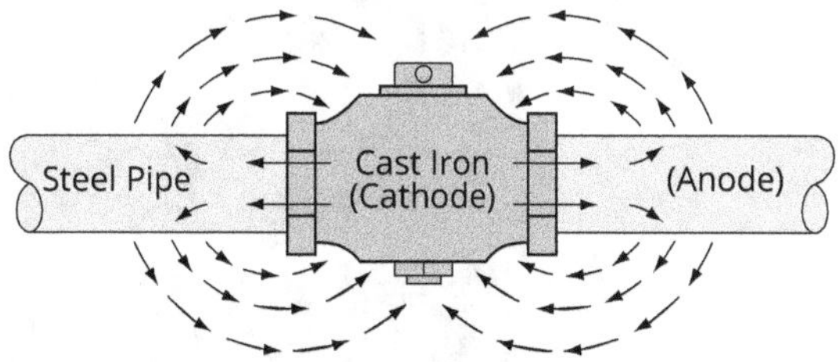

Cast-iron valve is the cathode (protected area), uncoated steel pipe is anode (corroding area), and surrounding earth is the electrolyte. As long as cathode is small in area relative to anode, corrosion is not ordinarily severe or rapid. If these area proportions are reversed, corrosion may be much more rapid.

Figure 10-7 Corrosion caused by dissimilar metals in contact on buried pipe

Table 10-1 Galvanic series of metals and alloys (in seawater)* at 77°F

Anodic, Active End at Top of Series (Read Down)†	
	Muntz metal
	Manganese bronze
Magnesium	Naval brass
Magnesium alloys	Nickel (active)
Zinc	Inconel—76% Ni, 16% Cr, 7% Fe (active)
Aluminum 52SH	Yellow brass
Aluminum 4S	Aluminum bronze
Aluminum 3S	Red brass
Aluminum 2S	Copper
Aluminum 53S-T	Silicon bronze
Alclad	Ambrac—5% Zn, 20% Ni, 75% Cu
Cadmium	70% Cu, 30% Ni
Aluminum 17S-T	Comp. G bronze—88% Cu, 2% Zn, 10% Sn
Aluminum 24S-T	Comp. M bronze—88% Cu, 4% Zn, 6.5% Sn, 1.5% Pb
Mild steel	Nickel (passive)
Wrought iron	Inconel—75% Ni, 16% Cr, 9% Fe (passive)
Gray iron and ductile iron	Monel—70% Ni, 30% Cu
Ni-resist	18-8 stainless steel, Type 304 (passive)
13% Cr stainless steel, Type 410 (active)	18-8, 3% Mo stainless steel, Type 316 (passive)
50–50 lead–tin solder	Titanium
18-8 stainless steel, Type 304 (active)	Silver
18-8, 3% Mo stainless steel, Type 316 (active)	Graphite
Lead	Gold
Tin	Platinum
	Cathodic, Noble End at Bottom (Read Up)†

Source: ASTM Standard G82-02, Guide for Development and Use of a Galvanic Series for Predicting Galvanic Corrosion Performance.

*Each environment has its own specific galvanic series. The relative positions of the various metals and alloys may vary slightly from environment to environment.

†In a galvanic cell of two dissimilar metals, the more active metal will act as the anode and be corroded, while the more noble metal will act as the cathode and be protected.

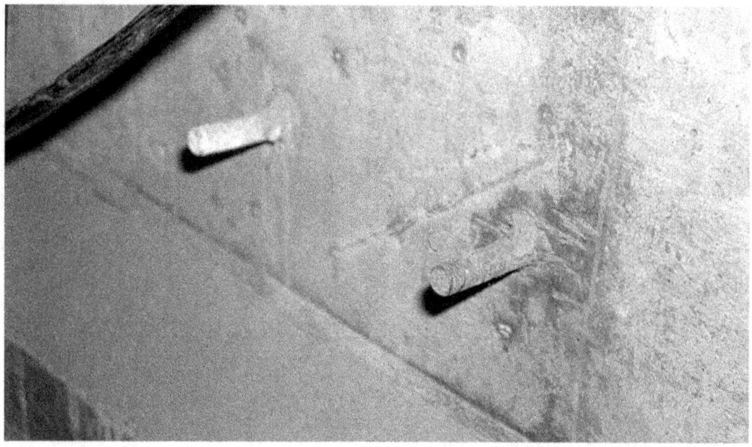

Figure 10-8 Corroding anchor bolt contacting reinforcement is subject to differential pH corrosion in water holding basin

Differential Oxygen Corrosion Cell

A differential oxygen concentration in soils is one reason underground corrosion of steel pipelines occurs. Differential oxygen concentration (or differential aeration) may be caused by unequal porosity of different soils, a rock against the pipe, saturated soil conditions caused by groundwater or river or drainage crossings, or restriction of air and moisture movement caused by the presence of buildings, roadways, or pavement.

Tubercles on the interior of unlined waterlines are caused by pitting corrosion that is commonly caused by differential oxygen cells.

Electrolytic Corrosion Cell

Electrolytic corrosion cells are one of the most common corrosion cells within water systems and can affect buried or submerged steel pipeline, unless properly mitigated. Concrete embedded or cement-mortar–coated steel is passivated by the highly alkaline environment. As a result, cement embedded steel will become more noble in the galvanic series in Table 10-1, to a potential near stainless steel or copper, making it a cathode relative to bare steel or iron.

Electrical contact with steel reinforcement in concrete structures is a major source of electrolytic corrosion cells. Figure 10-8 shows two galvanized steel anchor bolts that were immersed for 3 weeks in a concrete water basin where one bolt contacted the steel reinforcement and the other did not. Steel equipment that is connected to these anchor bolts would also be subject to electrolytic corrosion because of bolt contact with the reinforcement.

Common locations where electrolytic corrosion cells occur include bare pipe penetrations of a reinforced concrete wall or floor in vaults, manholes, and structures.

Microbiologically Influenced Corrosion

Microbiologically influenced corrosion (MIC) can be caused from a variety of living organisms including, but not limited to, a variety of bacteria, mussels, barnacles, and so on. These can be external or internal in the pipeline and supporting components.

Certain soil bacteria create chemicals—typically acids—that may result in corrosion. Bacterial corrosion, or anaerobicbacterial corrosion, is not so much a distinct type of corrosion as it is another cause of electrochemical corrosion. The bacteria cause changes in the physical and chemical properties of the soil (electrolyte) to produce active concentration corrosion cells. The bacterial action may remove the protective hydrogen film (polarization). Differential aeration may increase MIC by providing variable oxygen concentration cells. Differential aeration plays a major role in MIC.

While there are methods that provide a good indication that soil bacteria are present, the only certain way of determining the presence of anaerobic bacteria, the particular kind of microorganism responsible for this type of corrosion, is to secure a sample of the soil in the immediate vicinity of the pipe and develop a bacterial culture from that sample for analysis by a trained professional.

Crevice Corrosion

Crevice corrosion in a steel pipeline may be caused by a concentration cell formed where the dissolved oxygen of the water varies from one segment of the pipe to another. In an unprotected or uncoated crevice area, the dissolved oxygen is hindered from diffusion, creating an anodic condition that may cause metal to corrode inside the crevice. Crevice corrosion is commonly associated with stainless-steel alloys but can also occur with carbon steel.

Stress and Fatigue Corrosion

Stress corrosion cracking (SCC) is caused by tensile stresses that slowly build up in a corrosive atmosphere. Electrical potential and residual tensile stress have to be present with the correct electrolyte for SCC to occur. High temperatures can also accelerate SCC. Tensile stress, whether from residual stress or from stress caused by loading, is developed at the metal surfaces. At highly stressed points, accelerated corrosion can occur, resulting in increased tensile stress and potential failure when the metal's safe tensile strength is exceeded.

Fatigue corrosion occurs from cyclic loading. In a corrosive atmosphere, alternate loadings can cause fatigue corrosion substantially below the metal's failure in noncorrosive conditions.

Other Corrosion Cells

The electrochemical cells described previously demonstrate the fundamental principles of many kinds of electrochemical cells. Other common forms of corrosion encountered on unprotected buried pipelines are shown in Figures 10-9 through 10-14.

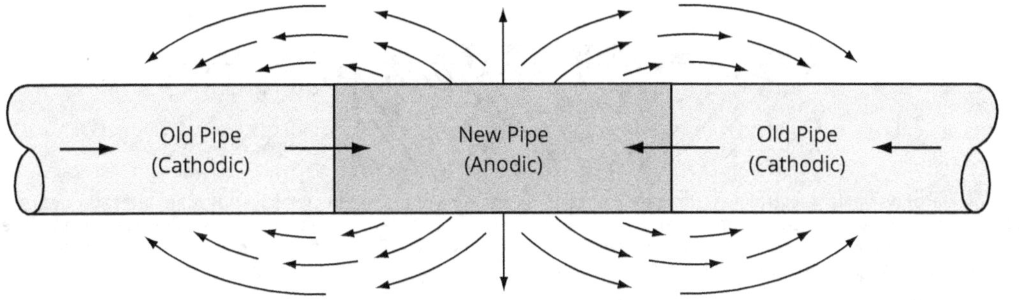

Although seldom considered, a galvanic cell is created by installing a piece of new uncoated pipe in an old line. New pipe always becomes the anode, and its rate of corrosion will depend on type of soil and relative areas of anode and cathode. Therefore, corrosion protection measures are essential.

Figure 10-9 Corrosion caused by new versus old steel pipe

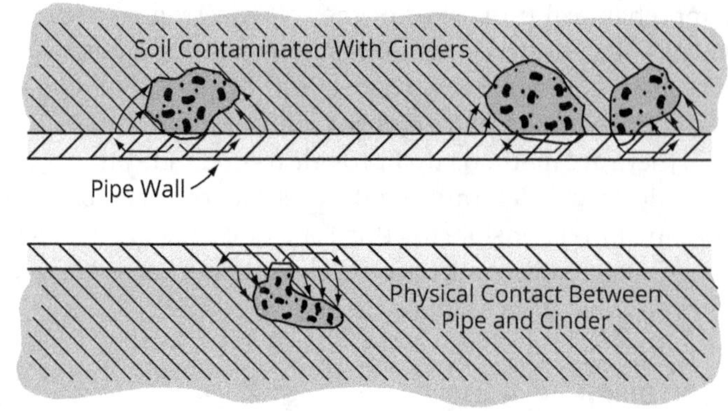

When uncoated metal pipe is laid in cinders, corrosive action is that of dissimilar metals. Cinders are one metal (cathode) and pipe the other (anode). Acid leached from cinders contaminates soil and increases its activity. No hydrogen collects on the cinder cathode, the cell remains active, and corrosion is rapid.

Figure 10-10 Corrosion caused by cinders

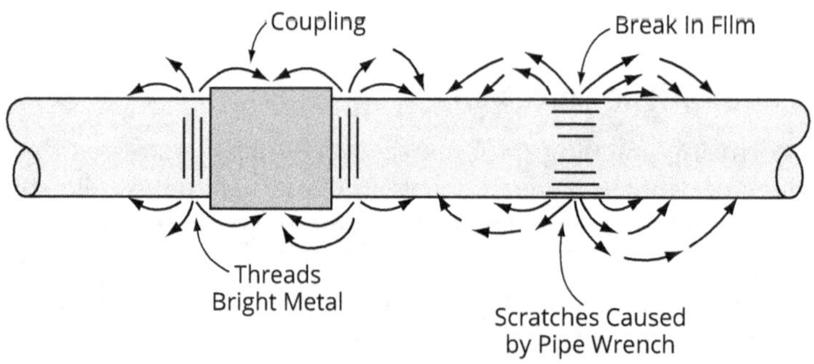

Bright scars or scratches of threads become anode areas in buried pipe, and the rest of the pipe is a cathodic area. In some soils, these bright areas can be very active and destructive because the small anode area and large cathode area produce the most unfavorable ratios possible.

Figure 10-11 Corrosion caused by dissimilarity of surface conditions

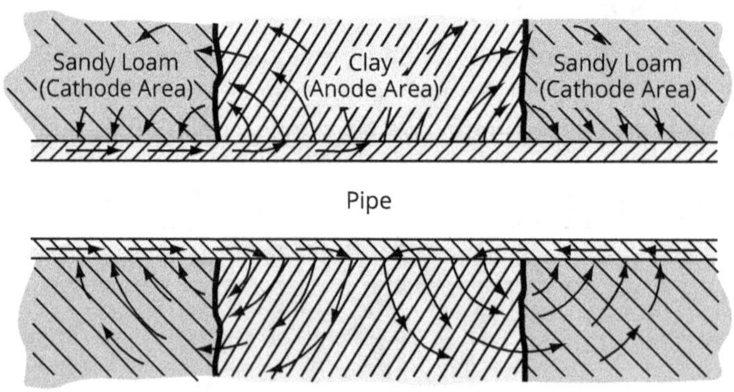

In this corrosion cell of dissimilar electrolytes (compare Figure 10-3), sections of uncoated pipe in sandy loam are cathodes (protected. areas), sections in clay are anodes (corroding areas), and soil is electrolyte. If resistance to electric-current flow is high in the electrolyte, the corrosion rate will be slow. If resistance to current flow is low, corrosion rate will be high. Thus, knowledge of soil resistance to electric-current flow becomes important in corrosion protection studies.

Figure 10-12 Corrosion caused by dissimilar soils

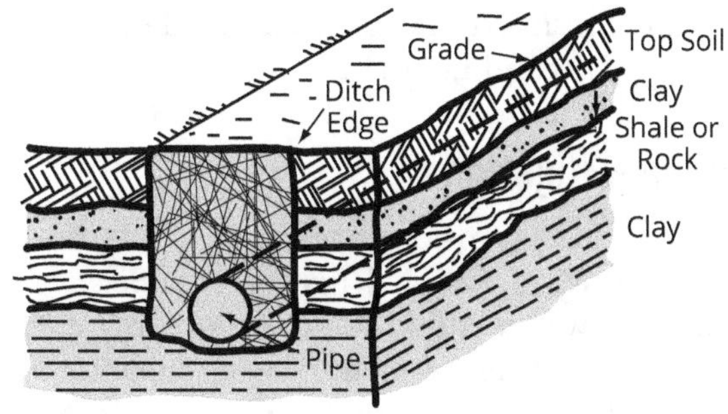

Dissimilarity of electrolytes, because of the mixture of soils, causes formation of a corrosion cell. If large clods of dirt, originally from different depths in the excavation, rest directly against uncoated pipe wall, contact area tends to become anodic (corroding area), and adjacent pipe cathodic. Small well-dispersed clods, such as result in trenching by machine, reduce cell-forming tendency. These corrosion cells having anode and cathode areas distributed around the circumference of pipe are often called short-path cells.

Figure 10-13 Corrosion caused by mixture of different soils

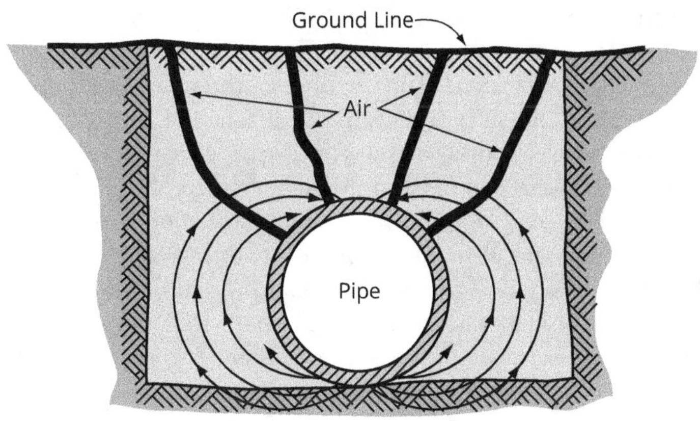

Moist or Saturated Soil, Poor or No Aeration

This is another galvanic cell of dissimilar-electrolyte type. Soil throughout the depth of the excavation is uniform, but a portion of the pipe rests on heavy, moist, undisturbed ground at the bottom of the excavation, while the remainder of the uncoated pipe is in contact with drier and more aerated soil backfill. Greatest dissimilarity—and the most dangerous condition—occur along a narrow strip at the bottom of the pipe, which is the anode of the cell.

Figure 10-14 Corrosion caused by differential aeration of soil

CORROSIVITY ASSESSMENT

Corrosion Survey

A corrosion survey should be conducted to review a proposed pipeline route for corrosive conditions that could affect long-term performance of the pipeline. Corrosion surveys commonly include field soil resistivity testing, chemical–physical analyses of soil samples, identification of stray current sources, and present and future land use and drainage that could affect corrosive conditions along the route. Soil sample laboratory analysis may include saturated soil box resistivity, soil pH, chlorides, sulfates, and sulfides.

Information from the corrosion survey can be used to determine pipeline coating requirements, corrosion monitoring system, the need for cathodic protection or electrical isolation, and the type of cathodic protection system required. In some situations a corrosion survey may indicate that cathodic protection will not be needed until proven necessary at a future date.

NACE SP0169, Control of External Corrosion on Underground or Submerged Metallic Piping Systems, or NACE SP0100, Cathodic Protection to Control External Corrosion of Concrete Pressure Pipelines and Mortar-Coated Steel Pipelines for Water or Waste Water

Service, should be consulted and used for guidance regarding corrosion surveys and assessment of corrosive conditions and risks.

Soil Corrosion Investigations

The first organized soil corrosion investigation was begun by the National Bureau of Standards (NBS) (now the National Institute for Science and Technology) in 1911. The purpose at that time was to study the effect of stray currents from street railway lines on buried metallic structures. In its initial investigation, the bureau found that in many instances where rather severe corrosion was anticipated, little damage was observed; whereas in others, more corrosion was found than seemed to be indicated by the electrical data associated with the corroded structure. These observations led to a second investigation, undertaken in 1921. Originally about 14,000 specimens were buried at 47 test sites, but the number was subsequently increased to 36,500 specimens at 128 test sites. The American Petroleum Institute and the American Gas Association collaborated in analyzing the results of the latter tests.

Burial sites were selected in typical soils representing a sampling of areas in which pipe was or might be buried. The purpose of the investigation was to determine whether corrosion would occur in pipelines in the absence of stray currents under conditions representative of those encountered by working pipelines. Tables 10-2 and 10-3 give summary data on the corrosivity of soils and the relationship of soil corrosion to soil resistivity.

The NBS soil corrosion tests were extensive, well-coordinated, and well analyzed. A final report on the studies made between 1910 and 1955, including over 400 references, was published (Romanoff 1957 [for the National Bureau of Standards]). An important finding was that in most soils the corrosion rate of bare steel decreased with time. This is largely because corrosion products, unless removed, tend to protect the metal.

Severity of Corrosion

Severity of corrosion in any given case will depend on many different factors, some of which may be more important than others. The factors most likely to affect the rate of corrosion are

- Conductivity or resistivity of electrolyte
- Uniformity of electrolyte
- Depolarizing conditions
- Type and composition of electrolyte
- Relative positions of metals in the galvanic series
- Size of anodic area with respect to cathodic area
- Relative location of anodic area with respect to cathodic area
- Resistance of metallic circuit

DC Stray Current Corrosion

Stray current corrosion, also known as *interference corrosion*, occurs when direct current (DC) in the earth collects on a pipeline and then discharges from that pipe, causing corrosion as the current returns to its source. Common sources of DC electricity in the earth are

- Impressed current cathodic protection systems on neighboring buried pipelines, which is mandated for oil and gas pipelines.
- Electrically discontinuous joints or joint bonds on cathodically protected pipes.
- DC transit or light rail transportation systems.

Table 10-2 Soils grouped in order of typical corrosive action on steel

Soil Group	Aeration and Drainage	Characterization	Soil Types
I—Lightly Corrosive	Good	Uniform color and no mottling anywhere in soil profile; very low water table.	Sands or sandy loams Light textured silt loams Porous loams or clay loams thoroughly oxidized to great depths
II—Moderately Corrosive	Fair	Slight mottling (yellowish brown and yellowish gray) in lower part of profile (depth 18–24 in.); low water table. Soils would be considered well drained in an agricultural sense, as no artificial drainage is necessary for crop raising.	Sandy loams Silt loams Clay loams
II—Severely Corrosive	Poor	Heavy texture and moderate mottling close to surface (depth 6–8 in.); water table 2–3 ft below surface. Soils usually occupy flat areas and would require artificial drainage for crop raising.	Clay loams Clays
IV—Unusually Corrosive	Very poor	Bluish-gray mottling at depths of 6–8 in.; water table at surface or extreme impermeability because of colloidal material contained.	Muck Peat Tidal marsh Clays and organic soils Adobe clay

Source: AWWA M27.

Table 10-3 Soil resistivity versus degree of corrosivity

Soil Group	Description	Resistivity, *ohm-cm*
I	Excellent	> 6,000
II	Good	6,000–4,500
III	Fair	4,500–2,000
IV	Bad	< 2,000

Source: AWWA M27.

The most common source of stray currents on water systems is from cathodic protection systems on neighboring pipelines. Impressed current cathodic protection systems use rectifiers that discharge DC electrical currents into the earth through ground beds to protect miles of pipe. The current returns to the rectifier through the wall of the protected pipe but may also take other parallel paths in proportion to path resistance. Another pipeline located within close proximity of a cathodic protection ground bed may collect some of the cathodic protection current, which can discharge from the pipe causing accelerated corrosion at pipe crossings or near the current source.

The diagram of a DC electricrified railway system shown in Figure 10-15 is another example of a DC electrical system that can cause stray currents. Many modern subway and light rail systems operate on the same principle. In Figure 10-15, direct current flows from the generator into the trolley wire, along this wire to the streetcar, and through the streetcar to the motors driving it. To complete the circuit, the return path for the current is

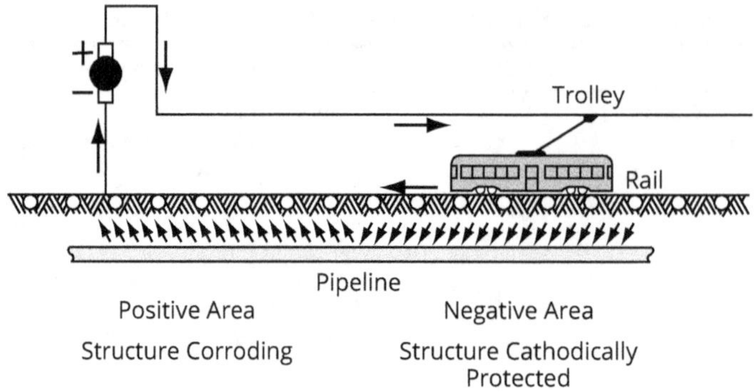

Figure 10-15 Stray-current corrosion caused by electrified railway systems

intended to be from the motors to the wheels of the streetcar, then through the rails to the generator. Because the earth is a parallel conductor, a portion of the current will leak from the rails and return to the generator through the earth. Current returning through the earth may use a buried metallic pipe as a secondary path where some of the earth return current enters the pipe and discharges from that pipe near the generator. Corrosion of the pipe occurs where current discharges.

The proper use of pipeline coatings and cathodic protection offers the best method for stray current corrosion mitigation. Cathodic protection (discussed in more detail later in the chapter) applies a current to the pipeline to reverse the effect of stray current discharging from the pipeline. Pipelines and appurtenances subject to DC stray currents can be subject to rapid metal losses if not properly protected from corrosion. Other methods for stray current mitigation include

- Proper cathodic protection design or pipeline route selection.
- Dielectric coatings to increase circuit resistance and minimize current collection.
- Pipeline coatings to decrease cathodic protection current densities at crossings.
- Galvanic anodes at current discharge locations (require monitoring and periodic replacement).

While bonds between a pipeline and transit system or another pipeline could provide a means to control stray current corrosion, this approach is the least desirable as it can increase corrosion and, in some situations, could create an unsafe operating condition on the pipeline. A corrosion engineer should be consulted to determine the most effective alternative for stray current mitigation.

AC Interference

AC interference primarily occurs on dielectrically coated steel pipe with a high coating resistance, but is minimal with cement-mortar–coated steel pipe due to its low resistance. Interference from AC extra high voltage (EHV) transmission lines is caused by the expansion and collapse of a magnetic field produced by current flow in EHV conductors. AC voltages are induced on coated metallic pipelines that parallel EHV transmission lines, with the magnitude of induced voltage proportional to the magnitude of current flow in the conductors and the length of the parallel.

Waterlines installed parallel to EHV transmission lines can cause the following issues on pipelines and issues for operating personnel:

- AC-induced voltage can produce unsafe touch potentials for operating personnel and at the highest levels can cause coating damage.

- AC faults under conductive conditions can result in pipe wall pitting damage or coating damage when the pipe is in close proximity to EHV towers (typically less than 50 ft).
- AC faults under inductive conditions can result in extremely high and unsafe touch potentials for operating personnel or coating damage during the fault, but risks are minimal as fault durations are less than one-half second.
- AC corrosion can occur on pipelines with induced AC voltages that are located in low to very low resistivity soils.

NACE SP0177 addresses many of the issues associated with AC interference, AC corrosion, the criteria for unsafe operating conditions, and mitigation alternatives. AC interference is best identified and mitigated with the assistance of computer modeling that can reasonably predict the magnitude and location of AC interference for the most cost-effective mitigation.

Personnel safety criteria presented in NACE SP0177 are based on an average resistance of 2,000 ohms. Safety criteria may need to be reviewed to account for wet conditions commonly associated with waterline condensation and moisture within vaults and manholes.

INTERNAL CORROSION PROTECTION

Corrosion of the internal surfaces of a pipe is principally caused by differential aeration and/or localized corrosion cells (Eliassen and Lamb 1953). The extent of corrosion on the interior of a pipe depends on the corrosivity of water or wastewater carried. Langelier (1936) developed a method for determining corrosive effect of different kinds of water on bare pipe interiors, and Weir (1940) extensively investigated and reported the effect of water contact on various kinds of early pipe linings.

Cement-mortar lining has been and continues to be the most common steel pipe lining and has provided excellent protection to the interior of steel pipe since the 1930s. Although unlined steel pipes have been pitted through by some waters, the principal result of interior corrosion is a reduction in flow capacity. This reduction is caused by a formation of tubercles of ferric hydroxide, a condition known as *tuberculation* (Linsey and Franzini 1979). Originally pipe linings were developed to maintain flow capacity, but today's lining materials also provide high levels of electrochemical corrosion protection. These modern linings have been developed (see chapter 11) to provide corrosion protection from aggressive water applications.

Where internal corrosion is allowed to persist, water quality deteriorates, pumping costs increase, hydraulic capacity decreases, leaks can occur, and costly replacement of the pipe may become inevitable. The occurrence of these problems can be eliminated by using quality protective linings as described in chapter 11.

ATMOSPHERIC CORROSION PROTECTION

Atmospheric corrosion of exposed metal pipelines can be significant, especially in industrial and seacoast areas. Protective coatings will protect against uniform corrosion in most atmospheric environments. Crevices, transitions from buried to exposed areas, insulated pipes, and other hidden areas can be significant areas for water pollutants and other contaminants to collect and accelerate corrosion. Each corrosion concern may require different methods of control. Where such corrosion is significant, the maintenance problem incurred is similar to that for bridges or other exposed steel structures.

EXTERNAL CORROSION PROTECTION

Due to the electrochemical nature of corrosion, common methods for corrosion protection for underground and underwater pipelines are as follows:

- Application of bonded dielectric pipe coatings that provides a chemically resistant barrier between the pipe and surrounding soil and water, thereby controlling corrosion; see chapter 11.
- Application of cement-mortar coatings that provide protection by producing an alkaline environment that passivates the steel surface; see chapter 11.
- Use of galvanic or impressed current cathodic protection in conjunction with a pipeline coating system.
- Electrical isolation between known corrosion cells to break the metallic path, such as dissimilar metal corrosion cell and differential pH corrosion cells.

At a minimum, steel pipes should be provided with a bonded dielectric or cement-mortar pipeline coating and a corrosion monitoring system that monitors the pipe for external corrosion. In addition, corrosion monitoring systems allow for the application of cathodic protection at the time of installation or at a future date depending on monitored corrosion activity. A corrosion monitoring system includes:

1. Bonding of all nonwelded or gasket pipe joints at the time of installation.
2. Corrosion monitoring stations installed for measuring current flow, pipe potentials, stray current influences, and pipeline electrical continuity testing.
3. Electrical isolation to control corrosion cells and isolate other facilities from the pipeline.

CP can be provided using a galvanic or an impressed current system, and can be applied to underground or submerged steel pipelines. CP is recommended for areas of identified active corrosion or for high risk installations where the longest possible service life is required. By judicious use of these methods, any required degree of corrosion protection can be economically achieved. The combination of pipeline coatings with a corrosion monitoring system and, if required, supplemental cathodic protection is the most cost-effective method for corrosion protection of buried and submerged steel water pipelines.

Methods of corrosion prevention utilizing coatings, corrosion monitoring, and/or cathodic protection make it unnecessary and uneconomical to require extra wall thickness or corrosion allowances as safeguards against corrosion and therefore these are not recommended.

Protective Coatings

Coatings and linings protect metal by providing chemical and/or water resistant barriers between the metal and the electrolyte. Pipeline coatings are the primary corrosion protection system and their performance ultimately determines the long-term service life of a pipeline. Proper coating specifications, use of qualified applicators, and inspection should be followed to ensure longest possible service life. Cathodic protection is most effective when used in conjunction with a pipeline coating to minimize current requirements and improve current distribution over the pipeline surface. Generally, the lower the current requirement of a coating, the further current will distribute over a pipeline. Improved current distribution reduces the quantity of galvanic anode or impressed current cathodic protection stations needed, making the overall cost of a cathodic protection system lower. Protective coatings suitable for steel pipelines are identified and discussed in chapter 11.

Bonding of Joints

When a pipeline is cathodically protected or when a pipeline is to be installed with a corrosion monitoring system, joint bonding of gasket joints is required in order to render the pipeline electrically continuous (Figures 10-16 through 10-18). It is desirable to bond all joints at the time of installation, as the practicality and cost to excavate and install joint bonding at a later time will be many times greater. Welded pipe joints are inherently electrically continuous and do not need joint bonds.

Joint bonds need to be properly designed and installed to provide long-term service and meet minimum electrical requirements. Joint bonds should meet the following requirements:

- Joint bond connections should be welded. Mechanical connections, such as compression or bolted connections, should be avoided as they are subject to corrosion and will become electrically discontinuous.
- Joint bonds and associated welds have sufficient strength and/or flexibility to withstand joint movement after backfilling.
- All bare metal surfaces at the bonding connection need to be properly coated to mitigate corrosion.

In addition to bonding, the pipeline should have corrosion monitoring stations installed at appropriate intervals to permit monitoring of the pipeline, whether under cathodic protection or not.

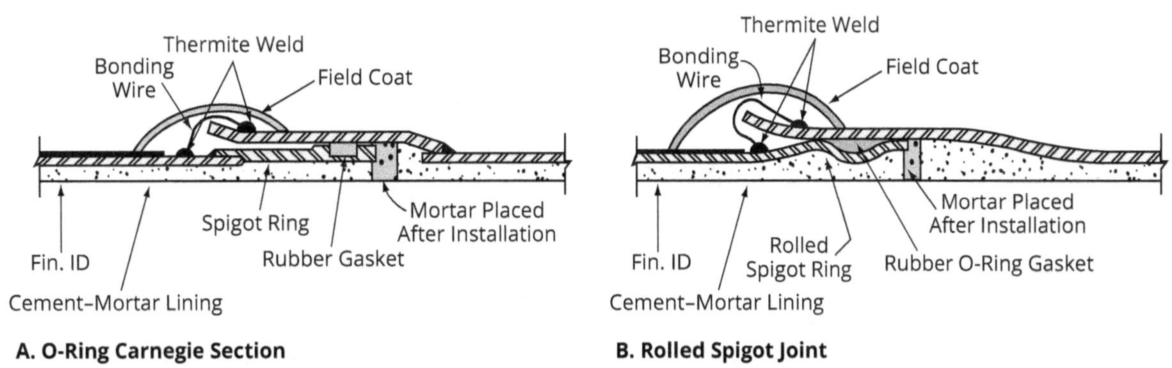

Figure 10-16 Bonding wire for bell-and-spigot rubber-gasketed joint

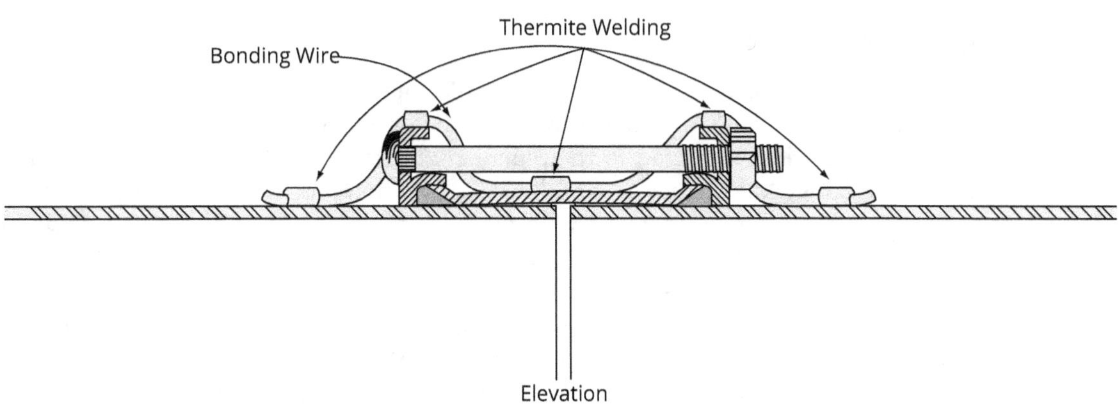

Figure 10-17 Bonding wires installed on sleeve-type coupling

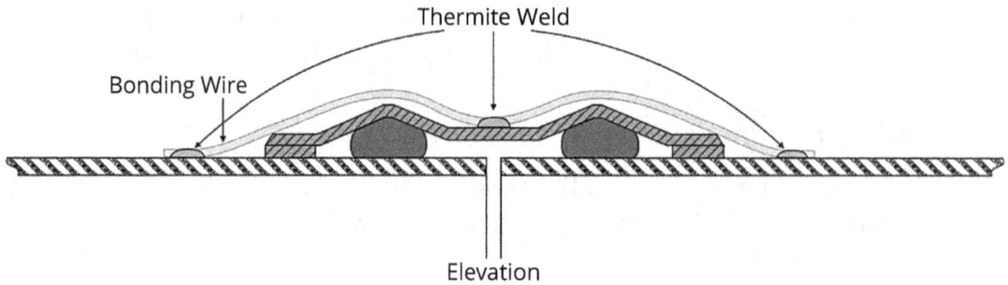

Figure 10-18 Bonding wires installed on split-sleeve–type coupling

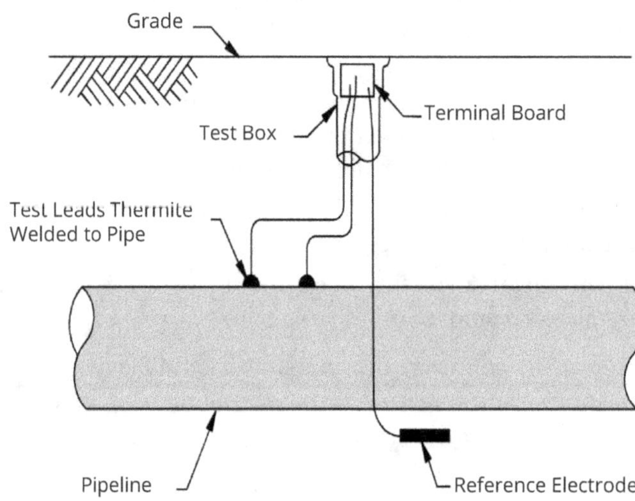

Figure 10-19 Corrosion monitoring station

Corrosion Monitoring Stations

Corrosion monitoring stations (Figure 10-19) are part of a corrosion monitoring system and are required with cathodic protection systems to allow evaluation of a pipeline for corrosion or stray current activity, electrical continuity testing of bonded pipelines, and corrosion protection monitoring. While cathodic protection can be applied at a future date, corrosion monitoring stations are most economical when installed during pipeline installation.

Electrical Isolation

Electrical isolation is used to mitigate dissimilar metal corrosion cells, to mitigate pH differential corrosion cells, and to control cathodic protection current losses to other structures and facilities connected to the pipeline. Electrical isolation can be achieved using insulating flange kits, insulated couplings, or isolation joints. Isolation at concrete structure penetrations can be achieved with modular rubber seals or dielectric coating of the concrete embedded pipe.

Testing of electrical isolation after installation is necessary to ensure that insulating materials were properly installed and are providing the electrical isolation needed. Secondary metallic paths need to be identified and electrically isolated where necessary. Secondary metallic paths can be bypass piping at valves, electrical conduits, electrical grounding, pipe supports, anchor bolts, or other connections to the pipe that may also contact concrete reinforcement or electrical grounding systems.

Cathodic Protection

Cathodic protection is an electrochemical corrosion cell that uses an auxiliary anode immersed in water or buried in the ground and makes the steel pipe a cathode, thereby protecting it from corrosion. Direct current in a corrosion cell is forced to collect on the pipeline (cathode) with rectifiers, using an external power source or using galvanic anodes, such as magnesium or zinc. Corrosion protection of a pipe is indicated when polarization occurs, where direct current collects on the steel, at sufficient magnitude to achieve a minimum potential change on protected surfaces.

Cathodic protection systems include impressed current and galvanic anode systems that are intended to be consumed and eventually replaced.

Galvanic Anode Systems

Galvanic anode systems use sacrificial-anode material such as magnesium or zinc to create a dissimilar metal corrosion cell. An electrical potential difference between the anode and pipe metals causes current to flow from the galvanic anode through the electrolyte to the pipe, returning to the anode through the anode lead wire (Figure 10-20 and Figure 10-21).

Galvanic anode systems use anodic metals with fixed voltages that restrict current output based on the total circuit resistance. While multiple ground beds can be provided to achieve higher current outputs and adequate current distribution, they may result in the galvanic system becoming uneconomical in comparison to an impressed current system.

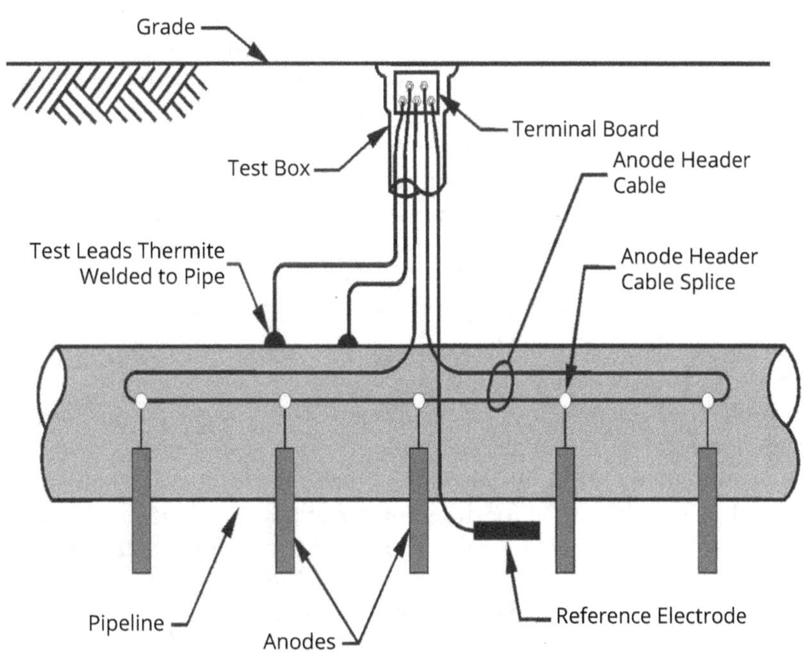

Figure 10-20 Galvanic anode cathodic protection

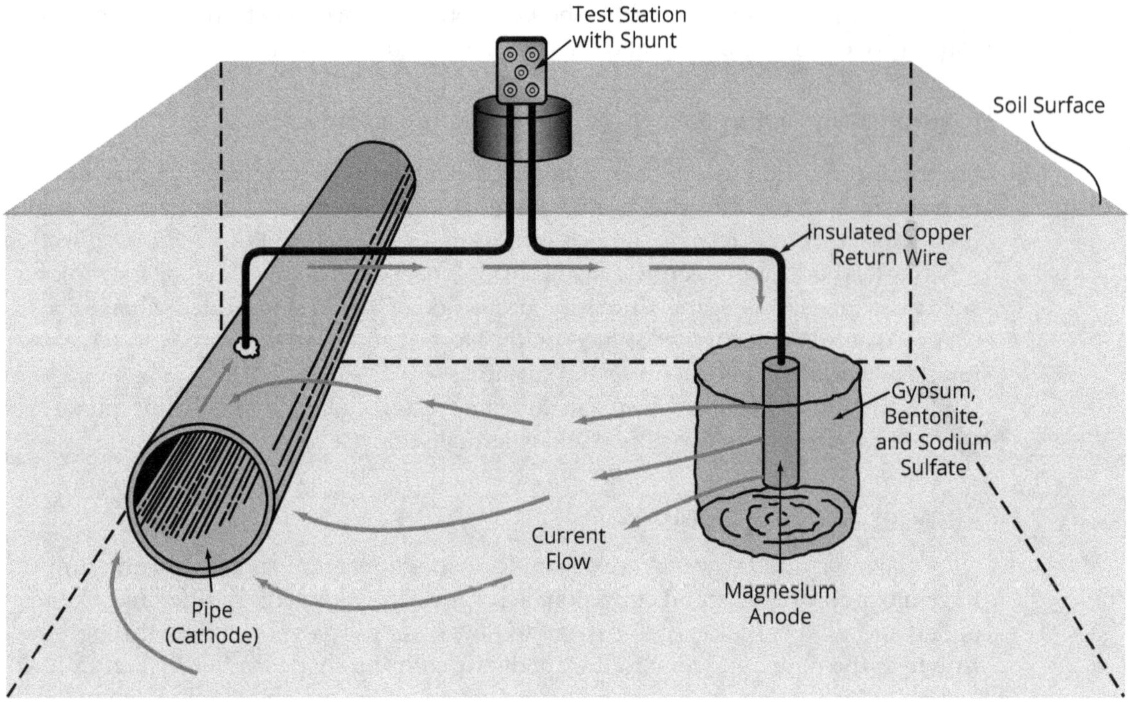

Source: Courtesy NACE International.

Figure 10-21 Cathodic protection—galvanic anode type

Impressed Current Systems

Impressed current systems operate similar to galvanic anodes but require an external DC power source for operation, which also provides greater voltage and current output control (see Figure 10-22). Rectifiers connected to an electrical grid source are a common power source that converts AC electric power to direct current. Other suitable DC power sources are solar panels, wind turbines, thermo-electric generators, inline pipe turbines, and combinations of these sources.

Anodes used in impressed current systems are inert anodes, such as graphite, silicon cast iron, and mix metal (platinum) oxide, because of their low consumption rates. Inert anodes are available in many styles that allow for their use in a multitude of applications and configurations.

An advantage of impressed current systems is their economical application on long pipelines or high current demand applications. A single impressed current station (rectifier) can protect several miles of well-coated pipe. Multiple stations can be combined to protect longer pipelines or pipelines with a higher current density coating, such as cement-mortar–coated pipe.

Impressed current systems offer greater flexibility than galvanic anodes because output voltage can be varied for different current outputs and circuit resistances and because inert anodes are available in more types and configurations. Although output voltages can be varied, the system needs to be monitored to make sure it is not creating stray current issues for neighboring metallic pipelines or structures.

Current Requirements

Pipeline dielectric coating systems reduce the magnitude of cathodic protection current needed to adequately protect a metallic pipe from corrosion. The magnitude of current

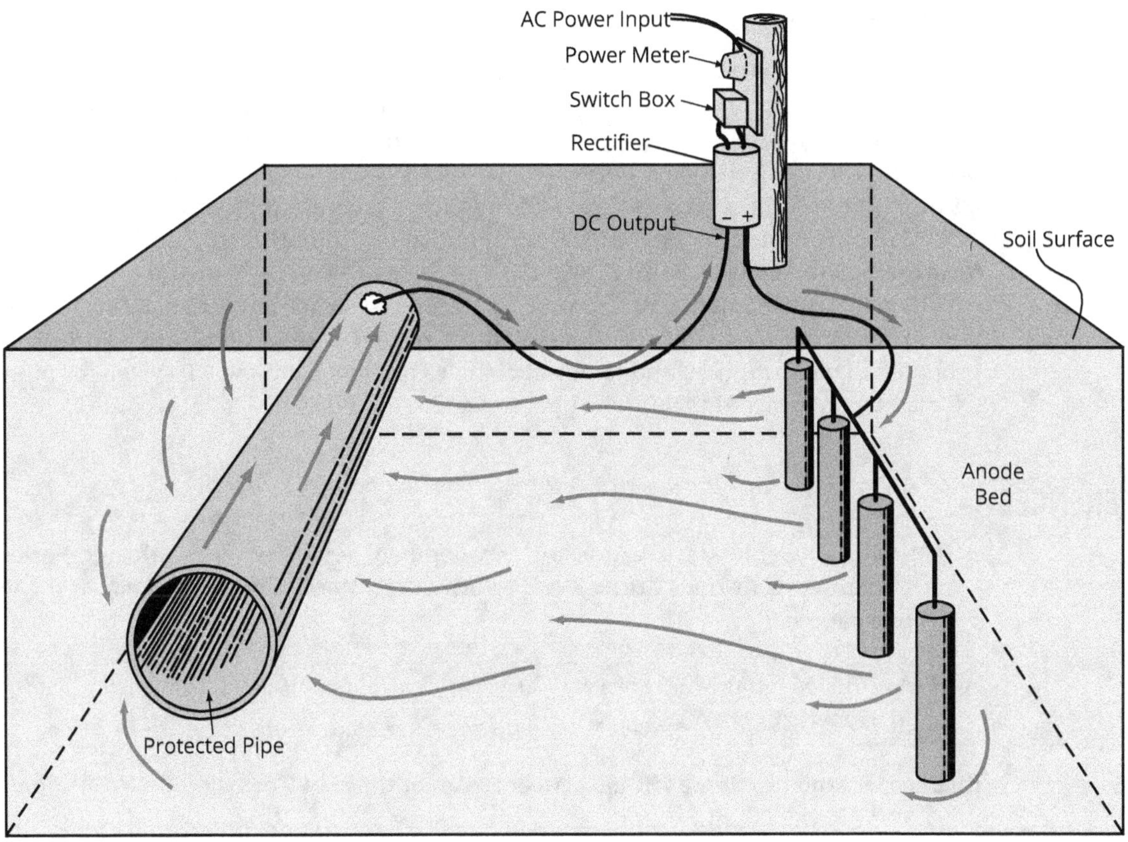

Source: Courtesy NACE International.

Figure 10-22 Cathodic protection—rectifier type

density changes with the coating system applied and will vary for pipe age, size, soil conditions, resistivity, and anticipated coating damage during construction.

Design of Cathodic Protection Systems

For design of cathodic protection systems, including determination of current requirements, it is recommended that the most current editions of NACE SP0169 or SP0100 be used for dielectric bonded or cement-mortar–coated pipe, respectively.

For any cathodic protection system to be effective and provide polarization, sufficient current must flow from the CP anode(s) through the electrolyte to the pipe to ensure that no part of the structure acts as an anode. This is normally achieved when the potential between steel pipe and a copper–copper sulfate reference electrode in contact with the soil and adjacent to the pipe meets or exceeds at least one of the criteria listed in NACE SP0169, Section 6 for bonded coatings, or NACE SP0100, Section 5, for cement-mortar coatings. Any combination of these NACE criteria can be used on a piping system at any particular test station location. Other criteria can be used if it is proven there is no corrosion occurring on the pipeline.

On activation of a cathodic protection system, pipe-to-soil potential measurements and stray current tests can be made along the pipeline to verify the cathodic protection is functioning properly for long-term protection of the pipeline as defined in the next section, Operational Testing.

Operational Testing

Cathodic protection system operation needs to be tested for proper operation and to ensure pipe-to-soil potentials are not below minimum criteria for corrosion protection or above maximum values for a pipeline coating system to prevent formation of stray current or potential coating damage. Impressed current system output can be adjusted to meet corrosion protection criteria, but galvanic anode systems are self-regulating.

Cathodic protection operational testing includes initial startup and ongoing monitoring and maintenance testing.

Periodic maintenance of cathodic protection systems is needed to verify continued corrosion protection of the pipe, to adjust system current output to compensate for ground-bed consumption, and to detect changes or damage to the CP system or pipeline. CP maintenance records should be updated and retained as historical records.

REFERENCES

ASTM G82. 2002. Standard Guide for Development and Use of a Galvanic Series for Predicting Galvanic Corrosion Performance. West Conshohocken, PA: ASTM International.

AWWA. 2013. Manual M27, *External Corrosion Control for Infrastructure Sustainability*, 3rd Ed. Denver, CO: AWWA.

Eliassen, R., and J.C. Lamb III. 1953. Mechanism of Internal Corrosion of Water Pipe. *Jour. AWWA*, 45(12):1281.

Langelier, W.F. 1936. The Analytical Control of Anticorrosion Water Treatment. *Jour. AWWA*, 28:1500.

Linsey, R.K., and J.B. Franzini. 1979. *Water Resources Engineering*. New York: McGraw-Hill Book Co.

Munger, C.G, and L.D. Vincent. 1999. *Corrosion Prevention by Protective Coatings*, 2nd Ed. Houston, TX: National Association of Corrosion Engineers.

NACE. 1975. *NACE Basic Corrosion Course*. Houston, TX: NACE.

NACE. 2006. *Corrosion Basics: An Introduction*, 2nd Ed. Houston, TX: NACE International.

NACE, SP0100. Standard Practice: Cathodic Protection to Control External Corrosion of Concrete Pressure Pipelines and Mortar-Coated Steel Pipelines for Water or Waste Water Service. Houston, TX: NACE International.

NACE, SP0169. Standard Practice: Control of External Corrosion on Underground or Submerged Metallic Piping Systems. Houston, TX: NACE International.

NACE, SP0177. Mitigation of Alternating Current and Lightning Effects on Metallic Structures and Corrosion Control Systems. Houston, TX: NACE International.

Romanoff, M. 1957. *Underground Corrosion*. National Bureau of Standards (NBS) Circular No. 579. Washington, DC: US Government Printing Office.

Weir, P. 1940. The Effect of Internal Pipe Lining on Water Quality. *Jour. AWWA*, 32:1547.

The following references are not cited in the text.

Columbia Gas System Service Corp. 1952. *Manual on Underground Corrosion*. New York.

Hertzberg, L.B. 1956. Suggested Non-technical Manual on Corrosion for Water Works Operators. *Jour. AWWA*, 48:719.

This page intentionally blank.

Chapter 11

Protective Coatings and Linings

Coatings (external) and linings (internal) for corrosion control are extremely effective when properly selected and installed. They are the primary line of defense against internal or external corrosion of underground, submerged, or aboveground steel piping systems. Coating and lining systems may be either passivating (cement mortar) or dielectric (isolation) type.

REQUIREMENTS FOR GOOD PIPELINE COATINGS AND LININGS

The requirements for a coating vary with the type of construction (buried versus submerged or exposed), the aggressiveness of the environment, and system operating conditions.

The effectiveness of a protective pipeline coating over its lifetime depends on many factors, some of which are

- Resistance to physical, chemical, and thermal degradation.
- Ability to retain physical characteristics over the anticipated life of the pipe.
- Physical resistance to soil stress during compaction and settling of the backfill.
- Compatibility with the type of bedding and backfill material.
- Resistance to damage and deterioration during handling, storage, transportation, and installation.
- Ease of repair.
- Resistance to outdoor weathering in aboveground applications.
- Application to pipe with a minimum of defects.
- Resistance to microorganisms.

The requirements for a lining also vary with the system and the environment. In addition to the factors considered for coatings, linings must be evaluated on their smoothness (flow resistance), and they must meet toxicological requirements when applied to pipes transporting potable water.

The effectiveness of a protective pipeline lining over its lifetime depends on many factors, some of which are

- Resistance to damage during handling, storage, and installation.
- Ease of repair.
- Resistance to chemical degradation.

SELECTION OF THE PROPER COATING AND LINING

Selection and recommendation of the coating and lining materials for use on underground, submerged, and aboveground steel pipelines are two of the most important responsibilities of the engineer.

Environmental conditions for coatings and linings are different; therefore, each should be considered separately according to the anticipated corrosion severity and project requirements. A variety of coating and lining system standards are issued by AWWA for pipeline applications as it has been determined that no single type is best for all conditions of exposure. Thus, the systems covered are not necessarily equivalent in terms of expected service life, nor are they equivalent in terms of initial cost. When selecting a system for a particular application, it is recommended that the purchaser establish conditions of exposure and then conduct an evaluation. For example, exposed piping that is subject to thermal cycling from sunlight or large strains from pressure fluctuations may require a flexible lining in lieu of a rigid cement-mortar lining.

Pipe surfaces covered with cement-mortar are protected by the alkaline cement environment that passivates the steel and prevents corrosion. The passivation of the steel by the cement mortar occurs quickly and is not destroyed by moisture and oxygen absorbed through the cement mortar. Dielectric coating and lining systems protect steel pipelines by electrically and chemically isolating the steel substrate from the environment, e.g., corrosive soils or waters and stray electrical currents. Selection of materials involves assessing the magnitude of potential corrosion, installation, and service hazards. ASTM International has developed test methods (ASTM standards G9, G20, and G11) to aid the engineer in evaluating and selecting the coating system that best meets a system's needs. NACE International and other organizations publish standard practices and other supporting documents.

Coating Selection

The corrosion potential for the exterior of steel pipe is difficult to evaluate because of the variety of environments encountered. Resistivity of the soil (see chapter 10) is the most important parameter for evaluating soil corrosivity. Soil chemical and physical analyses, pH, moisture content, presence of chloride or sulfate, fluctuating water table, and existence of stray electrical currents are also important factors in the evaluation process.

If the pipe is subjected to atmospheric conditions, the climate and environmental conditions as well as governmental or agency regulations must be evaluated to determine the proper coating system.

NACE SP0169 contains information to aid the engineer in evaluating considerations and factors for coatings selection. Information for the evaluation of cathodic protection for dielectric and cement-mortar coatings can be found in NACE SP0169 and NACE SP0100, respectively.

Coating performance depends on putting the pipeline into service with the least amount of coating damage. The coating system selected must not only meet the corrosion-control needs but must also allow for economical transportation, handling, storage, and pipeline construction with minimal coating damage or repair. To ensure precise control of the coating application and quality, most types of coatings are applied in a plant or shop. AWWA standards provide a guide to the proper protection during transportation, handling, and storage of pipe that has been coated in this manner. General guidelines are given in a later section of this chapter. There are several recognized testing procedures for evaluating coating system characteristics related to transportation, storage, and construction (ASTM test methods in ASTM standards G6, G10, and G11).

Lining Selection

The function of a lining is to prevent internal corrosion and to produce and maintain a smooth surface to enhance flow capacity. Dielectric and passivating lining materials both can prevent internal corrosion in water pipelines. When system operating conditions may result in cavitation or abrasion, the effects of these on the lining should be considered, regardless of the lining material selected.

AVAILABLE COATINGS AND LININGS

Current AWWA standards identify coatings and linings for steel water pipe that have proven to be the most reliable in water pipeline applications. However, the AWWA Steel Pipe Committee is alert to the possibilities of new developments, and additions to and modifications of existing standards will be made as deemed advisable. The current list of AWWA coating and lining standards for pipe protection is as follows:

ANSI/AWWA C203, Coal-Tar Protective Coatings and Linings for Steel Water Pipes. ANSI/AWWA C203 (latest edition) describes the material and application requirements for shop-applied, coal-tar protective coatings and linings for steel water pipelines intended for use under normal conditions. The standard describes coal-tar enamel applied to the interior and exterior of pipe, special sections, connections, and fittings. It also covers hot-applied coal-tar tape applied to the exterior of special sections, connections, and fittings.

Coal-tar enamel is applied over a coal-tar or synthetic primer. External coal-tar enamel coatings use bonded nonasbestos-felt and fibrous-glass mat to reinforce and cover the coal-tar enamel. The applied external coating is usually finished with either a coat of whitewash or a single wrap of kraft paper.

Internally, the coal-tar enamel is centrifugally applied to the pipe and does not contain reinforcement.

ANSI/AWWA C205, Cement-Mortar Protective Lining and Coating for Steel Water Pipe—4 In. (100 mm) and Larger—Shop Applied. ANSI/AWWA C205 (latest edition) describes the material and application requirements to provide protective linings and coatings for steel water pipe by shop application of cement mortar.

Cement mortar is composed of Portland cement, fine aggregate, and water, well mixed and of the proper consistency to obtain a dense, homogeneous lining or coating. Internally, the cement mortar is centrifugally applied to remove excess water and produce a smooth, uniform surface. Externally, the coating is a reinforced cement mortar, pneumatically or mechanically applied to the pipe surface. Reinforcement consists of spiral wire, wire fabric, or ribbon mesh. The standard provides a complete guide for the application and curing of the mortar lining and mortar coating. Cement-mortar overcoat can be used as a rock shield for pipe to which a dielectric flexible coating has been applied.

ANSI/AWWA C209, Cold-Applied Tape Coatings for Steel Water Pipe, Special Sections, Connections, and Fittings. ANSI/AWWA C209 (latest edition) describes the materials and application requirements for cold primer and cold-applied tape on the exterior of special sections, connections, and fittings for steel water pipelines installed underground. Tapes with polyvinyl chloride, polyethylene, polypropylene, and polyolefin backing are listed. The thicknesses of the tapes vary; however, all tapes may be sufficiently overlapped to meet changing performance requirements. If severe construction or soil conditions exist where mechanical damage may occur, a suitable overwrap of an extra thickness of tape or other wrapping may be required.

ANSI/AWWA C210, Liquid Epoxy Coatings and Linings for Steel Water Pipe and Fittings. ANSI/AWWA C210 (latest edition) describes the materials and application requirements for liquid epoxy coating system for the interior and exterior of steel water pipe, fittings, and special sections installed underground or underwater. The coating system consists of one coat of a two-part chemically cured inhibitive epoxy primer, and one or more coats of a two-part chemically cured epoxy finish coat. The coating system may alternately consist of one or more coats of the same epoxy coating without the use of a separate primer, provided the coating system meets the performance requirements of ANSI/AWWA C210.

The coating is suitable for use in potable and nonpotable water systems. The product is applied by spray application, preferably airless, but other application methods may be employed in accordance with the manufacturer's recommendations.

ANSI/AWWA C213, Fusion-Bonded Epoxy Coatings and Linings for Steel Water Pipe and Fittings. ANSI/AWWA C213 (latest edition) describes the material and application requirements for fusion-bonded epoxy protective coating for the interior and exterior of steel water pipe, special sections, welded joints, connections, and fittings of steel water pipelines installed underground or underwater under normal construction conditions. The coating is suitable for use in potable water systems. Currently ANSI/AWWA C213 coatings are commercially limited to sizes 48 in. and smaller.

Fusion-bonded epoxy coating is a heat-activated, chemically cured coating. The epoxy coating is supplied in powder form. Except for welded field joints, it is plant- or shop-applied to preheated pipe, special sections, connections, and fittings using fluid bed, air, or electrostatic spray.

ANSI/AWWA C214, Tape Coatings for Steel Water Pipelines. ANSI/AWWA C214 (latest edition) describes the materials, the systems, and the application requirements for prefabricated cold-applied tapes mechanically applied to the exterior of all diameters of steel water pipe. For normal construction conditions, this coating is applied as a three-layer system consisting of (1) primer, (2) corrosion preventive tape (inner layer), and (3) mechanical protective tape (outer layer). The primer is supplied in liquid form consisting of solid ingredients carried in a solvent or as a 100 percent solids material. The corrosion preventive tape and the mechanical protective tape are supplied in suitable thicknesses and in roll form. The standard describes the application at coating plants.

ANSI/AWWA C215, Extruded Polyolefin Coatings for Steel Pipe. ANSI/AWWA C215 (latest edition) describes the materials, systems, and application requirements for shop-applied extruded polyolefin coatings for the exterior of steel water pipe up to 144-in. diameter. The standard describes two types of extrusion, crosshead and side, and three types of applications as follows: Type A—crosshead-die extrusion, consisting of an adhesive and an extruded polyolefin sheath for pipe diameters from ½ in. through 36 in.; Type B—side extrusion, consisting of an extruded adhesive and an extruded polyolefin sheath for pipe diameters from 2 through 144 in.; and Type C–side extrusion, consisting of a liquid adhesive (primer) layer, extruded butyl rubber adhesive, and extruded polyolefin sheath for pipe diameters from 2 in. through 144 in.

ANSI/AWWA C216, Heat-Shrinkable Cross-Linked Polyolefin Coatings for Steel Water Pipe and Fittings. ANSI/AWWA C216 (latest edition) describes the material, application, and field-procedure requirements for protective exterior coatings consisting of heat-shrinkable cross-linked polyolefin coatings and their application to special sections, connections, and fittings to be used on underground and underwater steel water pipelines. These coatings may be field- or shop-applied in accordance with the provisions of the standard. Heat shrink sleeves are commonly used as a field joint coating system on dielectric coated pipe.

ANSI/AWWA C217, Microcrystalline Wax and Petrolatum Tape Coating Systems for Steel Water Pipe and Fittings. ANSI/AWWA C217 (latest edition) describes the materials and application requirements for field- or shop-applied exterior tape coatings that consist of cold-applied petrolatum or petroleum wax primer and saturant petrolatum or petroleum wax tape coatings and their applications to special sections, connections, and fittings to be used with buried or submerged steel water pipelines.

ANSI/AWWA C218, Liquid Coatings for Aboveground Steel Water Pipe and Fittings. ANSI/AWWA C218 (latest edition) describes several alternative coating systems for the protection of exterior surfaces of steel pipelines and associated fittings used by the water supply industry in aboveground locations. The coating systems described are not necessarily equivalent in terms of cost or performance, but are presented so that the purchaser can select the coating system that best meets the site-specific project requirements. Coating systems included are alkyds, epoxies, polyurethanes, silicones, and acrylics.

ANSI/AWWA C222, Polyurethane Coatings for the Interior and Exterior of Steel Water Pipe and Fittings. ANSI/AWWA C222 (latest edition) describes the materials and application processes for shop- and field-applied polyurethane linings and coatings for steel water pipe, special sections, welded joints, connections, and fittings installed underground or underwater. Polyurethanes adhering to this standard are suitable for use in potable and nonpotable water systems.

ANSI/AWWA C224, Nylon-11-Based Polyamide Coatings and Linings for Steel Water Pipe and Fittings. ANSI/AWWA C224 (latest edition) describes two-layer polyamide (Nylon-11-based) coating systems used for potable water-handling equipment installed aboveground, belowground, or underwater. Polyamide coating is thermoplastic and is ordinarily applied in a shop or manufacturing facility.

ANSI/AWWA C225, Fused Polyolefin Coatings for Steel Water Pipelines. ANSI/AWWA C225 (latest edition) describes the materials and application of fused polyolefin coating systems for buried service. Normally, these prefabricated polyolefin coatings are applied as a three-layer system consisting of a liquid adhesive, a corrosion-protection inner layer, and a mechanical-protection outer layer. This system is applied in pipe-coating plants, both portable and fixed.

ANSI/AWWA C229, Fusion-Bonded Polyethylene Coatings for Steel Water Pipe and Fittings. ANSI/AWWA C229 (latest edition) describes the minimum material and application requirements for fusion-bonded polyethylene (FBPE) coating to be factory-applied to the exterior of steel water pipe and fittings and the joint region (of rubber-gasketed field joints) of steel water and wastewater pipe.

ANSI/AWWA C602, Cement–Mortar Lining of Water Pipelines in Place—4 In. (100 mm) and Larger. ANSI/AWWA C602 (latest edition) describes the materials and application processes for the cement-mortar lining of pipelines in place, describing both newly installed pipes and existing pipelines. Detailed procedures are included for surface preparation and application, surface finishing, and curing of the cement mortar.

COATING AND LINING APPLICATION

This manual does not provide details on methods of coating or lining application or inspection, but the importance of obtaining proper application cannot be overemphasized. Effective results cannot be secured with any coating or lining material unless adequate care is taken in the preparation of surfaces, application, transportation, and handling of the pipe. Cement-mortar and dielectric coatings and linings have different surface preparation requirements because of the method each uses to protect the steel surface. Cement-mortar passivates the steel surface and dielectrics isolate the steel surface from the environment. AWWA standards provide the requirements for application, handling, and repair if needed. The user is encouraged to refer to the specific AWWA coating or lining standard for details.

Coating and Lining of Special Sections, Connections, and Fittings

The coating and lining of special sections, connections, and fittings are described in ANSI/AWWA standards C203, C205, C209, C210, C213, C214, C215, C216, C217, C218, C222, C224, C225, C229, and C602 (latest editions). The materials used are the same as those specified for use with steel water pipe. The methods of application may differ from those prescribed for pipe because of the variety of physical configurations encountered.

Interior pipe joints are normally lined in the field with materials similar to those used on the main body of the pipe. These materials are described in the appropriate AWWA lining standards. The exterior of joints for dielectrically coated exterior or buried pipe joints are typically coated with heat shrink sleeves per ANSI/AWWA C216 or materials similar to those used on the main body of the pipe. The exterior of joints for cement-mortar–coated pipe are field-coated with flowable cement mortar per ANSI/AWWA C205, using grout bands (diapers) to contain the material. Exposed pipe joints are typically coated with materials used on the main body of the pipe.

GOOD PRACTICE

The AWWA standards for protective coatings have been carefully prepared by experienced individuals and are based on the best current practice. They should be referred to in the job specification. Modification to the standards should be made only by experienced coating specialists after consultation with the fabricator, coating manufacturer, or applicator.

For AWWA standards to be complete for bidding purposes, the purchaser's job specifications should provide the Purchaser Options and Alternatives listed in each standard.

REFERENCES

ANSI/AWWA C203, Coal-Tar Protective Coatings and Linings for Steel Water Pipe. Denver, CO: American Water Works Association.

ANSI/AWWA C205, Cement-Mortar Protective Lining and Coating for Steel Water Pipe—4 In. (100 mm) and Larger—Shop Applied. Denver, CO: American Water Works Association.

ANSI/AWWA C209, Cold-Applied Tape Coatings for Steel Water Pipe, Special Sections, Connections, and Fittings. Denver, CO: American Water Works Association.

ANSI/AWWA C210, Liquid Epoxy Coatings and Linings for Steel Water Pipe and Fittings. Denver, CO: American Water Works Association.

ANSI/AWWA C213, Fusion-Bonded Epoxy Coatings and Linings for Steel Water Pipe and Fittings. Denver, CO: American Water Works Association.

ANSI/AWWA C214, Tape Coatings for Steel Water Pipe. Denver, CO: American Water Works Association.

ANSI/AWWA C215, Extruded Polyolefin Coatings. Denver, CO: American Water Works Association.

ANSI/AWWA C216, Heat-Shrinkable Cross-Linked Polyolefin Coatings for Steel Water Pipe and Fittings. Denver, CO: American Water Works Association.

ANSI/AWWA C217, Microcrystalline Wax and Petrolatum Tape Coating Systems for Steel Water Pipe and Fittings. Denver, CO: American Water Works Association.

ANSI/AWWA C218, Liquid Coating Systems for the Exterior of Aboveground Steel Water Pipelines and Fittings. Denver, CO: American Water Works Association.

ANSI/AWWA C222, Polyurethane Coatings for the Interior and Exterior of Steel Water Pipe and Fittings. Denver, CO: American Water Works Association.

ANSI/AWWA C224, Nylon-11-Based Polyamide Coating System for the Interior and Exterior of Steel Water Pipe, Connections, Fittings, and Special Sections. Denver, CO: American Water Works Association.

ANSI/AWWA C225, Fused Polyolefin Coatings for Steel Water Pipe. Denver, CO: American Water Works Association.

ANSI/AWWA C229, Fusion-Bonded Polyethylene Coatings for Steel Water Pipe and Fittings. Denver, CO: American Water Works Association.

ANSI/AWWA C602, Cement---Mortar Lining of Water Pipelines in Place—4 In. (100 mm) and Larger. Denver, CO: American Water Works Association.

ASTM Standard G6, Test for Abrasion Resistance of Pipeline Coatings. Philadelphia, PA: American Society for Testing and Materials.

ASTM Standard G9, Test for Water Penetration Into Pipeline Coatings. Philadelphia, PA: American Society for Testing and Materials.

ASTM Standard G10, Test for Bendability of Pipeline Coatings. Philadelphia, PA: American Society for Testing and Materials.

ASTM Standard G11, Test for Effects of Outdoor Weathering on Pipeline Coatings. Philadelphia, PA: American Society for Testing and Materials.

ASTM Standard G20, Test for Chemical Resistance of Pipeline Coatings. Philadelphia, PA: American Society for Testing and Materials.

NACE Standard SP0100. Cathodic Protection to Control External Corrosion of Concrete Pressure Pipelines and Mortar-Coated Steel Pipelines for Water or Waste Water Service. Houston, TX: NACE.

NACE Standard SP0169. Control of External Corrosion on Underground or Submerged Metallic Piping Systems. Houston, TX: NACE.

The following references are not cited in the text.

Steel Structures Painting Council. Steel Structures Painting Manual. Latest edition. *Good Painting Practice*—Vol. 1; *Systems and Specifications*—Vol. 2. Pittsburgh, PA: Steel Structures Painting Council.

US Bureau of Reclamation. 2002. *Guide to Protective Coatings: Inspection and Maintenance.* Washington, DC; US Bureau of Reclamation. (available at http://www.usbr.gov/pmts/materials_lab/pubs/Coatings.pdf).

Chapter 12

Transportation, Installation, and Testing

The detailed procedures for transporting, trenching, laying, backfilling, and testing any steel pipeline depend on many controlling factors, including the character and purpose of the line; its size, operating pressure, and operating conditions; its location—urban, suburban, or rural; and the terrain over which it is laid—flat, rolling, or mountainous. Procedures also are affected by trench depth, character of the soil, and backfill. This chapter briefly discusses several of the more common requirements of installation, omitting precise details that vary in individual installations. A more detailed explanation of installation practices can be found in ANSI/AWWA C604, Installation of Buried Steel Water Pipe—4 In. (100 mm) and Larger. Throughout the chapter, the importance of the engineering properties of the excavated soil and the backfill soil should be considered. The principles of soil mechanics properly applied to excavation and backfill practices lead to safer working conditions and to better and more economical pipeline installations.

TRANSPORTATION AND HANDLING OF COATED STEEL PIPE

Lined and coated steel pipe is typically transported by truck and can be shipped via rail or water with special precautions. Regardless of which mode of transportation is used, lined and coated steel pipe is valuable cargo and should be handled as such.

Modes of Transportation

Requirements for packaging, stowing, and restraining pipe during transit depend on the mode of transportation and the properties of the pipe itself, such as diameter, wall thickness, linings, and coatings.

Truck. Most coated pipe is carried on flatbed trucks and trailers directly to the job site. Actual dimensional limitations of loads vary by jurisdiction. Loads in excess of the allowed dimensions are considered oversize and involve the acquisition of permits, which increases the cost of transportation. A single oversize permitted load is not allowed if the material being shipped can be separated so that it will fit on two or more trailers within the general height, width, and length allowances. The shipper should caution the trucking firms against use of tie-down equipment that could damage the coating.

Rail. For rail transportation, pipe can be loaded in shipping containers or on flat railroad cars. Pipe can be restrained on the cars using stake pockets or made into floating loads in accordance with current rules of the Association of American Railroads. An inspector from the railroad will check each car for proper loading before accepting it for shipment. Additional consideration may be needed beyond the requirements of the railroad to properly protect pipe during shipment. Rail shipment generally involves pipe being offloaded at a transloader for temporary storage, followed by the pipe being loaded onto flatbed trailers for delivery to the job site.

Water. Constant pitching and rolling motions should be anticipated for pipe stowed aboard ships. Small pipe must be packaged, and large pipe must be stowed in such a manner to ride with or offset the pitching and rolling motion. Adequate padded timbers or similar barriers must be used to keep pipe from rubbing together or coating and linings from being damaged. Salt water shipments present the potential for corrosion and contamination from salt should be considered.

Air. Delivery of the pipe to distant sites can be expedited by airplane, and delivery into otherwise inaccessible locations may require cargo helicopters. The air carrier should be contacted to obtain maximum length, width, height, and weight limitations for the route involved. Generally, the carriers will require pipe to be strapped directly to pallets suitable for handling.

Loading and Unloading

Regardless of the mode of transportation, loads should be prepared and packaged in a manner that will protect the pipe and any associated lining and coating. Sufficient stringers or cradles should be used to layer the pipe without placing too much load on a single bearing point. When plain-end pipe is shipped, a pyramid load with the full length of pipe resting on adjacent pipe should be considered, providing due consideration is given to the protection of any applied coating. Interior bracing is discussed later in this chapter and may be necessary to maintain roundness during shipment and prevent damage to linings or coatings. Additionally, contoured blocks or bunks may be necessary to give proper support to some loads. Throughout the transportation process, pipe should not be allowed to roll or fall from the conveyance to the ground.

Handling Equipment. Both loading and unloading of coated pipe should be performed with equipment that is designed for such use and operated in a manner that will not damage the materials, their linings, coatings, or packaging. Approved equipment for handling coated pipe includes nylon straps, wide canvas or padded slings, wide padded forks, and skids designed to prevent damage to the pipe and coating. Unpadded chains, sharp edges on buckets, wire ropes, narrow forks, hooks, and metal bars are unacceptable.

Stringing. When pipe is to be distributed along the right-of-way, it should be supported in two places (at about one-quarter length from each end) using mounded soil, sandbags, or other suitable supports that protect the pipe and any associated coating. Padded wood blocks can be used as supports, but care needs to be taken to ensure that the pipe is not point loaded if the surface on which the blocks are placed is not substantially level. Regardless of the method of support used, sufficient room must be left under the pipe to allow access of the lifting mechanism without damage to the pipe or any associated coating.

Interior Bracing of Pipe

Temporary bracing may be needed during transportation, handling, and placement of pipe. If needed, bracing can help support the pipe shape until sufficient side soil support is in place to hold the circular shape of the pipe. Bracing will not support construction activity above the pipe. Flexible pipe relies on adequate soil side support and adequate soil cover to distribute the load from construction equipment and other loading above the pipe. Damage to the pipe lining, coating, and cylinder can occur from attempting to support excessive construction loads with the bracing in the pipe. Analysis of heavy construction loads can be found in chapter 5.

Table 12-1 and Figure 12-1 offer guidance for common bracing and stulling applications. For pipe with painted linings, felt, carpet, or padding is commonly attached to the ends of the stulls to prevent lining damage. For cement-mortar lining or pipe left bare on the inside, stulls are typically cut to the inside diameter less ¼ to 1 in. Wood shingles or wedges are then used to tighten the stull and provide a snug fit. Other bracing systems such as adjustable rods have also been successfully used.

INSTALLATION OF PIPE

Trenching

Depth. Trenches should be excavated to grade as shown in the profile. Where no profile is provided, the minimum cover should be generally selected to protect the pipe from external live loads but may be controlled by the depth of the frost line in freezing climates. The profile should be selected to minimize high points where air may be trapped. The depth of trench in city streets may be governed by the location of existing utilities, regulatory requirements for potable water pipelines, or other conditions.

Width. Where the sides of the trench will allow reasonable side support, the trench width that must be maintained at the top of the pipe, regardless of the depth of excavation, is the narrowest practical width that will allow proper densification of pipe-zone bedding, haunch zone, and backfill materials. If the pipe-zone bedding and backfill

Table 12-1 Pipe bracing

Diameter-to-Thickness Ratio (D/t)	Nominal Pipe Diameter / Minimum Stull Dimension					
	$D < 24$ in.	$D = 24$ in. to ≤ 30 in.	$D = 30$ in. to < 48 in.	$D = 48$ in. to < 60 in.	$D = 60$ in. to < 84 in.	$D = 84$ in. to 120 in.
	(2 in. × 4 in.)	(3 in. × 3 in.)	(4 in. × 4in.)	(4 in. × 4 in.)	(4 in. × 4 in.)	(4 in. × 4 in.)
$D/t \le 120$	No stulls	No stulls	No stulls	No stulls	No stulls	No stulls
$120 < D/t \le 160$	Brace between bunks	2 stulls—vertical	2 stulls—crossed	2 stulls—crossed	2 stulls—3 legs	2 stulls—3 legs
$160 < D/t \le 200$	Brace between bunks	2 stulls—vertical	2 stulls—crossed	2 stulls—3 legs	3 stulls—3 legs	3 stulls—3 legs
$200 < D/t \le 230$	—	2 stulls—vertical	2 stulls—crossed	3 stulls—3 legs	3 stulls—3 legs	3 stulls—3 legs
$230 < D/t \le 288$	—	2 stulls—crossed	2 stulls—3 legs	3 stulls—3 legs	3 stulls—3 legs	3 stulls—3 legs

Notes: D = Nominal, finished inside diameter; t = pipe wall thickness. Stulls should be placed 15 to 20% of the total pipe length from each end, but no less than 4 ft from the end. Shipping bunks are to be located near stulls.

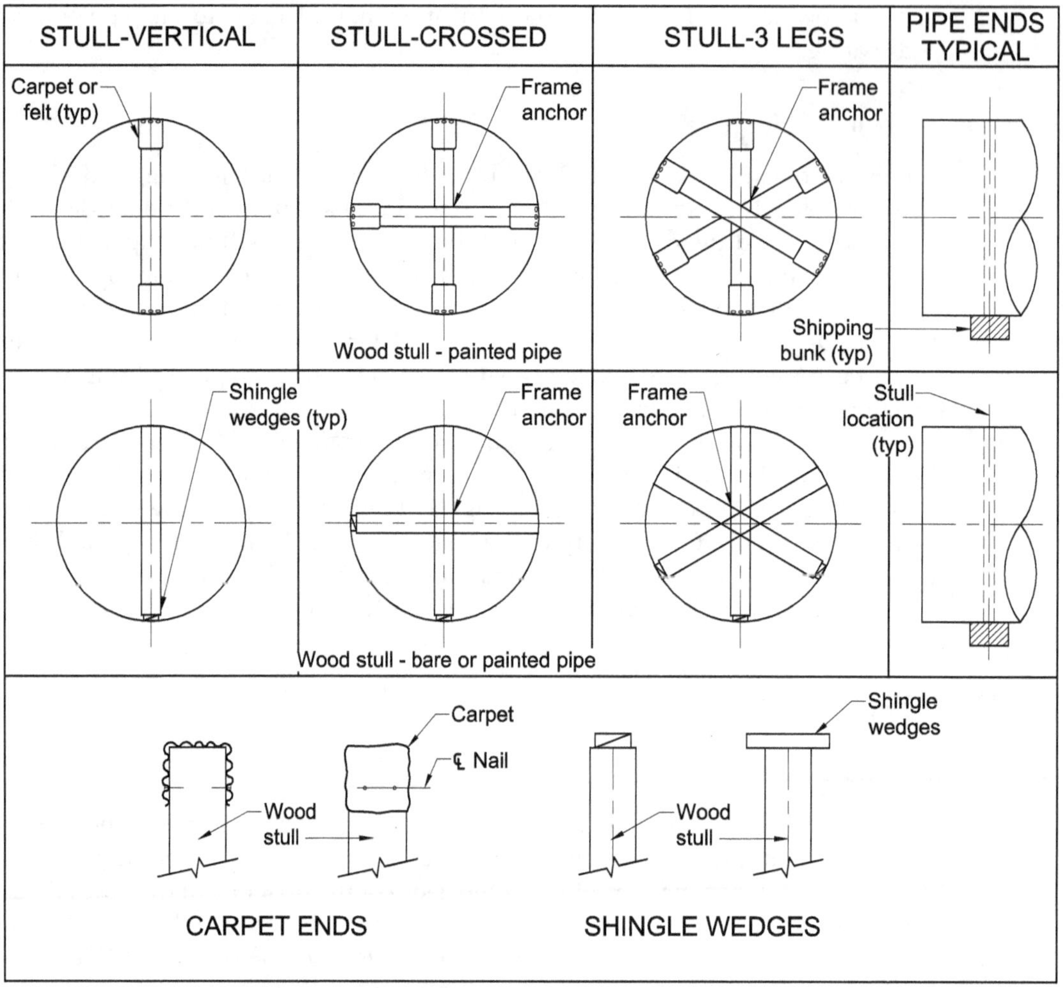

Figure 12-1 Pipe stulling and bracing configurations

require densification by compaction, the width of the trench at the bottom of the pipe should be determined by the space required for the proper and effective use of compaction equipment.

When joints are assembled in the trench, bell holes may be required at the joint to safely facilitate workers in the trench to complete the joint or install joint coatings. Holes excavated to permit removal of the slings without damage to the pipe coating may also be required.

Bottom Preparation. Flat-bottom trenches should be excavated to a depth of 2 in. to 6 in. below the established grade line of the outside bottom of the pipe. The excess excavation should then be filled with loose material from which all stones and hard lumps have been removed. The loose subgrade material should be graded uniformly to the established grade line for the full length of the pipe. Steel pipe should not be set on rigid blocks on the trench bottom that would cause concentration of the load on small areas of pipe coating or cause deformation of the pipe wall.

Where the bottom of the trench is covered with solid, hard objects that might penetrate the protective coating, a bedding of crushed rock or sand should be placed under the barrel of the pipe. Screened earth also has been used successfully for such bedding. It may be advantageous to shape the trench bottom under large steel pipe for full arc contact.

Regulations. All applicable local, state, and federal laws and regulations should be carefully observed including those relating to the protection of excavations, the safety of persons working therein, and provision for the required barriers, signs, and lights.

Handling and Laying

Care similar to that exercised during loading, transporting, unloading, and stringing should be observed during installation of the pipe in the trench.

Coated pipe should not be stored directly on rough ground, nor should it be rolled on rough surfaces.

While handling and placing pipe in the trench, fabric or nylon slings should be used. The pipe should not be dragged along the bottom of the trench or bumped. It should be supported by the slings while preparing to make the joint. The coating on the underside of the pipe should be inspected while it is suspended from the sling, and any visible damage to the coating should be repaired before lowering the pipe into the trench.

Pipe should be laid to lines and grades shown on the contract drawings and specifications, except where modified by the manufacturer's detailed layout drawings or laying schedule, as reviewed by the purchaser. All fittings and appurtenances should be at the required locations, and all valve stems and hydrant barrels should be plumb. The pipe trench should be kept free from water that could impair the integrity of bedding and joining operations. On grades exceeding 10 percent, the pipe should be laid uphill or otherwise held in place by approved methods.

Special means of supporting the pipe may be provided, but under no conditions should pipe sections be installed permanently on timbers, earth mounds, pile bents, or other similar supports unless specific pipe designs for these special conditions have been provided.

Slight deflections for horizontal and vertical angle points, long-radius curves, or alignment corrections may be made by nonsymmetrical closure of joints. The manufacturer can provide data to the purchaser and the contractor indicating maximum joint offsets and deflections for each type of joint supplied.

Assembly of Pipe

Pipe larger than 24 in. in diameter is normally assembled in the trench except under the most unusual conditions. Smaller-diameter pipe joined by welding, flanges, or couplings may be assembled aboveground in practicable lengths for handling and then lowered into the trench by suitable means allowing for progressive lowering of the assembled run of pipe. If the method of assembling pipe aboveground prior to lowering it into the trench is used, care must be taken to limit the degree of curvature of the pipe during the lowering operation so as to not exceed the yield strength of the pipe material or damage the lining or coating materials on the pipe. Pipe deflection at any joint during the lowering operation should be limited to the manufacturer's recommendation. Pipe that has O-ring rubber gaskets as seals should only be assembled section by section in the trench.

Trestle and ring-girder construction can be used for highway, river, and similar crossings.

When pipe is installed on the decks of highway bridges, saddles are generally used to support the pipe at proper intervals, and hold-down clamps provided as required. Expansion joints for restrained pipe are supplied as designed and are commonly installed at areas where the bridge contains an expansion joint in its construction. Steel pipe is also often suspended from or attached to the underside of highway bridges, with appropriate attention given to the flexibility of the bridge's structure. Exposed pipelines in any location should be protected against freezing in areas where such a possibility exists.

Field-Welded Joints

Technical requirements for field welding are contained in ANSI/AWWA C206. If pipe that has been lined and coated is to be field welded, a short length of the pipe barrel at either end must be left bare so that the heat of the welding operation will not adversely affect the protective coating. The length of the unprotected section may vary depending on the kind of protective coating and the pipe wall thickness. Care must be exercised when cutting and welding on pipes with combustible linings and coatings to avoid the risk of fire.

Following the completion of the weld, the gaps in the lining and coating must be completed, normally with similar material as that used for the pipe. The proper application and use of joint linings and coatings can be found in the applicable AWWA standards.

The use of welded joints results in a continuous pipeline. This stiffness provides a considerable advantage where long unsupported spans are required. It is also advantageous in restraining steel pipe. Welded joints are capable of resisting thrusts caused by closed valves or by changes in the direction of a pipeline. Welded joints may be provided to transmit such thrusts over a sufficient distance to absorb the force through skin friction provided by the backfill material against the pipe. In such cases, determination of the thrust and strength of the weld must be made, particularly for larger pipe under high pressures, to determine if the weld is sufficiently strong to transmit the force from one pipe section to the next. Calculations for weld design can be found in chapter 6. Calculations for thrust can be found in chapter 8.

Except during the construction period when an open trench exists, pipe with welded joints will usually have no problems with excessive thermal expansion and contraction. Where immediate shading or backfill of welded-joint steel pipe is impractical, it is advisable to weld the pipe in sections of approximately 400 to 500 ft and leave the final joint unwelded. If the final open joints are then welded in the early morning hours when the pipe is typically coolest, a minimum of temperature stress will occur in the pipeline.

Weld-after-backfill is also an acceptable method for field-welded joints. This method involves assembling the joint, applying the appropriate exterior protective coating to the joint, and backfilling around the joint prior to welding the interior joint. Welding is done on the inside of the pipe after the backfill is complete. With inside single-welded joints, weld-after-backfill can reduce or eliminate thermal stress issues associated with welding pipe exposed to the sun or atmosphere and eliminates the need for special thermal expansion joints. Once interior welding and/or weld testing are complete, the interior lining at the joint may be completed. Project-specific conditions and materials can vary, such as the exterior protective joint coating, backfill materials, use of lighter-gage cylinders, and welding procedures, which may warrant testing to verify the method at the start of construction.

Pipe laid on piers above the ground can be continuously welded; however, it is necessary to account for thermal expansion and contraction when the pipe is not rigidly fixed.

Bell-and-Spigot Rubber-Gasket Joints

Under normal laying conditions, work should proceed with the bell end of the pipe facing the direction of laying. Before setting the spigot in place, the bell should be thoroughly cleaned and then lubricated.

After the O-ring rubber gasket has been placed in the spigot groove, it should be adjusted so the tension on the rubber is uniform around the circumference of the joint. Following assembly, the pipe gasket should be checked around the full circumference of the joint with a feeler gauge to ensure the gasket has not rolled out of the groove or "fish mouth." Joint harness can be installed after installation of gasket joint if restraint against thrust and/or thermal forces is needed.

Installation Practice for Flanged Joints

The method of flange assembly and sequence of tightening bolts and controlling torque are very important and become more critical with larger-diameter flanges and bolts. A detailed description of flange assembly can be found in ANSI/AWWA C604.

Bolted joint closure depends principally on three factors: applying the correct amount of gasket load, achieving a uniform load around the joint, and the condition of the flange components, especially the flange faces and gasket.

Gasket Load. Gaskets seal joints as they are compressed to fill imperfections and spaces in the joint faces. Insufficient compression can result in imperfections and spaces not being filled and can also result in insufficient resistance to hydrostatic pressure. Excessive compression can result in pinching or crushing of the gasket and consequently possible leakage. Desirable gasket load is generated by tightening (stretching) the bolts in a controlled and uniform manner, usually to a given percent of their yield. By convention, most bolting of gasketed flanges is carried out to achieve a stress of approximately 50 percent of bolt material yield strength, although values may vary according to gasket material and the design specifications. The sum of all the bolt loads in the joint divided by the contact area of the gasket with the flanges equals the gasket load, in psi. In addition to seating the gasket, bolts also resist movement at the joint.

Bolt Load. It is bolt load that compresses the gasket and seals the joint. However, it is not just how tight the bolts are, but also how consistently they are loaded that ensures a proper seal. The intent is not just to achieve a sufficient load but also to minimize variations or "scatter" among the bolts. A complication to achieving uniform tightness is that bolts act like interactive springs in a joint. Tightening any one bolt will change the effective load of the nearby bolts on either side. This interaction is referred to as bolt "crosstalk." No matter how accurately torque or tension is initially applied to any single bolt, a subsequent retightening of all the bolts must be done to account for crosstalk to even out the load. A further complication is that although bolts react quickly and completely as loads shift during tightening, soft gasket materials do not exhibit similar resilience. Care must be taken to not overcompress the gasket because it may not spring back to fill the void if the load under any given bolt is reduced. That is the reason for gradually increasing the pressure in stages to final load. In order to even out these loads, bolts are tightened in a specific sequence using gradually increasing passes or steps to the final tightness.

ANSI/AWWA C604 gives guidance on various tightening patterns that can be utilized. There is no single "right" tightening pattern for a joint. Whatever pattern is chosen, it should accomplish several interrelated goals to

- Apply sufficient bolt stress to maintain a leak-free connection
- Achieve uniform bolt stress and therefore uniform gasket stress around the flange
- Maintain parallel closure of the flanges during tightening
- Minimize excessive loading/unloading of the gasket during tightening
- Avoid overcompression or crushing of the gasket
- Reduce tool movements to improve efficiency
- Be simple to implement for the assemblers

Since the load applied to the joint by the bolts is difficult to measure directly, torque is used as a convenient way to approximate the desired load. Load is achieved by a controlled turning force (torque) applied to the bolt head or nut. Torque is the product of a force times the distance over which it is applied, usually expressed in foot-pounds. Torque is developed by the turning of threaded nuts and bolts that require surfaces to slide past one another. To accurately convert torque into load, the friction that exists between those sliding surfaces must be known. Friction is reduced by the application of a lubricant. It may take three times the torque to achieve the same bolt load with a dry fastener as with one that has been lubricated.

Nut-and-bolt assemblies have two sliding surfaces: (1) the mating threads and (2) the face of the nut where it contacts the flange or washer. Therefore the amount of force necessary to tighten a nut/bolt to any given load value depends on the friction between the two surfaces.

The relationship between torque and load in this context may be expressed by the following formula cited in ASME PCC-1, appendix J (ASME 2013):

$$T = F\left[0.16\,p + 0.58\,\mu_t d_2 + \frac{D_e \mu_n}{2}\right] \qquad \text{(Eq 12-1)}$$

Torque to Stretch the Bolt (0.16 p) — ***Torque to Overcome Thread Friction*** (0.58 $\mu_t d_2$) — ***Torque to Overcome Face Friction*** ($D_e \mu_n / 2$)

Where:

T = total tightening torque, ft-lb
F = bolt preload, lb
μ_t = coefficient of friction for the threads
n = number of threads per in.
d_1 = thread major diameter, in.
d_2 = basic pitch diameter of the thread, in (for inch threads $d_2 = d_1 - 0.6495/n$)
p = pitch of the thread; normally threads/ inch, i.e., $p = 1/n$
μ_n = coefficient of friction for the nut face or bolt head
D_e = effective bearing diameter of the nut, in. = $(d_o + d_i)/2$
d_o = outer bearing diameter of the nut, in.
d_i = inner bearing diameter of the nut face, in.

Notice in the above formula that the applied torque has to overcome three resistant forces: bolt stretching, thread friction, and face friction.

To simplify this calculation, a formula that combines these three resisting forces into a single variable, *K*, referred to as a "nut factor" when used in the following simplified torque/load formula, is presented in ASME PCC-1, appendix K (ASME 2013).

$$T = \frac{(K \times D_s \times F)}{12} \qquad \text{(Eq 12-2)}$$

Where:

T = torque—the turning force required, ft-lb
K = nut factor—how hard it is to turn, expressed as a decimal
D_s = nominal diameter of the stud, in.
F = load desired in the stud, lb

K is an empirical value related to the total resistance, derived experimentally by applying tightening torque to a bolt of a given nominal diameter in a scale device and observing the resultant load.

K is dependent on a number of factors such as temperature, quantity and application of lubricant to sliding surfaces, bolt diameter, thread pitch, fastener condition, and the applied load. History has shown the use of a nut factor to be as reliable and accurate as the more complex torque formula, and it can be relied on to produce acceptable results if consistently applied. Once determined, it is then applied generally to bolts of the same grade and diameter to relate torque to load. At normal assembly temperatures with standard bolts, the derivation of K can be simplified by adding 0.04 to the lubricant's coefficient of friction, μ (Bickford 1995). Coating or plating of bolts will also result in different friction conditions and therefore different torque requirements than for uncoated fasteners.

Suggested Values for *K*. Tests conducted by Brown et al. (2006) showed that average K values for a number of copper-, nickel-, molybdenum-, and graphite-based antiseize lubricants on standard ASTM A193 B7 studs were in a fairly consistent range from 0.16 to 0.18, regardless of assembly temperatures ranging from 23°F to 105°F. Consensus figures from multiple sources place the K value for petroleum-based lubricants such as SAE 20 oil at approximately 0.19 and that for machine oil at approximately 0.21. Although dry alloy steels exhibit a rather wide scatter of values, a K value of 0.30 has been successfully used.

Testing by Cooper and Heartwell (2011) demonstrated that the presence of a through-hardened washer under the nut or bolt head has as great a positive effect on reducing overall friction as the use of a lubricant. The conclusion being that a hardened washer should be used under all turning nuts both to reduce required torque and also to improve consistency of load among the bolts.

Example Calculation. Assume a flanged joint requires 1-in. diameter ASTM A193 B7 studs, coated with a graphite antiseize having a coefficient of friction, μ, of 0.13 to be loaded to 50 percent of their minimum yield of 105,000 psi.

Using Eq 12-2:

$$T = \frac{(K \times D_S \times F)}{12}$$

$K = \mu + 0.04 = 0.13 + 0.04 = 0.17$

D_S = nominal bolt diameter = 1 in.

F = desired bolt load in pounds $= \pi\left(\frac{D_S}{2}\right)^2 \times 0.5(105{,}000\text{ psi}) = 41{,}234\text{ lb}$

Solving for T:

$T = (0.17 \times 1 \times 41{,}234)/12 \approx 580$ ft-lb

Table 12-2 shows recommended torque values in foot-pounds for A193 B7 bolts at 50 percent of their minimum yield values using various common lubricants. Tightening bolts to approximately 50 percent of their yield strength is common rule of thumb for many joints, but the actual load required may be higher or lower, based on the gasket load requirements for each specific joint. An interactive tool for torque calculations can be found at www.awwa.org/TorqueCalc.

Gasket stresses can change after tightening takes place. Please see the discussion of gasket "creep relaxation" in chapter 6. Joints may have to be retightened in order to ensure leak-free service.

Installation Practice for Mechanical Couplings

Requirements for both mechanical couplings and pipe end preparation can be found in applicable AWWA standards, such as ANSI/AWWA C200, C219, C227, and C606. These should be read in conjunction with the manufacturer's installation instructions. A mechanical coupling relies on uniform contact of the gaskets with the pipe surface. It is therefore important to ensure that the pipe ends in the areas where the coupling gaskets will seat comply with the requirements of the coupling standard, otherwise the full pressure capability of the coupling may not be achieved.

Mechanical couplings are used to introduce flexibility or restraint that will accommodate pipe or ground movement over the life of the pipeline and they can provide access to the pipeline. Manufacturers specify the setting gap between adjacent pipe ends

Table 12-2 Bolt torque

AWWA Torque Guide for ASTM A193 Grade B7 Bolts Bolt Tension Based on 50 Percent Yield* and AWWA C207 Class B, D, or E Flanges

							Required Torque (ft/lbs)				
Nominal Pipe Size	Number of Bolts	Bolt Size Diameter Inches	Hex Nut Size	Nominal Area (in^2)	Minimum Yield Stress (*psi*)	Bolt Tension F (*lbs*)	Moly/Copper/ Nickel or Graphite Anti-Sieze 1000 $K = 0.17$	Petroleum-Based Grease $K = 0.19$	Machine Oil $K = 0.21$	Dry Steel $K = 0.30$	Custom Plug In *K* Value† 0.185
4	8	5/8	1 1/16"	0.307	105,000	16,107	143	159	176	252	155
6	8	3/4	1 1/4"	0.442	105,000	23,194	246	275	304	435	268
8	8	3/4	1 1/4"	0.442	105,000	23,194	246	275	304	435	268
10	12	7/8	1 7/16"	0.601	105,000	31,569	391	437	483	691	426
12	12	7/8	1 7/16"	0.601	105,000	31,569	391	437	483	691	426
14	12	1	1 5/8"	0.785	105,000	41,234	584	653	722	1,031	636
16	16	1	1 5/8"	0.785	105,000	41,234	584	653	722	1,031	636
18	16	1 1/8	1 13/16"	0.994	105,000	52,186	832	930	1,027	1,468	905
20	20	1 1/8	1 13/16"	0.994	105,000	52,186	832	930	1,027	1,468	905
22	20	1 1/4	2"	1.227	105,000	64,427	1,141	1,275	1,409	2,013	1,242
24	20	1 1/4	2"	1.227	105,000	64,427	1,141	1,275	1,409	2,013	1,242
30	28	1 1/4	2"	1.227	105,000	64,427	1,141	1,275	1,409	2,013	1,242
36	32	1 1/2	2 3/8"	1.767	105,000	92,775	1,971	2,203	2,435	3,479	2,145
42	36	1 1/2	2 3/8"	1.767	105,000	82,775	1,971	2,203	2,435	3,479	2,145
48	44	1 1/2	2 3/8"	1.767	105,000	92,775	1,971	2,203	2,435	3,479	2,145
54	44	1 3/4	2 3/4"	2.405	105,000	126,278	3,131	3,499	3,867	5,525	3,407
60	52	1 3/4	2 3/4"	2.405	105,000	126,278	3,131	3,499	3,867	5,525	3,407
66	52	1 3/4	2 3/4"	2,405	105,000	126,278	3,131	3,499	3,867	5,525	3,407
72	60	1 3/4	2 3/4"	2.405	105,000	126,278	3,131	3,499	3,867	5,525	3,407
78	64	2	3 1/8"	3.142	105,000	164,934	4,673	5,223	5,773	8,247	5,085
84	64	2	3 1/8"	3.142	105,000	164,934	4,673	5,223	5,773	8,247	5,085
90	68	2 1/4	3 1/2"	3.976	105,000	208,745	6,654	7,437	8,219	11,742	7,241
96	68	2 1/4	3 1/2"	3.976	105,000	208,745	6,654	7,437	8,219	11,742	7,241
102	72	2 1/2	3 7/8"	4.909	105,000	257,709	9,127	10,201	11,275	16,107	9,933
108	72	2 1/2	3 7/8"	4.909	105,000	257,709	9,127	10,201	11,275	16,107	9,933
114	76	2 3/4	4 1/4"	5.940	95,000	282,130	10,991	12,284	13,578	19,396	11,961
120	76	2 3/4	4 1/4"	5.940	95,000	282,130	10,991	12,284	13,578	19,396	11,961
126	80	3	4 5/8"	7.069	95,000	335,759	14,270	15,949	17,627	25,182	15,529
132	80	3	4 5/8"	7.069	95,000	335,759	14,270	15,949	17,627	25,182	15,529
144	84	3 1/4	5 3/8"	9.621	95,000	457,005	22,660	25,326	27,992	39,988	24,659

*50% of yield is a generally accepted working goal in that it fits within most gaskets sealing parameters. Caution should be taken with softer gasket materials such as rubber. Other special conditions may also apply and either lower or higher values may be required. Consult the gasket manufacturer for more specific values.

†At normal assembly temperatures, *K* is equal to a lubricant's coefficient of friction plus 0.04.

to facilitate proper installation and movement allowed by the design after the pipe is in service.

Mechanical couplings use elastomeric material in their gaskets, which due to its nature will relax over the design life of the product. To ensure the couplings continue to offer leaktight performance over the life of the pipeline, it is critical that the correct bolt torque or installation instruction as recommended by the manufacturer is achieved on all bolts at installation.

Field Coating of Joints

Acceptable procedures for coating of field joints are described in applicable AWWA standards. More information on coatings can be found in chapter 11.

Pipe-Zone Bedding and Backfill

The following discussion on pipe bedding and backfill is somewhat general in nature. A foundation study should be performed to provide more precise design criteria for large projects or those with unusual conditions. More detailed bedding and backfill information can be found in ANSI/AWWA C604.

Backfill. A typical trench detail can be found in Figure 5-2 in chapter 5. A critical area of pipe backfill installation is located in the haunch of the pipe. This area provides the majority of support for the installed pipe but can be the most difficult area in which to achieve proper compaction levels.

Trench backfill should not be placed until confirmation that compaction of pipe-zone, haunch, and backfill complies with the specified compaction. Native backfill material above the pipe zone up to the required backfill surface should be placed to the density required in the contract specifications. To prevent excessive live loads on the pipe, sufficient densified backfill should be placed over the pipe before power-operated hauling or rolling equipment is allowed directly over the pipe.

Compaction Methods. Regardless of the densification method used, materials must be brought up at relatively the same rate on both sides of the pipe. Backfill materials must be placed such that the haunch area under the pipe will be completely filled and that no voids exist in the backfill zone. Care also should be taken so that the pipe is not floated or displaced before backfilling is complete.

Cohesive soils should be densified by compaction using mechanical or hand tamping. Care must be taken to not damage coatings during compaction. Equipment with suitably shaped tamping feet for compacting the material will generally provide soil density as required by the specifications.

Soils identified as free draining are usually densified by mechanical or hand tamping. Methods using water for consolidation are less frequently used, such as water jets, immersion-type vibrators, bulkheading, and flooding or sluicing. Consolidation of earth backfill by hydraulic methods should be used only if both the backfill and the native soil are free draining. The thickness of layers should not exceed the penetrating depth of the vibrators if consolidation is performed by jetting and internal vibration.

ANCHORS AND THRUST BLOCKS

Anchors or thrust blocks may be used at angle points, side outlets, valves, and on steep slopes when using unrestrained pipe. The type of pipe joint (restrained or nonrestrained) used determines the necessary anchoring at these points.

All-welded (restrained) pipelines laid in trenches will ordinarily not need anchors or thrust blocks except on extremely steep slopes since they are fully restrained. An all-welded pipeline laid aboveground on piers may be stable when filled and under pressure but may require heavy anchorage at angle points and particularly on steep slopes to resist stresses resulting from temperature changes when the pipe is empty.

When unrestrained joints are used that have little or no ability to resist tension, all of the previously mentioned critical points must be adequately blocked or anchored. In order to provide resistance to thrust at angles in large-diameter pipelines, whether buried or exposed, welded joints or otherwise restrained joints should be provided on each side of the angle point for a distance sufficient to resist the thrust. See chapter 8 for more information.

Where pipe is laid on piers, antifriction material should separate the pipe from the supporting structure; 90° to 120° of the pipe surface should be made to bear on the pier.

Pipelines laid on slopes, particularly aboveground, always have a tendency to creep downhill. It may be necessary to provide anchor blocks placed against undisturbed earth at sufficiently frequent intervals on a long, steep slope to reduce the weight of pipe supported at each anchorage to a safe value. When disturbance of the trench is unlikely, concrete thrust blocks may be used to resist the lateral thrust. Vertical angles causing downward thrust require no special treatment if the pipe is laid on a firm and carefully trimmed trench bottom, but vertical angles causing upward thrust should be properly anchored.

Soil Resistance to Thrust

A force caused by thrust against soil, whether applied horizontally or vertically downward, may cause consolidation and shear strains in the soil, allowing a thrust block to move. The safe load that a thrust block can transfer to a given soil depends on the consolidation characteristics and the passive resistance (shear strength) of that soil, the amount of block movement permissible, the area of the block, and the distance of force application below ground line. For general guidelines for thrust block design refer to chapter 8.

STEEL TUNNEL LINERS AND CASING PIPE

In tunnel applications, steel tunnel liners are installed to maintain the tunnel opening and to prevent tunnel leakage caused by unfavorable geological conditions of the surrounding rock or soils. The liner may be used as the conveyance pipe or a carrier pipe may be installed inside the liner. Joints may be lap welded, butt welded with backing, gasketed, or mechanically interlocked depending on the design of the liner and the installation methods. The annular space between the tunnel liner and the soil or rock walls is typically grouted, as is the space between the liner and carrier pipe when a dual system is used. The steel tunnel liner is usually designed to withstand all internal pressures and external loads. For rock tunnels, load sharing may be applicable (*ASCE MOP 79, Steel Penstocks*). Depending on the tunneling method, the steel liners may be installed with special pipe carriers or specially designed conveyance systems, or pushed and/or pulled in as the tunneling operation advances. The steel tunnel liner pipe is commonly supplied bare on the outside when the design includes cementious grouting between the exterior of the pipe and the tunnel wall. Lining requirements are determined by the project design and installation methods. As an example, cement-mortar lining may be plant-applied or applied in the field after the liner has been installed in the tunnel.

REHABILITATION OF PIPELINES

With large-diameter water transmission pipelines reaching the end of their intended service lives or needing repair, steel cylinder–based fully structural rehabilitation solutions have been embraced by municipal agencies throughout North America since the 1980s. In particular, large-diameter pipelines have been rehabilitated by the process of internal relining and sliplining with steel cylinders. These renewal methods eliminate the need to remove and replace structurally deficient large-diameter pipe sections with new pipe.

Rehabilitation of water or wastewater pipelines offers an alternative solution to replacement of structurally deficient pipelines by traditional cut, remove, and replace methods. As pipelines reach the end of their intended service lives or deficiencies in design, manufacture, installation, or operation impact the serviceability of a pipeline, rehabilitation can be a viable option. The following trenchless rehabilitation methods result in a fully structural carrier pipe that can be manufactured to the desired design pressures of existing transmission line or "upgraded" to handle higher pressures and flow rates. Although there is some loss in the internal diameter with these methods, the benefit of trenchless construction, especially in busy urban areas with their inherent risks, is of value.

Relining With Steel Cylinders

Relining involves the insertion of designed and fabricated steel relining cylinders into the host pipe. The cylinders are provided rolled but with the longitudinal seam left unwelded and the cylinder strapped with the diameter reduced for ease of insertion. Once in place the straps on the cylinders can be released allowing the cylinder to expand and the cylinder precisely fit inside the host line and completed by longitudinal and circumferential field welds. Applicable nondestructive tests (NDTs) are used to ensure weld quality. Grout ports fabricated on the reliner sections are utilized to fill the annular space between the liner and host pipe with a lightweight cellular grout. Finally, lining is field-applied to the

Figure 12-2 Steel reliner section

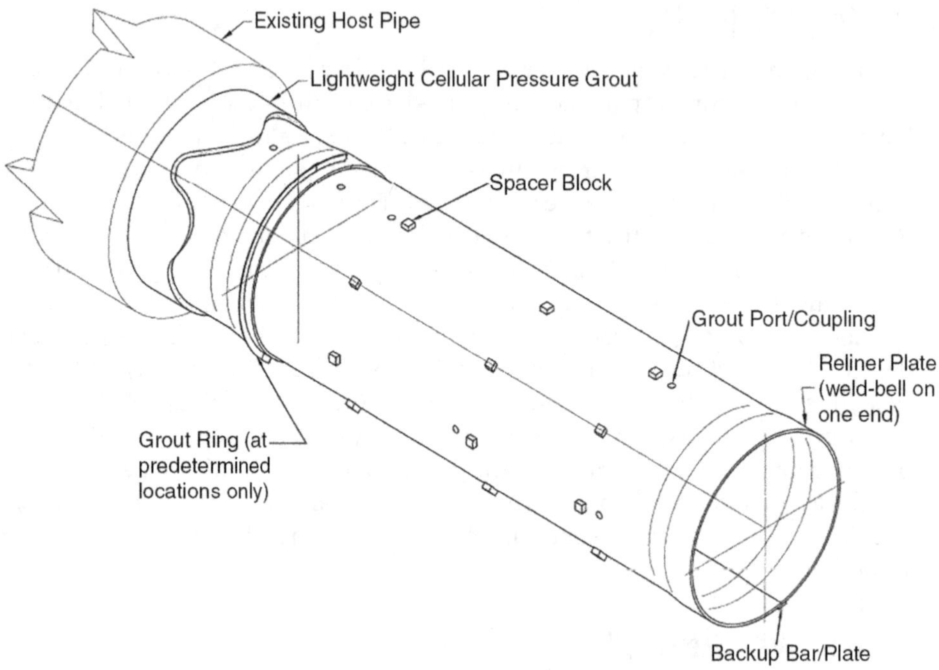

Figure 12-3 Steel reliner assembly view (not to scale)

inside of the reliner sections for corrosion protection. Relining results in minimal internal diameter loss in the host pipe, typically less than 5 in. Figure 12-2 shows the unwelded steel reliner strapped with a reduced diameter ready for insertion. Figure 12-3 shows an assembly view of a typical reliner section and its components.

Sliplining

Sliplining consists of inserting complete sections of steel cylinder into the host pipe, connecting the adjoining slipliner cylinder sections by either internal lap welding or using O-ring gasket joints, then typically filling the annular space between the host pipe and the slipliner sections with a lightweight cementious grout. Grout ports may be fabricated on each slipliner section during manufacture. For internal corrosion protection, lining may be shop-applied or field-applied for larger diameters. Although cylinder design, fabrication, and installation are less specialized for the sliplining process in comparison to relining, the loss in internal flow area of the host pipe with sliplining is typically as high as 6 to 8 in. for a host pipe of 60-in. diameter. Figure 12-4 shows a slipliner section being inserted into a host pipe.

Structurally Independent Systems

Both relining and sliplining provide a structurally independent rehabilitation solution with a long-term internal burst strength, when independently tested from the host pipe, equal to or greater than the maximum allowable operating pressure (MAOP) of the host pipe. Relined and sliplined systems are designed to withstand any dynamic loading or other short-term effects associated with a complete failure of the host pipe, usually due to deteriorating soil conditions and further corrosion of various components of the deficient composite host pipe. These two capabilities place these renewal methods into the Class IV Linings, which is a fully structural system and is the highest category as described in AWWA M28, *Rehabilitation of Water Mains*.

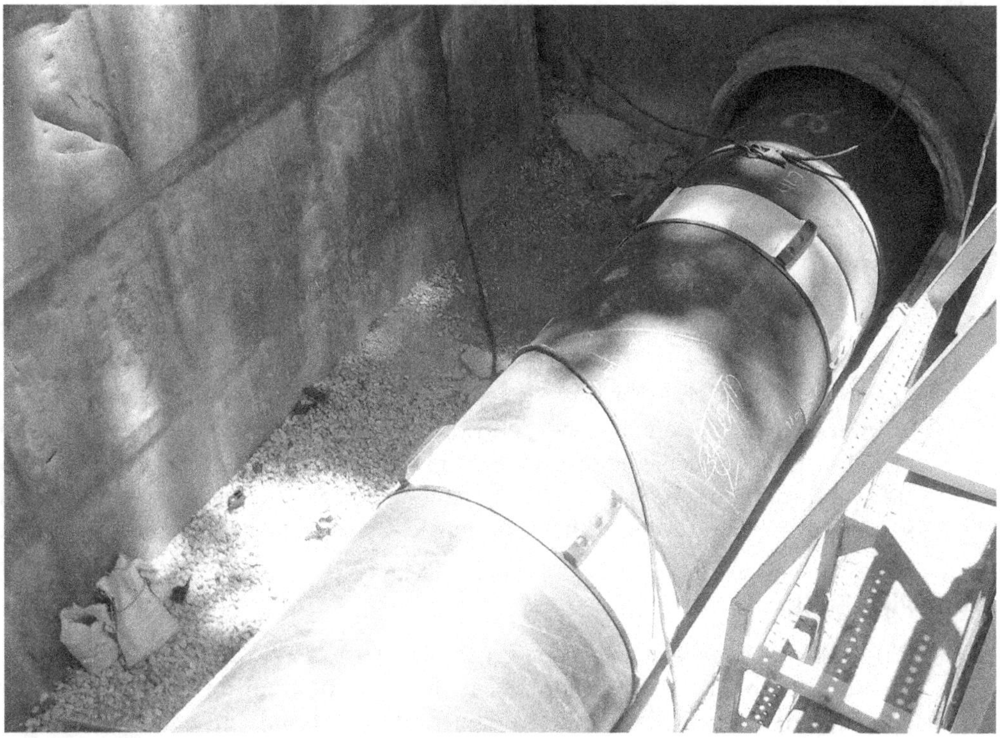

Figure 12-4 Steel slipliner section being inserted into host pipe with casing spacers

HORIZONTAL DIRECTIONAL DRILLING

Horizontal directional drilling, or HDD, is a steerable trenchless method of installing underground pipes, conduits, and cables in a shallow arc along a prescribed bore path by using a surface launched drilling rig, with minimal impact on the surrounding area. HDD is used when trenching or excavating is not practical. It is suitable for a variety of soil conditions and jobs including road, wetland, and river crossings. HDD has become a widely accepted form of trenchless construction in the water industry, although the process itself was developed for, and grew out of, the petroleum pipeline industry. More steel pipe has been used in HDD applications worldwide than any other pipe materials because in the petroleum industry, the use of steel pipe was dictated by high-pressure service. Some of the longest length and largest diameter HDD projects with considerable challenges in water systems have been successfully completed with steel pipe.

During an HDD installation, the bore is generally reamed 12 in. larger than the pipe. Depending on its size and weight, spiral-weld pipe will either float or sink in the drilling fluid that occupies the excess space. In either case, the pipe is "dragged" through the drilling fluid and cuttings. For diameters 36 in. and larger, the pipe is normally filled with water to reduce buoyancy and resulting pulling loads. Tensile stress due to bending is generally limited to 90 percent of the yield strength of the steel.

Design of an HDD steel pipe installation differs from the design of a buried water transmission line because of the high tension loads, bending stresses, and the external fluid pressures acting on the pipeline during the installation. In normal transmission lines, a designer is concerned primarily with internal pressures and external live and dead loads. HDD pipe installation load requirements are normally far in excess of those required for typical open trench applications. For design of HDD projects, reference the American Gas Association manual *Installation of Pipelines by Horizontal Directional Drilling: an Engineering Design Guide* (Hair and PRCI 2008) or ASCE MOP 108, *Pipeline Design for Installation by Horizontal Direction Drilling.*

SUBAQUEOUS PIPELINES

There are basically two systems for constructing subaqueous pipelines: pipe-laying systems, and pipe-pulling systems.

Pipe Laying

In a pipe-laying system, the pipe is transported by water to the laying platform, which is a barge equipped primarily with a heavy crane and possibly a horse. The horse is a winch capable of moving on skid beams in two directions with cables extending vertically downward into the water. On arrival at the job site, sections of pipe are often assembled on the barge into sections of 100 to 150 ft in length. The crane picks up the assembled pipe segment and holds it while the horse is centered above it. The pipe, once attached to the horse, is lowered to the bottom. Divers report the position of the segment in relation to the completed section before it, and the horse is moved up and down, forward and backward, and sideways until the spigot end lines up with the bell end of the completed section. The pipe sections are then pulled together with harness rods, winches, or vacuum pull devices.

Pipe Pulling

Pipe pulling has been used for crossing rivers, bays, and in the open ocean. The pipe-pulling method requires pipe capable of withstanding the tensile stresses developed during the pulling operation. The method is usually used with steel pipe because of these high tensile stresses.

A steel-pipe–pulling operation begins on assembly ways established ashore. To prevent floating, the pipe may be allowed to fill with water as it leaves the assembly way. Alternatively, the pipe may be capped to exclude water, then concrete weighted or coated to overcome its buoyancy. The pipe lengths are welded in continuous strings. The completed pipe string is transferred to launchways (Figure 12-5), which lead to the submerged placement area. Once shore assembly is complete, the reinforced head of the pipe string is attached to a pull barge by wire rope and pulled along the bottom by a winch until it is in position (Figure 12-6).

A variation of the bottom-pull method is the floating-string method of pipe installation. The line is initially assembled in long segments and transferred to the launchways. It is then pulled off the launchway by a tugboat, floated out to location, and sunk

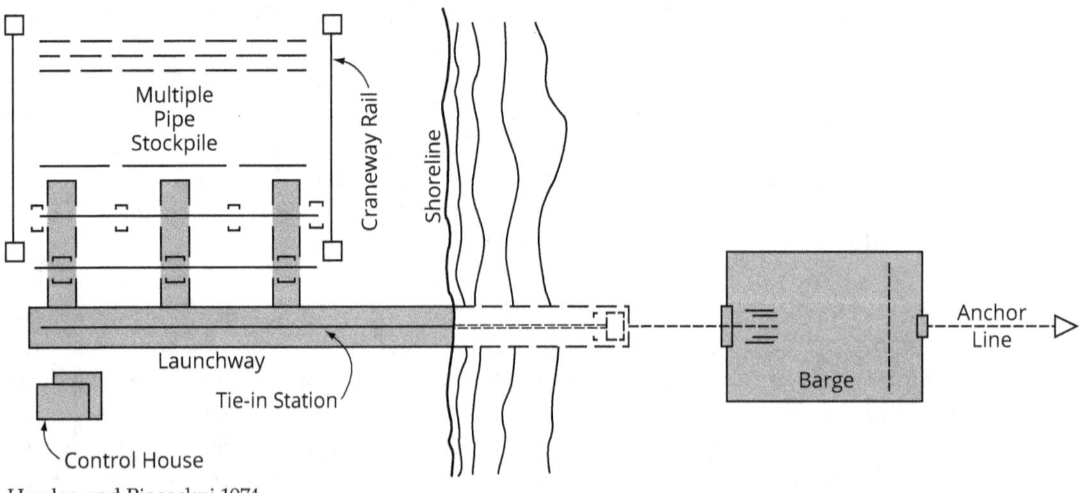

Source: Hayden and Piaseckyi 1974.

Figure 12-5 Subaqueous pipeline—assembly and launching

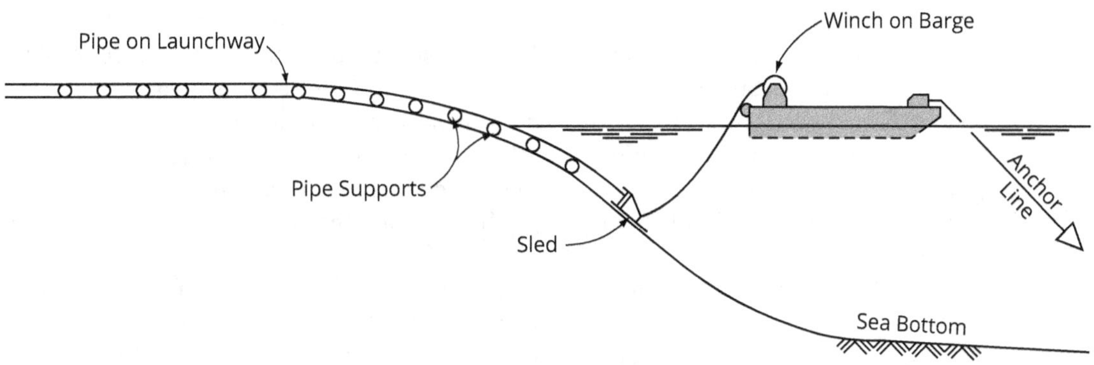

Source: Hayden and Piaseckyi 1974.

Figure 12-6 Subaqueous pipeline—positioning by barge

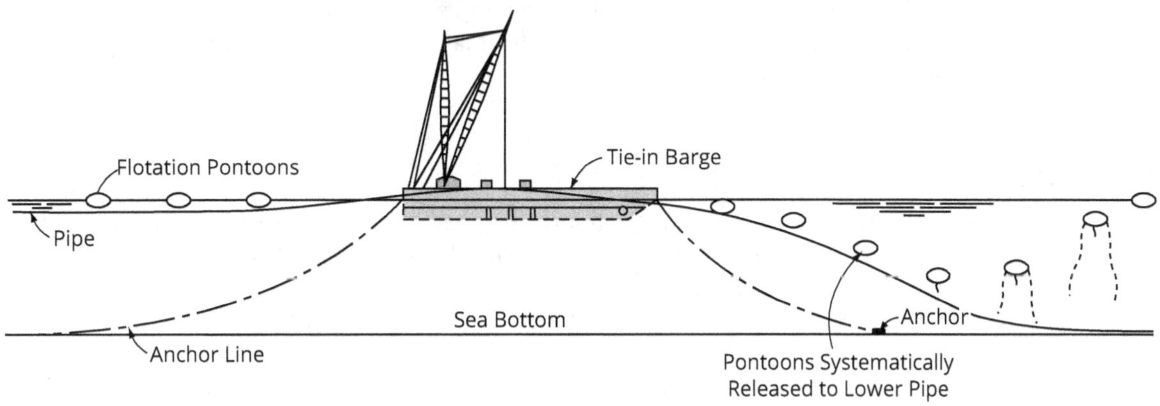

Source: Hayden and Piaseckyi 1974.

Figure 12-7 Subaqueous pipeline—floating string positioning

(Figure 12-7). Individual strings are connected by divers, as in the pipe-laying method, or strings are joined by picking up the end of the last piece installed and putting it on a deck of a special tie-in platform, where the connection to the beginning of the next string is made.

Lay Barge

Smaller-diameter pipelines are sometimes laid at sea or across rivers from a lay barge, which has onboard facilities for welding pipe sections together. The pipe string is fed over the end of the barge as the barge moves along the route of the pipeline, adding pipe as it goes. The pipe undergoes bending stresses as it is laid, so the barge should include quality-control facilities for checking the soundness of the circumferential welds.

HYDROSTATIC FIELD TEST

The purpose of the hydrostatic field test is primarily to determine if the field joints are watertight. The hydrostatic test is usually conducted after backfilling is complete; some areas of a pipeline may need to be left exposed for inspection during the hydrotest, such as bolted flexible joints. It is performed at a fixed pressure not exceeding 125 percent of the design working pressure unless a higher pressure is taken into consideration in the design. If thrust resistance is provided by concrete thrust blocks, the blocking must be allowed to cure before the test is conducted. Some general guidelines are provided here; detailed procedures and requirements for field hydrotesting are described in ANSI/AWWA C604.

Cement-mortar-lined pipe to be tested should be filled with water and allowed to stand for at least 24 hours to permit maximum absorption of water by the lining. Additional makeup water should be added to replace water absorbed by the cement-mortar lining. (Pipe with other types of lining may be tested without this waiting period.) Pipe to be cement-mortar lined in place may be hydrostatically tested before or after the lining has been placed.

If the pipeline is to be tested in segments and valves are not provided to isolate the ends, the ends must be provided with bulkheads for testing. A conventional bulkhead usually consists of a section of pipe 2-ft to 3-ft long, with a flat plate or dished plate bulkhead welded to the end and containing the necessary outlets for accommodating incoming water and outgoing air.

The pipeline should be filled slowly to prevent possible water hammer, and care should be exercised to allow all of the air to escape during the filling operation. After filling the line, a pump and makeup water may be necessary to raise and maintain the desired test pressure.

REFERENCES

American Society of Civil Engineers (ASCE) MOP 108. *Pipeline Design for Installation by Horizontal Direction Drilling*. Latest Edition. Reston VA: ASCE.

ASCE MOP 79. *Steel Penstocks*. Latest Edition. Reston VA: ASCE.

American Society of Mechanical Engineers (ASME) PCC-1-2013. Guidelines for Pressure Boundary Bolted Flange Joint Assembly. New York: ASME.

American Water Works Association. M28, *Rehabilitation of Water Mains*. 3rd ed. Denver, CO: AWWA.

ANSI/AWWA C200. Steel Water Pipe, 6 In. (150 mm) and Larger. Latest Edition. Denver, CO: AWWA.

ANSI/AWWA C205. Cement-Mortar Protective Lining and Coating for Steel Water Pipe 4 In. (100 mm) and Larger—Shop Applied. Latest edition. Denver, CO: AWWA.

ANSI/AWWA C206. Field Welding of Steel Water Pipe. Latest edition. Denver, CO:

ANSI/AWWA 219. Bolted, Sleeve-Type Couplings for Plain-End Pipe. Denver, CO: AWWA.

ANSI/AWWA C227. Bolted, Split-Sleeve Restrained and Nonrestrained Couplings fior Plain-End Pipe. Denver, CO: AWWA.

ANSI/AWWA C602. Cement-Mortar Lining of Water Pipelines in Place—4 In. (100 mm) and Larger. Latest edition. Denver, CO: AWWA.

ANSI/AWWA C604. Installation of Buried Steel Water Pipe—4 In. (100 mm) and Larger. Latest edition. Denver, CO: AWWA.

ANSI/AWWA C606, Grooved and Shouldered Joints. Denver, CO: AWWA.

ASTM A193/193M. Standard Specification for Alloy-Steel and Stainless Steel Bolting for High Temperature or High Pressure Service and Other Special Purpose Applications. Latest edition. West Conshohocken, PA: ASTM.

Bickford, J.H. 1995. *An Introduction to the Design and Behavior of Bolted Joints*. Boca Raton, FL: CRC Press.

Brown, W., L. Marchand, and T. LaFrance. 2006. *Bolt Anti-Seize Performance in a Process Plant Environment*. New York. ASME Pressure Vessel Research Council, PVP2006-ICPVT11-93072.

Cooper, W., and T. Heartwell. 2011. *Variables Affecting Nut Factors for Field Assembled Joints*. New York: ASME Pressure Vessels and Piping Conference, PVP2011-57197.

Hair, J.D. and Associates, and Pipeline Research Council International. 2008. *Installation of Pipelines by Horizontal Directional Drilling: an Engineering Design Guide*. Houston, TX: Technical Toolboxes.

Hayden, W.M., and P.J. Piaseckyi. 1974. Economic and Other Design Considerations for a Large Diameter Pipeline. In *Proc. Sixth International Harbour Congress*. Antwerp, Belgium: K. Vlaam Ingenieursver.

For more information on bolted joints, the reader is referred to the following articles:

Bickford, J. 1998. *Gaskets and Gasketed Joints*. New York: Marcel Dekker.

Brown, W. 2004. *Efficient Assembly of Pressure Vessel Joints*. New York: ASME Pressure Vessel Research Council, PVRC 2365.

This page intentionally blank.

Nomenclature

α = Angular influence factor [chapter 7]
α = Slope angle, degrees [chapter 8]
β = Shallow cover factor [chapter 8]
β = Contact angle, degrees, [chapter 9]
Δ = Deflection angle of elbow or outlet, deg [chapters 7, 8]
Δ = Angular deflection per pipe section, deg [chapter 8]
Δ_p = Plan angle of elbow, deg [chapter 7]
Δx = Predicted horizontal pipe deflection, in. [chapter 5]
ε = Coefficient of thermal expansion/contraction of steel [chapter 6]
γ = Specific weight of fluid [chapter 3]
γ = Unit weight of backfill [chapter 8]
γ_w = Unit weight of water, lb/ft^3 [chapter 5]
θ = Elbow segment deflection angle, deg [chapter 7]
θ = Total deflection angle of curve, deg [chapter 8]
φ = Angle of internal friction, degrees [chapter 8]
φ_s = Factor to account for variability in stiffness of compacted soil [chapter 5]
σ = Soil bearing capacity, psf [chapter 8]
σ_1 = Longitudinal stress in cylinder at anchor, ksi [chapter 7]
σ_2 = Circumferential or hoop stress in pipe, ksi [chapter 7]
σ_a = Average bearing stress of anchor ring on concrete, psi [chapter 7]
σ_A = Allowable design stress for an elbow with reduced radius, psi [chapter 7]
σ_{A1} = Allowable stress in mainline steel cylinder, psi [chapter 7]
σ_{A2} = Allowable stress in outlet steel cylinder, psi [chapter 7]
σ_b = Bending stress resulting from ring restraint, psi [chapter 9]
σ_b = Secondary bending stress in steel cylinder resulting from concrete encasement, psi [chapter 7]
σ_{cs} = Local bending stress at saddle, psi [chapter 9]
σ_e = Effective stress in restrained elbow, psi [chapter 7]
σ_e = Equivalent stress , psi [chapter 9]
σ_{eq} = Equivalent stress, psi [chapter 7]
σ_h = Hoop stress, psi or ksi [chapters 6, 7, 9]
σ_l = Longitudinal stress, ksi [chapters 6, 7, 9]
σ_{ls} = Longitudinal stress at saddle support, psi [chapter 9]
σ_L = Longitudinal stress in pipe shell, ksi [chapter 7]
σ_r = Bending stress in anchor ring at connection to pipe, ksi [chapter 7]

σ_t = Total longitudinal tension stress, psi [chapter 9]
σ_T = Thermal stress, ksi [chapters 6, 9]
$\sigma_{T+\nu}$ = Allowable combination of Poisson's and thermal stress, ksi [chapter 6]
σ_U = Minimum ultimate tensile strength of material, psi or ksi [chapters 6, 9]
σ_ν = Poisson ratio of hoop stress [chapter 6]
σ_w = Stress in weld, psi [chapter 6]
σ_w = Minimum tensile strength of weld electrode, ksi [chapter 7]
σ_x = Principal stress in span of pipe, psi [chapter 9]
σ_y = Flexural stress in pipe, psi [chapter 9]
σ_y = Principal stress in span of pipe, psi [chapter 9]
σ_Y = Minimum yield strength of material, psi or ksi [chapters 4, 6, 7]
μ = Coefficient of friction of bolt lubricant [chapter 12]
μ = Friction coefficient between pipe and soil [chapter 8]
μ_n = Coefficient of friction for the nut face or bolt head [chapter 12]
μ_t = Coefficient of friction for the threads [chapter 12]
ν = Kinematic viscosity of fluid, ft^2/s [chapter 3]
ν_c = Poisson's ratio for cement mortar [chapter 4]
ν_s = Poisson's ratio for soil [chapters 5, 7]
ν_s = Poisson's ratio for steel [chapters 4, 7]
a = Wave velocity, ft/s [chapter 3]
A = Area of pipe [chapters 3, 8]
A = Dimension between front and back harness lug plates or rings, in. [chapter 7]
A = Anchor ring height, in. [chapter 7]
A_1 = Cross sectional area of main pipe, in.^{2} [chapter 8]
A_2 = Cross sectional area of wye branch or reducer, in.^{2} [chapter 8]
A_a = Excess area available for outlet reinforcement, in.^{2} [chapter 7]
A_o = Cross sectional area of tee or wye outlet, in.^{2} [chapter 8]
A_r = Theoretical outlet reinforcement area, in.^{2} [chapter 7]
A_T = Half the width of tire pattern print, ft [chapter 5]
A_w = Required reinforcement area for outlet, in.^{2} [chapter 7]
B = Anchor ring thickness, in. [chapter 7]
B = Saddle support width, in. [chapter 9]
B' = Anchor ring minimum thickness, in. [chapter 7]
B_d = Horizontal width of trench width at top of pipe, ft [chapter 5]
B_c = Outside diameter of pipe, ft [chapter 5]
B_T = Half the length of tire pattern print, ft [chapter 5]
C = Hazen-Williams Coefficient [chapter 3]
C_d = Load coefficient based on H_c/B_d [chapter 5]
C_Δ = Factor for pipe deflection limit [chapter 5]
C_n = Scalar calibration factor [chapter 5]
d = Initial crotch plate depth, in. [chapter 7]
d = Inside diameter of pipe, ft or in. [chapters 3, 7]
d'_b = Base depth of crotch plate for unequal diameter, in. [chapter 7]
d'_t = Top depth of crotch plate for unequal diameter pipes, in. [chapter 7]
d'_w = Crotch depth of crotch plate for unequal diameter pipes, in. [chapter 7]

d_1 = Crotch plate depth, in. [chapter 7]
d_1 = The thread major diameter, in. [chapter 12]
d_2 = The basic pitch diameter of the thread, in. [chapter 12]
d_b = Base depth of crotch plate for equal diameter pipes, in. [chapter 7]
d_i = The inner bearing diameter of the nut face, in. [chapter 12]
d_n = Nominal diameter of pipe, in. [chapter 3]
d_o = Outside diameter of outlet pipe, in. [chapter 7]
d_o = The outer bearing diameter of the nut, in. [chapter 12]
d_t = Top depth of crotch plate for equal diameter pipes, in. [chapter 7]
d_w = Crotch depth of crotch plate for equal diameter pipes, in. [chapter 7]
D = Nominal diameter of pipe, in. [chapters 4, 12]
D = Inside diameter of pipe, ft. [chapters 3, 7]
D = Nominal diameter of harness rod, in. [chapter 7]
D'_r = Anchor ring minimum outside diameter, in. [chapter 7]
D_c = Outside diameter of coated pipe, in. [chapters 4, 5]
D_e = The effective bearing diameter of the nut, in. [chapter 12]
D_I = Inside diameter of steel cylinder, in. [chapter 4]
D_l = Deflection lag factor [chapter 5]
D_o = Outside diameter of steel cylinder, in. [chapters 4, 5, 7]
D_o = Outside diameter of ellipsoidal head, in. [chapter 7]
D_r = Anchor ring actual outside diameter, in. [chapter 7]
D_s = Nominal diameter of the stud, in. [chapter 12]
D_x = Inside horizontal diameter, in. [chapter 5]
D_y = Inside vertical diameter, in. [chapter 5]
e = Absolute roughness [chapter 3]
E = Modulus of elasticity, general, psi [chapters 4, 5, 9]
E = Harness lug back plate/ring hole height dimension, in. [chapter 7]
E' = Modulus of soil reaction, psi [chapter 5]
E_C = Modulus of elasticity for cement mortar, psi [chapters 4, 5, 6]
E_S = Modulus of elasticity for steel, psi [chapters 4, 5, 6, 7]
EI = Pipe wall stiffness, lb-in. [chapters 5, 6]
f = Darcy friction factor [chapter 3]
f = Desired load in stud, lb [chapter 12]
f'_c = Minimum 28-day compressive strength of concrete, psi [chapter 7]
f_μ = Frictional resistance per linear ft. [chapter 8]
f_b = Unit shear force in fillet weld to resist ring moment, lb/in. or kip/in. [chapter 7]
f_r = Resultant shear force in fillet weld, lb/in. or kip/in. [chapter 7]
f_v = Unit shear force in fillet weld to resist direct shear from ring, lb/in. or kip/in. [chapter 7]
F = Bolt preload, lbs [chapter 12]
F = Desired load in stud, lb [chapter 12]
F = Unbalanced axial thrust, lb [chapter 8]
F_1 = Unbalanced axial thrust, lb [chapter 8]
F_2 = Unbalanced axial thrust, lb [chapter 8]
FS = Factor of safety [chapter 5]
g = Acceleration due to gravity, 32.2 ft/s^2 [chapter 3]

G = Gradient, ft/ft [chapter 9]
h = Pressure head rise above normal, ft [chapter 3]
h = Height of ground above bottom of thrust block, ft [chapter 8]
h_L = Head loss, ft [chapter 3]
H = Head loss in 1,000 units of length of pipe, ft/1,000 ft [chapter 3]
H = Depth of cover, ft. [chapter 8]
H_b = Thrust block height, ft [chapter 8]
H_c = Height of fill above pipe, ft [chapter 5]
H_w = Height of water above top of pipe, ft. [chapter 5]
HB = Harness lug back plate/ring height, in. [chapter 7]
HF = Harness lug front plate/ring height, in. [chapter 7]
I = General case, transverse moment of inertia per unit length of individual pipe wall components= $t^3/12$, where t is in in., in.^{3} [chapter 5]
I = Slope of incoming leg of elbow, deg [chapter 7]
I_C = Moment of inertia per unit length of CMC, in.^{3} [chapter 5]
I_L = Moment of inertia per unit length of CML, in.^{3} [chapter 5]
I_S = Moment of intertia per unit length of steel cylinder, in.^{3} [chapter 5]
I_s = Moment of inertia of pipe wall cross section, in.^{4} [chapter 9]
k = Bulk modulus of compressibility of liquid, lb/in.^{2} [chapter 3]
k = Saddle support contact angle factor [chapter 9]
k_v = Soil modulus correction factor for Poisson's ratio [chapter 5]
K = Bedding constant [chapter 5]
K = PDV multiplier, [chapter 7]
K = Resistance coefficient [chapter 3]
K = Nut factor - how hard it is to turn, expressed as a decimal [chapter 12]
K_o = Coefficient of lateral soil pressure [chapter 8]
K_s = Scobey constant [chapter 3]
L = Length or span of pipe, ft [chapters 3, 9]
L = Length between fixed points, in. [chapter 6]
L = Restraint length, ft [chapter 8]
L_b = Thrust block length, ft [chapter 8]
L_p = Calculated centerline pipe laying length [chapter 8]
L_r = Length of increased steel cylinder thickness or reinforcement at saddle support, in. [chapter 9]
L_R = Length of reinforcement extension past concrete encasement, in. [chapter 7]
m = A_T/H_C [chapter 5]
m = Radius to thickness ratio [chapters 3, 9]
M = Beam bending moment, in.- lb or ft- lb [chapters 7, 9]
M = Factor for outlet reinforcement design [chapter 7]
M_1 = Unit bending moment in pipe at anchor ring connection, in.-kip /in. [chapter 7]
M_l = Longitudinal bending moment at mid-span, in.- lb [chapter 9]
M_r = Unit bending moment in harness lug or anchor ring, in.- lb/in. or in.-kip/in.[chapter 7]
n = B_T/H_C [chapter 5]
n = Manning coefficient [chapter 3]
n = Number of threads per inch [chapter 12]

n_e = Diameter multiplier for elbow radius calculation, in. [chapter 7]
N = Number of bolt threads per inch, [chapter 7]
N_b = Angle factor for crotch plate design [chapter 7]
N_L = Number of harness lugs [chapter 7]
N_w = Angle factor for crotch plate design [chapter 7]
O = Slope of outgoing leg of elbow, deg [chapter 7]
p = Internal design pressure, psi [chapters 4, 5, 6, 7, 8, 9]
p = Pitch of the thread. Normally threads/ in. [chapter 12]
p_c = Collapsing pressure, psi [chapters 4, 9]
p_{cr} = Critical collapse pressure, psi [chapters 4, 9]
p_{cvx} = Design pressure on convex side of head, psi [chapter7]
p_s = Surge allowance [chapter 4]
p_t = Transient pressure or test pressure, psi [chapters 4, 7, 8, 9]
p_w = Working pressure, psi [chapters 4, 7]
P = Pressure, psi [chapter 3]
P = Pressure rise above normal, ft. of water [chapter 3]
P = Total saddle reaction, lb. [chapter 9]
P_v = Internal vacuum pressure, psi [chapter 5]
PC = Point of curvature [chapter 8]
PDV = Pressure-diameter value lb/in. [chapter 7]
PI = Point of intersection [chapter 8]
PT = Point of tangent [chapter 8]
q_a = Allowable buckling pressure, psi [chapter 5]
Q = Volume of flow, ft^3/sec [chapter 3]
Q_b = Unequal factor for crotch plate design [chapter 7]
Q_w = Unequal factor for crotch plate design [chapter 7]
r = Hydraulic radius of pipe, ft [chapter 3]
r = Mean radius of steel cylinder, in. [chapter 5]
r_o = Outside radius of steel cylinder, in. [chapters 4, 5, 7, 9]
R = Pipeline centerline radius [chapter 8]
R = Centerline radius of elbow, in. [chapter 7]
R_B = Radius of mainline for crotch plate design, in. [chapter 7]
R_e = Reynolds number [chapter 3]
RF = Saddle stress reduction factor [chapter 9]
R_H = Correction factor for depth of fill [chapter 5]
R_S = Radius of outlet for crotch plate design, in. [chapter 7]
R_w = Water buoyancy factor [chapter 5]
s = Slope of hydraulic gradient [chapter 3]
s = Allowable design stress, [psi or ksi] [chapters 4, 7]
s_{r1} = Strength reduction factor between outlet and mainline material [chapter 7]
s_{r2} = Strength reduction factor between reinforcement and mainline material [chapter 7]
S = Segment length along inside of elbow, in. [chapter 7]
SF = Straight flange on ellipsoidal head [chapter 7]
$Sp\ gr$ = Specific gravity of fluid [water= 1.0] [chapter 3]
t = Pipe cylinder wall thickness, in. [chapters 3, 4, 5, 9, 12]

t = New/final crotch plate thickness, in. [chapter 7]
t_1 = Existing crotch plate thickness, in. [chapter 7]
t_2 = Crotch plate third plate thickness, in. [chapter 7]
t_a = Adjusted wall thickness, in. [chapters 4, 9]
t_C = Cement-mortar coating thickness, in. [chapters 4, 5]
t_g = Harness ring gusset attachment to front and back plates/rings fillet weld attachment size, in. [chapter 7]
t_L = Cement mortar lining thickness, in. [chapters 4, 5]
t_r = Required outlet cylinder thickness, in. [chapter 7]
t_w = Anchor ring, harness ring and harness lug gusset fillet weld attachment size, in. [chapter 7]
t_{w1} = Outlet attachment [no collar/wrapper] corner cap fillet weld size, in. [chapter 7]
t_{w2} = Outlet attachment [with collar/wrapper] corner cap fillet weld size, in. [chapter 7]
t_{w3} = Outlet reinforcement collar/wrapper to cylinder attachment fillet weld size, in. [chapter 7]
t_{ww} = Anchor ring wrapper reinforcement fillet weld attachment size, in. [chapter 7]
t_y = Outlet nominal steel cylinder thickness, in. [chapter 7]
T = Total tightening torque [ft-lbs] [chapter 12]
T = Closing time, s [chapter 3]
T = Thrust force, lb [chapter 8]
T = Temperature, °F [chapters 6, 9]
T_{alt} = Alternate collar/wrapper thickness, in. [chapter 7]
T_c = Minimum collar/wrapper thickness, in. [chapter 7]
T_h = Minimum thickness of head after forming, in. [chapter 7]
T_{max} = Maximum collar/wrapper thickness, in. [chapter 7]
T_r = Required mainline cylinder thickness, in. [chapter 7]
T_s = Plate thickness of harness lugs and rings/gussets, in. [chapter 7]
T_w = Wrapper thickness at anchor ring, in. [chapter 7]
T_y = Mainline or elbow nominal steel cylinder thickness, in. [chapters 4, 3, 7]
T_{ymin} = Minimum cylinder thickness under harness and anchor rings, in. [chapter 7]
V = Mean velocity of flow, ft/s [chapter 3]
w = Unit weight of fill, lb/ft^3 [chapter 5]
w = Combined weight of steel, water and additional loads, lb/linear in. [chapter 9]
w = Collar/wrapper edge width, in. [chapter 7]
w_{alt} = Alternate collar/wrapper edge width, in. [chapter 7]
w_{min} = Minimum collar/wrapper reinforcement edge width, in. [chapter 7]
W = Load per unit of pipe, lb/lin in. [chapter 5]
W = Total load on span, lb [chapter 9]
W = Overall collar/wrapper reinforcement width, in. [chapter 7]
W = Harness lug gusset separation, in. [chapter 7]
W_{alt} = Alternate collar/wrapper overall width, in. [chapter 7]
W_c = Prism load, dead load on a conduit, lb/in. [chapter 5]
W_d = Trench load on conduit [Marston trench load] lb/lin ft [chapter 5]
W_e = Weight of soil over pipe, lb/linear ft [chapter 8]
W_f = Weight of fluid in pipe, lb/linear ft [chapter 8]
W_L = Live load on a conduit, lb/ft^2 [chapter 5]
W_p = Weight of pipe, lb/linear ft [chapter 8]

X = Harness lug front plate width, in. [chapter 7]
y = Maximum deflection at center of span, in. or ft [chapter 9]
Y = Harness lug back plate width, in. [chapter 7]

CONSTANTS

ε = Coefficient of thermal expansion/contraction of steel = 6.5×10^{-6}/°F
γ_c = Unit weight of cement mortar = 144 lb/ft^3
γ_s = Unit weight of steel = 490 lb/ft^3
γ_w = Unit weight of water = 62.4 lb/ft^3
ν_c = Poisson's ratio for cement mortar = 0.25
ν_s = Poisson's ratio for steel = 0.30
π = Pi = 3.14

E_C = Modulus of elasticity of cement mortar = 4×10^6 psi
E_S = Modulus of elasticity of steel = 30×10^6 psi

This page intentionally blank.

Glossary of Pressure Terms

The following is a listing of pressure terms and associated definitions used in M11.

Design Pressure: A pressure for which a design analysis is performed. Design pressure is a purchaser-defined pressure to which a system will be exposed. Three common pressure analysis cases are: working pressure, transient pressure, and test pressure. Typically, material stress limitations will vary for each design case based on the specific pressure for which the analysis is being performed.

External Radial Pressure: Outside pressure, either atmospheric or hydrostatic, both of which are uniform and act radially.

Field-Test Pressure: A pressure that is utilized to verify the integrity and soundness of an installed piping system or a zone within the system.

Fluid Pressure: A force per unit area applied either internally or externally.

Internal Pressure: The pipe internal pressures used for design that include working pressure and transient or field-test pressures to which the pipe may be subjected during its lifetime.

Operating Pressure: Synonymous with working pressure.

Pipe Class: Synonymous with pressure class.

Pressure Class: Pressure class is generally not applicable or used for steel pipe. This is due to the wide range of steel pipe applications and the ability to design the pipe with variable wall thicknesses and steel grades. If used, pressure class is a numerical designation representing a working pressure for a piping system or a pressure zone within the system.

Rated Pressure: Synonymous with working pressure.

Shop Test Pressure: A pressure that is utilized to verify the integrity and soundness of pipe.

Surge Allowance: Transient pressure less working pressure

Surge Pressure: Synonymous with transient pressure.

Transient Pressure: The purchaser-determined unsteady state maximum pressure induced upon a system resulting from a rapid change in flow velocity. Transient pressure as defined herein represents the total pressure level achieved during the transient event and is not an incremental pressure increase above the working pressure.

Vacuum Pressure: An internal pressure less than atmospheric pressure.

Working Pressure: The maximum pressure under which a system or a zone within the system operates in a steady state or static condition, whichever is greater.

Appendices

A Pipe Deflection-Improving Embedment Versus Increasing Cylinder Thickness

B Harness Ring Assembly Design Example

C Harness Rod Placement When Using Multiple Couplings to Accommodate Vertical Differential Settlement

D Design of Steel Water Pipelines in Geohazard Areas

E Useful Equations and Conversions

This page intentionally blank.

Appendix A

Pipe Deflection-Improving Embedment Versus Increasing Cylinder Thickness

As noted in chapter 5, increasing the *E′* of the pipe embedment is more effective for controlling pipe deflection than increasing the steel cylinder thickness. This fact can be demonstrated for any given pipe diameter by comparing the pipe's deflection resistance capacity, as measured by calculated maximum height of fill above the top of pipe, based on two conditions. The first condition involves improved pipe embedment conditions resulting in increased *E′* values. The second condition involves increased steel cylinder thicknesses. To illustrate this, the calculated maximum heights of cover are shown below for a 48-in. nominal diameter pipe. The graph starts with an assumed initial condition, defined below, and indicates the resulting maximum associated height of fill from either increasing *E′* values by adjusting the pipe embedment condition (increasing embedment compaction or changing embedment material), or increasing the pipe wall thickness.

The maximum height of fill is based on analyses using Eq 5-4 and the following criteria:

- Initial pipe wall thickness—based on either 170 psi operating pressure (σ_Y = 42 ksi) or D/t = 240.
- Initial pipe embedment soil—SC3 stiffness category (CL or ML), compaction = 85 percent, minimum depth of cover 6 ft, maximum to be determined.
  - From Table 5-3, initial E' = 600 psi.
- E_S = 30,000,000 psi, E_C = 4,000,000 psi, D_l = 1, K = 0.1, and unit weight of soil = 120 lb/ft^3.

- Pipe has cement-mortar lining and a flexible coating.
  - From chapter 5, recommended maximum predicted (calculated) deflection for pipe with cement-mortar lining and flexible coating, $\Delta x = 3$ percent.

As is evidenced by Figure A-1, a minimal increase in the E' of less than 200 psi yields an increase in the maximum height of fill (to approximately 20 ft) similar to that achieved by doubling the wall thickness of the cylinder (to approximately 19 ft). The magnitude of and relative difference in maximum cover values reflected below are representative of any pipe diameter of similar stiffness as the one depicted, with the initial condition maximum fill height being the only variable. For 30-in.-diameter pipe, the initial condition maximum fill height would be about 17 ft, and for 96-in. pipe it would be about 14 ft. The graph (Figure A-1) clearly shows the more significant benefit of increased E' values from improving the pipe embedment versus increasing pipe wall thickness for better deflection control.

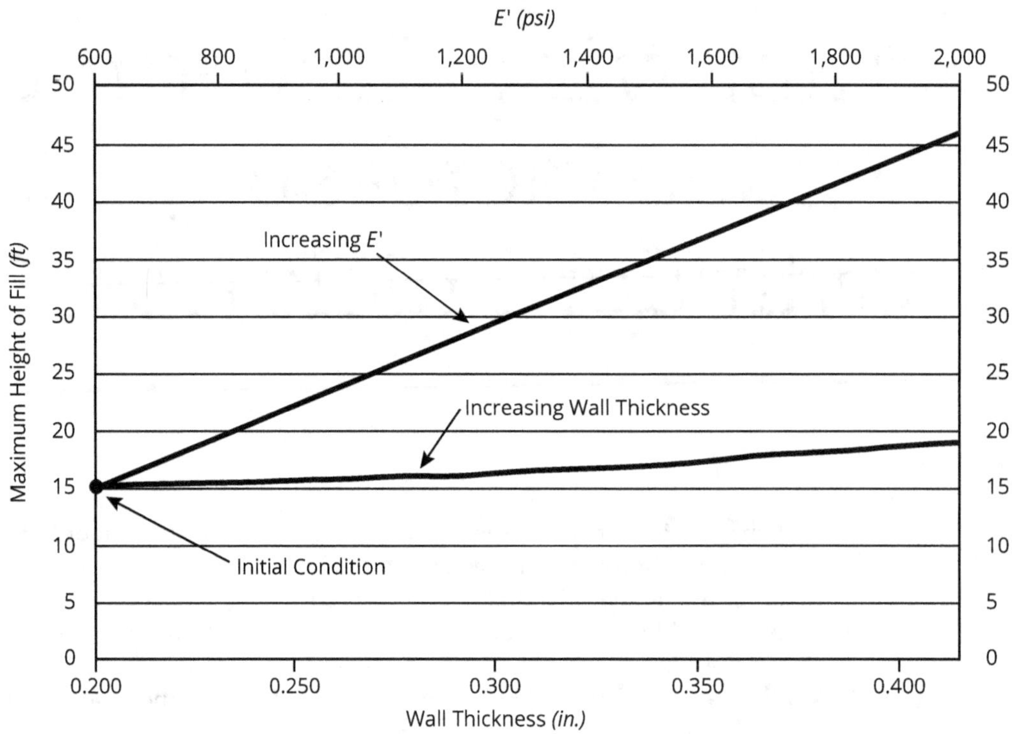

Figure A-1 Comparison of improving pipe embedment versus increasing wall thickness for 48-in. pipe

Appendix B

Harness Ring Assembly Design Example

The following not only presents the fully detailed step-by-step procedure for designing a harness ring assembly, but also presents the process used to generate the information for the 48-in. nominal diameter pipe with a maximum service pressure of 300 psi as shown in Table 7-3. Front and back ring stresses are calculated based on the applied analysis of Brockenbrough (1988), which is based on Case 7 in Table 9.2 of *Roark's Formulas for Stress and Strain* (7th ed., Young and Budynas 2001). The example is based on using gusset plates with a 15° flare but includes final stress values for gusset plates with a 0° flare for comparison purposes. The pipe and ring/gusset material are assumed to have a minimum yield strength of 36 ksi.

Note: Throughout this example the precision of the values shown is intentionally excessive relative to the practical accuracy of the final answer(s) for several reasons: (1) Truncation of each value in a written presentation of multistep calculations can yield a progressively increasing error if not defined to a sufficient precision throughout the process; (2) values subject to exponential operators can vary significantly if an insufficient base level of precision is not maintained in the written form, compared to common electronic calculations used to generate or verify this example; and (3) written presentation requires an increased level of accuracy so that values are consistent within and between calculations as presented. The final values at the end of the example have been properly rounded to reflect the practical level of precision against which to measure acceptability.

1. **Calculate the outside diameter of the pipe.**

   The wall thickness is based on the larger of that calculated based on the working pressure, D_O/T_y = 288 or a minimum functional manufacturing thickness of 0.135 in. Given that the information in Table 7-3 is based on the maximum service pressure, the maximum working pressure is (maximum pressure)/1.5.

   a. Calculate required wall thickness based on working pressure, at an allowable cylinder stress equal to 50 percent of the 36-ksi minimum yield strength, or 18 ksi. The outside diameter is 50.00 in. and is based on the diameter guidelines for Table 7-3 as outlined in chapter 7. Given that Table 7-3 is based on a maximum service pressure, the pressure used to calculate the steel cylinder thickness will be the maximum service pressure of 300 psi divided by 1.5.

$$T_y = \frac{pD_o}{2\sigma_A} = \left(\frac{(300/1.5)50.00}{2(18{,}000)}\right) = 0.2778 \text{ in.}$$

Therefore, the diameter is 50.00 in. and the wall thickness is 0.2778 in.

Note: The steel cylinder thickness at harness lug type-RR assemblies in this design analysis is based on the hoop stress design procedures shown in chapter 4. This practice is deemed acceptable, based on empirical data of successful service and test performance dating back to the early 1960s. In both scenarios, calculation of the cylinder stress is impractical as the evaluation is strain based and not stress based. Local deformation and stress redistribution in a cylinder properly designed for pressure have proven to allow successful performance in both service and proof-of-design test conditions without additional thickness. For reference, a documented test conducted at Thompson Pipe & Steel (TP&S) in 1994 involved strain gauges mounted externally on the cylinder of a 72-in.-nominal-diameter O-ring joint proof-of-design test assembly. Linear adjustment of TP&S test stress values based on the difference in the maximum pressure of the TP&S test and the procedure presented herein for a 72-in. pipe at a maximum design pressure of 275 psi resulted in maximum stresses well within the cylinder and harness assembly design stress limit of 27 ksi defined in this manual for material with a minimum yield strength of 36 ksi.

2. **Calculate the number of bolts, *N*, that will be required for the joint.**

   a. Calculate the maximum thrust, *T*, for the assembly.

   $$T = pA_p = p\left(\frac{\pi D_o^2}{4}\right) = 300\left(\frac{\pi 50.00^2}{4}\right) = 589{,}049 \text{ lb}_f$$

   Select an appropriate harness rod size, which is an iterative process to balance number of rods and size. Try 1.75-in.-diameter rods.

   b. From Table 7-6, the maximum load for a 1.75-in.-diameter bolt is 83,286 lb. Based on this allowable load, the minimum number of harness bolts is equal to

   $$\frac{589{,}049}{83{,}286} = 7.1$$

   To maintain symmetry of the restraint system, round up to the next even number of bolts; $N = 8$ rods.

3. **Calculate the angular geometry of the harness lug. See Figure B-1.**

   a. Calculate the linear distance, *L*1, between the centerline of the harness lug and the centerline of the gusset plate intersection with the front ring:

   Reference Figure B-1. For a 1.750-in. harness bolt, the lug dimensions from Table 7-3 are as follows:

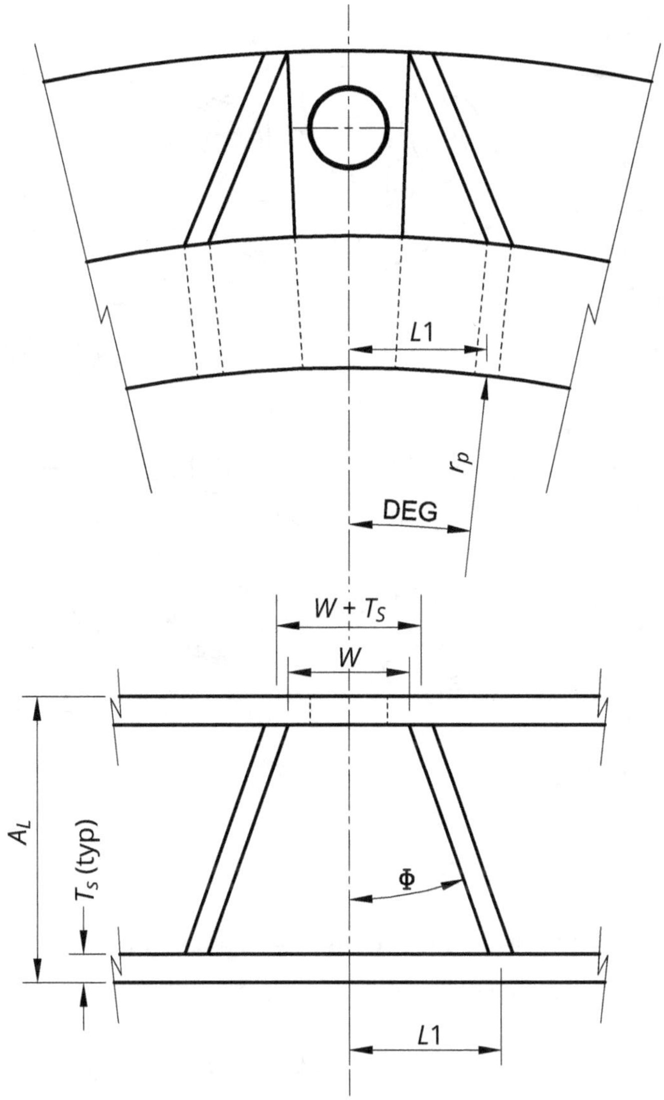

Figure B-1 Harness lug geometry

$$T_s = 0.875 \text{ in.}$$
$$H_B = 6.50 \text{ in.}$$
$$H_F = 2.50 \text{ in.}$$
$$A_L = 12 \text{ in.}$$
$$W = 2.5 \text{ in.}$$
$$\Phi = 15° = 0.2618 \text{ rad}$$

$$L1 = \frac{W + T_s}{2} + (A_L - 2T_s)\tan \Phi$$

$$L1 = \frac{2.5 + 0.875}{2} + [12 - 2(0.875)] \tan (0.2618) = 4.4340 \text{ in.}$$

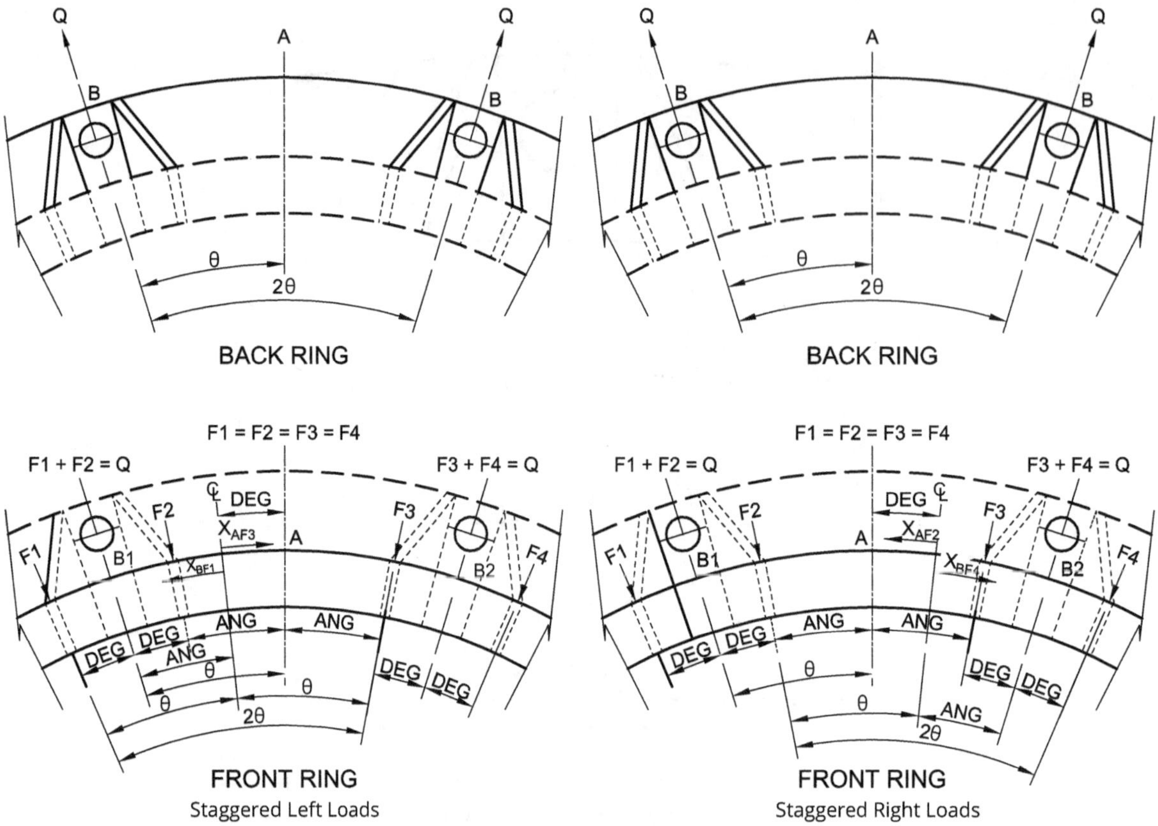

Figure B-2 Harness lug assembly geometry

b. Calculate half of the angle between harness lugs, θ:

Reference Figure B-2.

$$\theta = \frac{\pi}{N} = \frac{3.14159}{8} = 0.39270 \text{ rad}$$

c. Calculate the angle, *DEG*:

$$DEG = \sin^{-1}\left(\frac{L1}{r_p}\right) = \sin^{-1}\left(\frac{4.4340}{50.00/2}\right) = 0.17830 \text{ rad}$$

d. Calculate the angle, *ANG*:

$$ANG = \theta - DEG = 0.39270 - 0.17830 = 0.21440 \text{ rad}$$

4. **Calculate the ring coefficients, where angles are in radians.**

The fundamental analysis of ring loads is from Roark (Young and Budynas 2001), Table 9.2, case 7, for a ring under any number of radial forces, equally spaced.

The associated equations for the analysis are

$$\text{For } 0 < x < \theta \qquad M = \frac{Qr_p(\mu/s - k_2/\theta)}{2}, \text{ and } N = \frac{Q\mu}{2s}$$

The maximum positive moment occurs between the loads, at position A.

$$\text{Max + moment} = M_A = \frac{Qr_p(1/s - k_2/\theta)}{2}$$

The maximum negative moment occurs at each load, at position B.

$$\text{Max − moment} = M_B = \frac{-Qr_p}{2}\left(\frac{k_2}{\theta} - \frac{c}{s}\right)$$

N = the direct, or hoop, load in the ring due to Q.

Where:

$s = \sin\theta$

$c = \cos\theta$

$\mu = \cos x$

r_p = pipe radius

k_2 = 1 (neglecting shear)

Note: θ and x are in radians.

Front Ring Load Coefficients

Since the front ring has two applied loads at each lug location, the loading is not equally spaced. To apply the Roark analysis, which is based on equally spaced loads, independent analysis of the constants relative to the left-side loads and the right-side loads is required. The resulting values are respectively added together to yield load constant values at the centerline of a given lug, Position B, and at the location between the lugs, position A. The left-side loads are symmetric around the ring, as are the right-side loads. The geometry of the lug assembly and the associated loads are shown in Figure B-2, based on the applied analysis by Brockenbrough (1988). This analysis accounts for the angular spread of the gusset plates and the resulting asymmetric loads applied to the front ring by each of the gussets. The staggered left loads are identified as *F1* and *F3*, and the staggered right loads are identified as *F2* and *F4*. All staggered loads are equal and have a magnitude of $Q/2$, where Q equals the radial load on the back ring at the lug centerline. Each analysis, staggered left or staggered right, is oriented such that the circumferential midpoint between any two loads is offset from position A by the angle *DEG*. At location B1, the influence of *F1* and *F2* is found through symmetry by calculating the influence of *F1* on B1 (staggered left analysis with x = *ANG*), and *F4* on B2 (staggered right analysis with x = *ANG*), and summing the results. Similarly, for location A, the influence of *F2* (staggered right analysis with x = *DEG*) and *F3* (staggered left analysis with x = *DEG*) is found, and the results summed. The resulting equations are based on the Brockenbrough application of Roark.

$K1$ = front ring direct, or hoop, load coefficient, at the centerline of lugs (position B)

$$N = \frac{-Q\mu}{2s} = Q(K1)$$

$$\text{therefore } K1 = \frac{-\mu}{2s} = \frac{-\cos x}{2\sin\theta}$$

At the load, use x = *ANG*.

Therefore,

$$K1 = \frac{-\cos(ANG)}{2\sin\theta} = \frac{-\cos(0.21440)}{2\sin(0.39270)} = -1.27665$$

$K11$ = front ring direct, or hoop, load coefficient, midpoint between the centerline of lugs (position A)

$$N = \frac{-Q\mu}{2s} = Q(K11)$$

$$\text{therefore } K11 = \frac{-\mu}{2s} = \frac{-\cos x}{2\sin\theta}$$

Between the loads, use x = DEG.

Therefore,

$$K11 = \frac{-\cos(DEG)}{2\sin\theta} = \frac{-\cos(0.17830)}{2\sin(0.39270)} = -1.28585$$

$K2$ = front ring bending coefficient at the centerline of lugs (position B)

$$M_B = \frac{-Qr_p(k_2/\theta - \mu/s)}{2} = \frac{Qr_p(\mu/s - k_2/\theta)}{2} = Qr_p(K2)$$

$$\text{therefore } K2 = 0.5\left(\frac{\mu}{s} - \frac{k_2}{\theta}\right) = 0.5\left(\frac{\cos x}{\sin\theta} - \frac{1}{\theta}\right)$$

For analysis at the lug load centerline, $x = ANG$.

Therefore,

$$K2 = 0.5\left(\frac{\cos(ANG)}{\sin\theta} - \frac{1}{\theta}\right) = 0.5\left(\frac{\cos(0.21440)}{\sin(0.39270)} - \frac{1}{0.39270}\right) = 0.00341$$

$K22$ = front ring bending coefficient midpoint between the centerline of lugs (position A)

$$M_A = \frac{Qr_p(\mu/s - k_2/\theta)}{2} = Qr_p(K22)$$

$$\text{therefore } K22 = 0.5\left(\frac{\mu}{s} - \frac{k_2}{\theta}\right) = 0.5\left(\frac{\cos x}{\sin\theta} - \frac{1}{\theta}\right)$$

For analysis between the lug loads, $x = DEG$.

Therefore,

$$K22 = 0.5\left(\frac{\cos(DEG)}{\sin\theta} - \frac{1}{\theta}\right) = 0.5\left(\frac{\cos(0.17830)}{\sin(0.39270)} - \frac{1}{0.39270}\right) = 0.01261$$

Back Ring Load Coefficients

$KK1$ = back ring direct, or hoop, load coefficient, at the loads (position B)

$$N = \frac{Q\mu}{2s} = Q(KK1)$$

$$\text{therefore } KK1 = \frac{\mu}{2s} = \frac{\cos x}{2\sin\theta}$$

At the load, $x = \theta$.

Therefore,

$$KK1 = \frac{\cos\theta}{2\sin\theta} = \frac{\cos(0.39270)}{2\sin(0.39270)} = 1.20710$$

$KK11$ = back ring direct, or hoop, load coefficient, between the loads (position A)

$$N = \frac{Q\mu}{2s} = Q(KK11)$$

$$\text{therefore } KK11 = \frac{\mu}{2s} = \frac{\cos x}{2\sin\theta}$$

Between the loads, $x = 0$.

Therefore,

$$KK11 = \frac{\cos(0)}{2\sin\theta} = \frac{1}{2\sin(0.39270)} = 1.30656$$

$KK2$ = back ring bending coefficient, at the loads (position B)

$$M_B = \frac{-Qr_p(k_2/\theta - \mu/s)}{2} = \frac{Qr_p(\mu/s - k_2/\theta)}{2} = Qr_p(KK2)$$

$$\text{therefore } KK2 = 0.5\left(\frac{\mu}{s} - \frac{k_2}{\theta}\right) = 0.5\left(\frac{\cos x}{\sin\theta} - \frac{1}{\theta}\right)$$

The constant is necessary to calculate the bending stress at a given point, but the bending stress is related to the bending moment. The bending moment is constant at any point on the ring, but the bending stress varies in magnitude and sign from the outside to the inside of the ring. As the bending stress is the desired value, the constant can have the appropriate sign applied so that the correct stress values are achieved in the final calculation. For the analysis to achieve the correct sign of the bending stress in the final calculation, let

$$KK2 = -0.5\left(\frac{\mu}{s} - \frac{k_2}{\theta}\right) = -0.5\left(\frac{\cos x}{\sin\theta} - \frac{1}{\theta}\right)$$

At the load, $x = \theta$.

Therefore,

$$KK2 = -0.5\left(\frac{\cos\theta}{\sin\theta} - \frac{1}{\theta}\right) = -0.5\left(\frac{\cos(0.39270)}{\sin(0.39270)} - \frac{1}{0.39270}\right) = 0.06613$$

$KK22$ = back ring bending coefficient, between the loads (position A)

$$M_A = \frac{Qr_p(\mu/s - k_2/\theta)}{2} = Qr_p(KK22)$$

$$\text{therefore } KK22 = -0.5\left(\frac{\mu}{s} - \frac{k_2}{\theta}\right) = -0.5\left(\frac{\cos x}{\sin\theta} - \frac{1}{\theta}\right)$$

For the analysis to achieve the correct sign in the final calculation, let

$$KK22 = -0.5\left(\frac{\mu}{s} - \frac{k_2}{\theta}\right) = -0.5\left(\frac{\cos x}{\sin\theta} - \frac{1}{\theta}\right)$$

Between the loads, $x = 0$.

Therefore,

$$KK22 = -0.5\left(\frac{\cos(0)}{\sin(0.39270)} - \frac{1}{0.39270}\right) = -0.03332$$

5. **Calculate the effective length of shell, *LE*, that contributes to ring section.**

   a. $LE = 1.56\sqrt{r_p T_y} + T_s$, but not greater than $A_L - T_s$

   $$= 1.56\sqrt{\frac{50.00}{2}(0.2778)} + 0.875 = 4.986 \text{ in.}$$

   b. Check to verify that the calculated *LE* is not greater than $A_L - T_s$.

   $A_L - T_s = 12 - 0.875 = 11.125$ in.

   $LE = 4.986$ in. ≤ 11.125 in. O.K.

   Note: If *LE* is > than $A_L - T_s$ then $LE = (\text{initial } LE)/2 + (A_L - T_s)/2$

6. **Calculate front ring/shell centroid locations and section properties. Reference Figure B-3.**

   a. Calculate centroid locations:

   (1) Area of ring, *AST,*

   $$AST = H_F T_S = 2.5(0.875) = 2.1875 \text{ in.}^2$$

   (2) Area of effective length of shell, *ASH,*

   $$ASH = LE(T_y) = 4.986(0.2778) = 1.3851 \text{ in.}^2$$

   (3) Area of ring/shell combination, *EFA,*

   $$EFA = AST + ASH = 2.1875 + 1.3851 = 3.5726 \text{ in.}^2$$

   (4) Distance from shell ID to ring centroid, *YST,*

   $$YST = T_y + \frac{H_F}{2} = 0.2778 + \frac{2.5}{2} = 1.5278 \text{ in.}$$

   (5) Distance from shell ID to shell centroid, *YSH,*

   $$YSH = \frac{T_y}{2} = \frac{0.2778}{2} = 0.1389 \text{ in.}$$

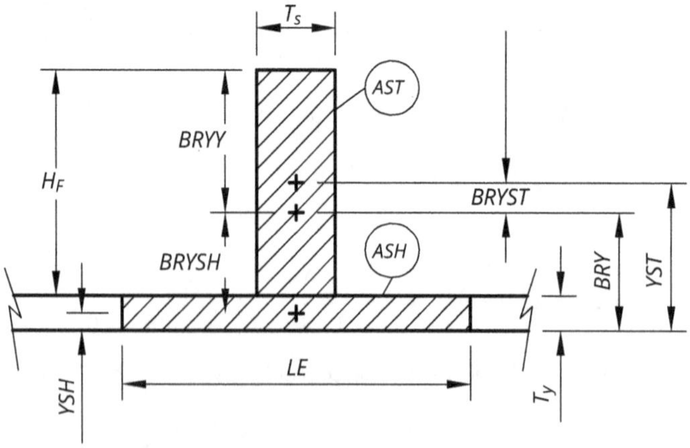

Figure B-3 Front harness ring/shell section

(6) Area of ring times centroid distance, *AYST*,

$$AYST = AST(YST) = 2.1875(1.5278) = 3.3421 \text{ in.}^3$$

(7) Area of shell times centroid distance, *AYSH*,

$$AYSH = ASH(YSH) = 1.3851(0.1389) = 0.1924 \text{ in.}^3$$

(8) Sum of areas times centroids, *SAY*,

$$SAY = AYST + AYSH = 3.3421 + 0.1924 = 3.5345 \text{ in.}^3$$

(9) Centroid distance of combined ring/shell section, *BRY*,

$$BRY = \frac{SAY}{EFA} = \frac{3.5345}{3.5726} = 0.9893 \text{ in.}$$

b. Calculate distances between centroids:

(1) Distance from section centroid to ring centroid, *BRYST*,

$$BRYST = YST - BRY = 1.5278 - 0.9893 = 0.5385 \text{ in.}$$

(2) Distance from section centroid to shell centroid, *BRYSH*,

$$BRYSH = YSH - BRY = 0.1389 - 0.9893 = -0.8504 \text{ in.}$$

(3) Distance from ring OD to section centroid, *BRYY*,

$$BRYY = H_F + T_y - BRY = 2.5 + 0.2778 - 0.9893 = 1.7885 \text{ in.}$$

c. Calculate moments of inertia:

(1) Ring moment of inertia, *STI*,

$$STI = \frac{T_S H_F{}^3}{12} = \frac{0.875(2.5)^3}{12} = 1.1393 \text{ in.}^4$$

(2) Shell moment of inertia, *SHI*,

$$SHI = \frac{LE(T_y)^3}{12} = \frac{4.986(0.2778)^3}{12} = 0.0089 \text{ in.}^4$$

(3) Effective moment of inertia of combined section, *EFI*,

$$\begin{aligned} EFI &= STI + AST(BRYST)^2 + SHI + ASH(BRYSH)^2 \\ &= 1.1393 + 2.1875(0.5385)^2 + 0.0089 + 1.3851(-0.8504)^2 \\ &= 2.7842 \text{ in.}^4 \end{aligned}$$

d. Calculate section moduli:

(1) Section modulus from ring OD, *SO*,

$$SO = \frac{EFI}{BRYY} = \frac{2.7842}{1.7885} = 1.5567 \text{ in.}^3$$

(2) Section modulus from ring ID, *SI*,

$$SI = \frac{EFI}{BRY} = \frac{2.7842}{0.9893} = 2.8143 \text{ in.}^3$$

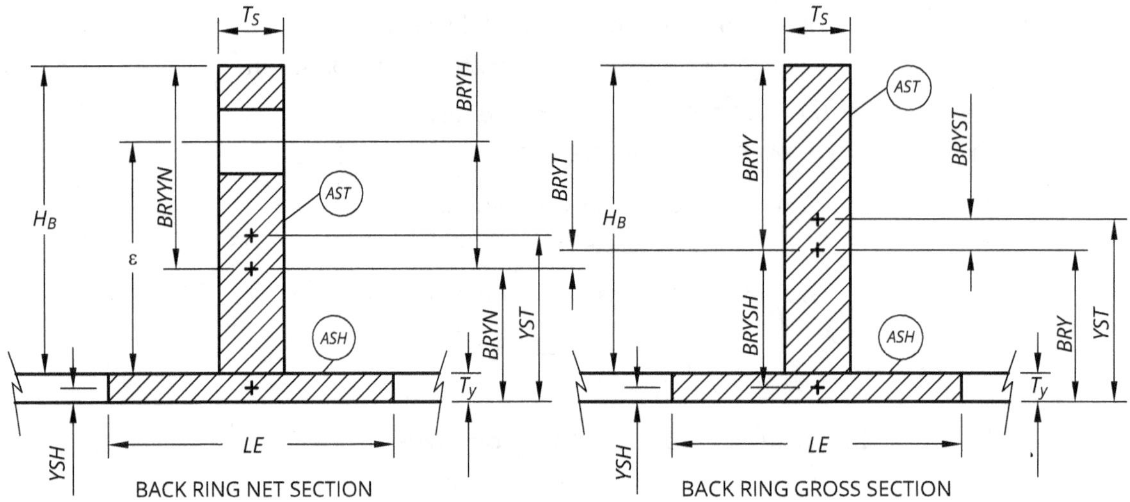

Figure B-4 Back harness ring/shell sections

7. **Calculate back ring/shell centroid locations and section properties. Reference Figure B-4.**

Gross Ring Section—Between Loads (Position A)

a. Calculate centroid locations for gross section:

(1) Area of ring, *AST,*

$$AST = H_B T_S = 6.50(0.875) = 5.6875 \text{ in.}^2$$

(2) Area of effective length of shell, *ASH,*

$$ASH = LE(T_y) = 4.986(0.2778) = 1.3851 \text{ in.}^2$$

(3) Effective area of ring/shell combination, *EFA,*

$$EFA = AST + ASH = 5.6875 + 1.3851 = 7.0726 \text{ in.}^2$$

(4) Distance from shell ID to ring centroid, *YST,*

$$YST = T_y + \frac{H_B}{2} = 0.2778 + \frac{6.50}{2} = 3.5278 \text{ in.}$$

(5) Distance from shell ID to shell centroid, *YSH,*

$$YSH = \frac{T_y}{2} = \frac{0.2778}{2} = 0.1389 \text{ in.}$$

(6) Area of ring times centroid distance, *AYST,*

$$AYST = AST(YST) = 5.6875(3.5278) = 20.0644 \text{ in.}^3$$

(7) Area of shell times centroid distance, *AYSH,*

$$AYSH = ASH(YSH) = 1.3851(0.1389) = 0.1924 \text{ in.}^3$$

(8) Sum of areas times centroids, *SAY,*

$$SAY = AYST + AYSH = 20.0644 + 0.1924 = 20.2568 \text{ in.}^3$$

(9) Centroid distance of combined ring/shell section, *BRY,*

$$BRY = \frac{SAY}{EFA} = \frac{20.2568}{7.0726} = 2.8641 \text{ in.}$$

b. Calculate distances between centroids:

(1) Distance from section centroid to ring centroid, *BRYST,*

$$BRYST = YST - BRY = 3.5278 - 2.8641 = 0.6637 \text{ in.}$$

(2) Distance from section centroid to shell centroid, *BRYSH,*

$$BRYSH = YSH - BRY = 0.1389 - 2.8641 = -2.7252 \text{ in.}$$

(3) Distance from ring OD to section centroid, *BRYY,*

$$BRYY = H_B + T_y - BRY = 6.50 + 0.2778 - 2.8641 = 3.9137 \text{ in.}$$

c. Calculate gross section moment of inertia:

(1) Ring moment of inertia, *STI,*

$$STI = \frac{T_S H_B{}^3}{12} = \frac{0.875(6.50)^3}{12} = 20.0247 \text{ in.}^4$$

(2) Shell moment of inertia, *SHI,*

$$SHI = \frac{LE(T_y)^3}{12} = \frac{4.986(0.2778)^3}{12} = 0.0089 \text{ in.}^4$$

(3) Effective moment of inertia of gross combined section, *EFI,*

$$EFI = STI + AST(BRYST)^2 + SHI + ASH(BRYSH)^2$$

$$= 20.0247 + 5.6875(0.6637)^2 + 0.0089 + 1.3851(-2.7252)^2$$

$$= 32.8257 \text{ in.}^4$$

d. Calculate gross section moduli:

(1) Section modulus from ring OD, *SO,*

$$SO = \frac{EFI}{BRYY} = \frac{32.8257}{3.9137} = 8.3874 \text{ in.}^3$$

(2) Section modulus from ring ID, *SI,*

$$SI = \frac{EFI}{BRY} = \frac{32.8257}{2.8641} = 11.4611 \text{ in.}^3$$

Net Ring Section—at Loads (Position B)

a. Calculate centroid locations for net section:

(1) Net effective area of ring/shell combination, *EFAN,*

$$EFAN = EFA - T_S(D_R + 0.125) = 7.0726 - 0.875(1.75 + 0.125) = 5.4320 \text{ in.}^2$$

(2) Net sum of areas times centroids, *SAYN,*

$$SAYN = SAY - T_S(D_R + 0.125)(E + T_y)$$

$$= 20.2568 - 0.875(1.75 + 0.125)(4 + 0.2778) = 13.2385 \text{ in.}^3$$

(3) Centroid of net section from shell ID, *BRYN,*

$$BRYN = \frac{SAYN}{EFAN} = \frac{13.2385}{5.4320} = 2.4371 \text{ in.}$$

(4) Centroid of net section from ring OD, *BRYYN,*

$$BRYYN = H_B + T_y - BRYN = 6.50 + 0.2778 - 2.4371 = 4.3407 \text{ in.}$$

(5) Centroid distance of combined ring/shell section, *BRY*,

$$BRY = \frac{SAY}{EFA} = \frac{20.2568}{7.0726} = 2.8641 \text{ in.}$$

b. Calculate distances between centroids:

(1) Distance from BRY to *BRYN*, *BRYT*,

$$BRYT = BRY - BRYN = 2.8641 - 2.4371 = 0.4270 \text{ in.}$$

(2) Distance from *E* to *BRYN*, *BRYH*,

$$BRYH = E + T_y - BRYN = 4 + 0.2778 - 2.4371 = 1.8407 \text{ in.}$$

c. Calculate net section moment of inertia:

(1) Effective moment of inertia of net combined section, *EFI*,

$$EFIN = EFA(BRYT)^2 - (D_R + 0.125)(T_S)BRYH^2 + EFI - \frac{T_S(D_R + 0.125)^3}{12}$$

$$= 7.0726(0.4270)^2 - (1.75 + 0.125)(0.875)(1.8407)^2 + 32.8257 - \frac{0.875(1.75 + 0.125)^3}{12}$$

$$EFIN = 28.0759 \text{ in.}^4$$

d. Calculate net section moduli:

(1) Section modulus from ring OD, *SON*,

$$SON = \frac{EFIN}{BRYYN} = \frac{28.0759}{4.3407} = 6.4681 \text{ in.}^3$$

(2) Section modulus from ring ID, *SIN*,

$$SIN = \frac{EFIN}{BRYN} = \frac{28.0759}{2.4371} = 11.5202 \text{ in.}^3$$

8. Calculate hoop stresses in front and back ring sections due to internal pressure.

a. Average pressure stress at front ring, *FCAP*,

$$FCAP = \frac{pD_o(LE)}{2EFA} = \frac{300(50.00)(4.986)}{2(3.5726)} = 10{,}467 \text{ psi}$$

b. Average pressure stress at back ring between loads (position A), *FCAP*,

$$FCAP = \frac{pD_o(LE)}{2EFA} = \frac{300(50.00)(4.986)}{2(7.0726)} = 5{,}287 \text{ psi}$$

c. Average pressure stress at back ring at loads (position B), *FCAPN*,

$$FCAPN = \frac{pD_o(LE)}{2EFAN} = \frac{300(50.00)(4.986)}{2(5.4320)} = 6{,}884 \text{ psi}$$

9. Calculate radial force applied to ring sections.

a. Radial force at front ring section, *FR*: The radial forces at the lug are determined, by static analysis, in terms of the force on the restraint rod. For simplicity, the thicknesses of the front and back rings are assumed to be negligible, and the moment resisting the rod force is equal to $FR(A_L - T_s)$. See Figure B-5.

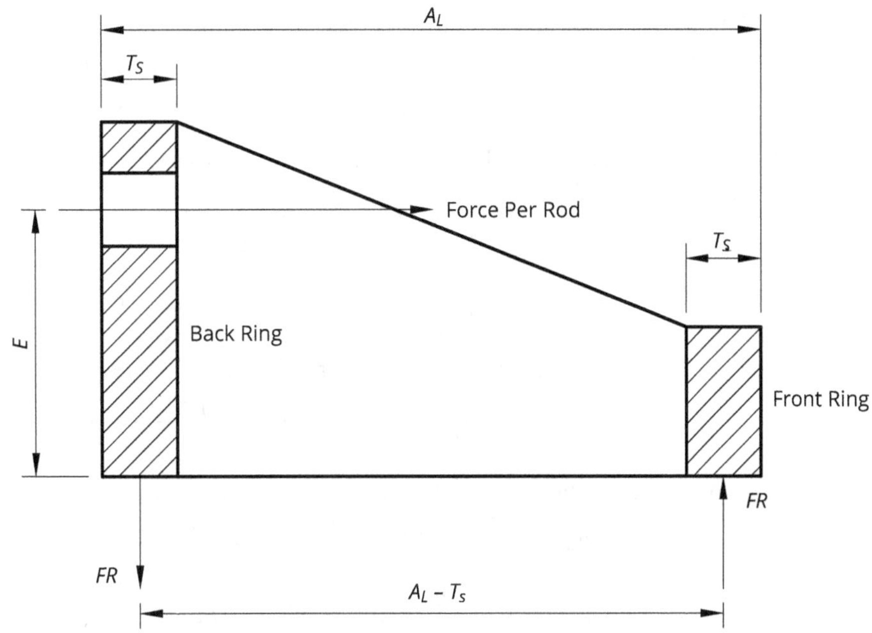

Figure B-5 Simplified lug free body diagram

Summing moments around either the front or back ring, and solving for *FR*,

$$FR = \left(\frac{(\text{Force per Rod})E}{A_L - T_S}\right) = \frac{pA_p}{N}\left(\frac{E}{A_L - T_S}\right) = \frac{p[\pi D_o{}^2/4]}{N}\left(\frac{E}{A_L - T_S}\right)$$

$$= \frac{300[\pi(50)^2/4]}{8}\left(\frac{4}{12 - 0.875}\right) = 26{,}474 \text{ lb}_f$$

10. Calculate stresses in the front ring, at centerline of lugs (position B), due to the radial force, *FR*.

a. Direct stress in ring at the centerline of a lug, *FCARALS*,

$$FCARALS = \frac{K1(FR)}{EFA} = \frac{-1.27665(26{,}474)}{3.5726} = -9{,}460 \text{ psi}$$

b. Outer surface bending stress at the centerline of a lug, *FCBRALSO*,

$$FCBRALSO = \frac{K2(FR)r_p}{SO} = \frac{0.00341(26{,}474)(50/2)}{1.5567} = 1{,}450 \text{ psi}$$

c. Inner surface bending stress at the centerline of a lug, *FCBRALSI*,

$$FCBRALSI = \frac{-K2(FR)r_p}{SI} = \frac{-0.00341(26{,}474)(50/2)}{2.8143} = -802 \text{ psi}$$

The final step is to sum the calculated stresses. Note that the total stress is the sum of the stress due to internal pressure (tension), the direct stress (compression), and the bending stress (tension or compression).

d. Total outer surface stress at the centerline of a lug, *TSALSO*,

$$TSALSO = FCAP \text{ (tension)} + FCARALS \text{ (compression)} + FCBRALSO \text{ (tension)}$$

$$= 10{,}467 + (-9{,}460) + 1{,}450 = 2{,}457 \text{ psi}$$

e. Total inner surface stress at the centerline of a lug, *TSALSI*,

$$TSALSI = FCAP \text{ (tension)} + FCARALS \text{ (compression)} + FCBRALSI \text{ (compression)}$$
$$= 10{,}467 + (-9{,}460) + (-802) = 205 \text{ psi}$$

11. Calculate stresses in the front ring, between lugs (position A), due to the radial force, *FR*.

a. Direct stress in ring between lugs, *FCARBLS*,

$$FCARBLS = \frac{K11(FR)}{EFA} = \frac{-1.28585(26{,}474)}{3.5726} = -9{,}529 \text{ psi}$$

b. Outer surface bending stress between the lugs, *FCBRBLSO*,

$$FCBRBLSO = \frac{K22(FR)r_p}{SO} = \frac{0.01261(26{,}474)(50/2)}{1.5567} = 5{,}361 \text{ psi}$$

c. Inner surface bending stress between the lugs, *FCBRBLSI*,

$$FCBRBLSI = \frac{-K22(FR)r_p}{SI} = \frac{-0.01261(26{,}474)(50/2)}{2.8143} = -2{,}966 \text{ psi}$$

d. Total outer surface stress between the lugs, *TSBLSO*,

$$TSBLSO = FCAP + FCARBLS + FCBRBLSO$$
$$= 10{,}467 + (-9{,}529) + 5{,}361 = 6{,}299 \text{ psi}$$

e. Total inner surface stress at the centerline of the of a lug, *TSBLSI*,

$$TSBLSI = FCAP + FCARBLS + FCBRBLSI$$
$$= 10{,}467 + (-9{,}529) + (-2{,}966) = -2{,}028 \text{ psi}$$

12. Calculate stresses in the back ring, at centerline of lugs (position B), due to the radial force, *FR*.

a. Direct stress in ring at the centerline of a lug, *FCARALS*,

$$FCARALS = \frac{KK1(FR)}{EFAN} = \frac{1.20710(26{,}474)}{5.4320} = 5{,}883 \text{ psi}$$

b. Outer surface bending stress at the centerline of a lug, *FCBRALSO*,

$$FCBRALSO = \frac{KK2(FR)r_p}{SON} = \frac{0.06613(26{,}474)(50/2)}{6.4681} = 6{,}767 \text{ psi}$$

c. Inner surface bending stress at the centerline of a lug, *FCBRALSI*,

$$FCBRALSI = \frac{-KK2(FR)r_p}{SIN} = \frac{-0.06613(26{,}474)(50/2)}{11.5202} = -3{,}799 \text{ psi}$$

d. Total outer surface stress at the centerline of a lug, *TSALSO*,

$$TSALSO = FCAPN + FCARALS + FCBRALSO$$
$$= 6{,}884 + 5{,}883 + 6{,}767 = 19{,}534 \text{ psi}$$

e. Total inner surface stress at the centerline of the of a lug, *TSALSI*,

$$TSALSI = FCAPN + FCARALS + FCBRALSI$$
$$= 6{,}884 + 5{,}883 + (-3{,}799) = 8{,}968 \text{ psi}$$

13. Calculate stresses in the back ring, between lugs (position A), due to the radial force, *FR*.

a. Direct stress in ring between lugs, *FCARBLS*,

$$FCARBLS = \frac{KK11(FR)}{EFA} = \frac{1.30656(26{,}474)}{7.0726} = 4{,}891 \text{ psi}$$

b. Outer surface bending stress between the lugs, *FCBRBLSO,*

$$FCBRBLSO = \frac{KK22(FR)r_p}{SO} = \frac{-0.03332(26{,}474)(50/2)}{8.3874} = -2{,}629 \text{ psi}$$

c. Inner surface bending stress between the lugs, *FCBRBLSI,*

$$FCBRBLSI = \frac{-KK22(FR)r_p}{SI} = \frac{0.03332(26{,}474)(50/2)}{11.4611} = 1{,}924 \text{ psi}$$

d. Total outer surface stress between the lugs, *TSBLSO,*

$$TSBLSO = FCAP + FCARBLS + FCBRBLSO$$
$$= 5{,}287 + 4{,}891 + (-2{,}629) = 7{,}549 \text{ psi}$$

e. Total inner surface stress between the lugs, *TSBLSI,*

$$TSBLSI = FCAP + FCARBLS + FCBRBLSI$$
$$= 5{,}287 + 4{,}891 + 1{,}924 = 12{,}102 \text{ psi}$$

14. Summary of harness ring calculated stresses.

Location		Stress* (*ksi*)
Back Ring at Rod—Outside	*TSALSO*	19.5
Back Ring at Rod—Inside	*TSALSI*	9.0
Back Ring Between Rods—Outside	*TSBLSO*	7.5
Back Ring Between Rods—Inside	*TSBLSI*	12.1
Front Ring at Rod—Outside	*TSALSO*	2.5 (–13.6)*
Front Ring at Rod—Inside	*TSALSI*	0.2 (9.6)*
Front Ring Between Rods—Outside	*TSBLSO*	6.3 (13.7)*
Front Ring Between Rods—Inside	*TSBLSI*	–2.0 (–6.3)*

*The parenthetical stresses for the front ring are shown for informational comparison only, and are based on analysis with straight gusset plate orientation, without the 15° flare.

Given that harness ring assemblies are designed at the maximum pressure to which they will be subjected, the allowable calculated stress is limited to 75 percent of the minimum specified yield strength of the steel. For steel with a minimum yield strength of 36 ksi, the allowable stress is 0.75(36) = 27 ksi. The calculated stress values are below the allowable stress and are therefore acceptable.

REFERENCES

Brockenbrough, R.L. 1988. Personal Communication. Pittsburgh, PA: R.L. Brockenbrough & Assoc. Inc.

Thompson Pipe & Steel (TP&S). 1994. "Pressure Vessel Test Results."

Young, W.C., and R.G. Budynas. 2001. *Roark's Formulas for Stress and Strain*, 7th ed. New York: McGraw-Hill.

This page intentionally blank.

Appendix C

Harness Rod Placement When Using Multiple Couplings to Accommodate Vertical Differential Settlement

When an alternate method is desired to accommodate anticipated vertical differential settlement across multiple harnessed joints, history has shown that some designers have grouped harness rods nearer to the springline on each side of the joints. When this type of rod configuration has been employed, no more than 6 rods have been used. In such cases, the grouping has typically been limited to 3 rods on each side of the assembly, with 1 rod being placed at springline, 1 rod no more than 30° above and 1 rod no more than 30° below the springline rod. See Figure C-1. To properly accommodate this nonuniform placement of harness rods, it is suggested that a full design analysis be performed in accordance with the procedure outlined in appendix B. Indiscriminate, noncircumferentially uniform relocation of the rod quantities shown in Table 7-3 is not recommended.

The design procedure detailed in appendix B is based on symmetric loading of the restraint assembly. Therefore, when performing the analysis for this special case, the quantity of rods must equal 360 divided by the angular separation of the rods at springline. For instance, if the angular separation of the rods on each side of the joint is chosen to be 30°, the analysis should be performed using a quantity of 12 rods. Increasing the number of rods in this fashion fulfills the necessary symmetry required by the design procedure and provides proper representation of the actual applied loads at each rod location. (*Note:* when decreasing the angular separation between harness rods from that shown in Table 7-3, the designer is cautioned to verify that sufficient circumferential distance exists between the

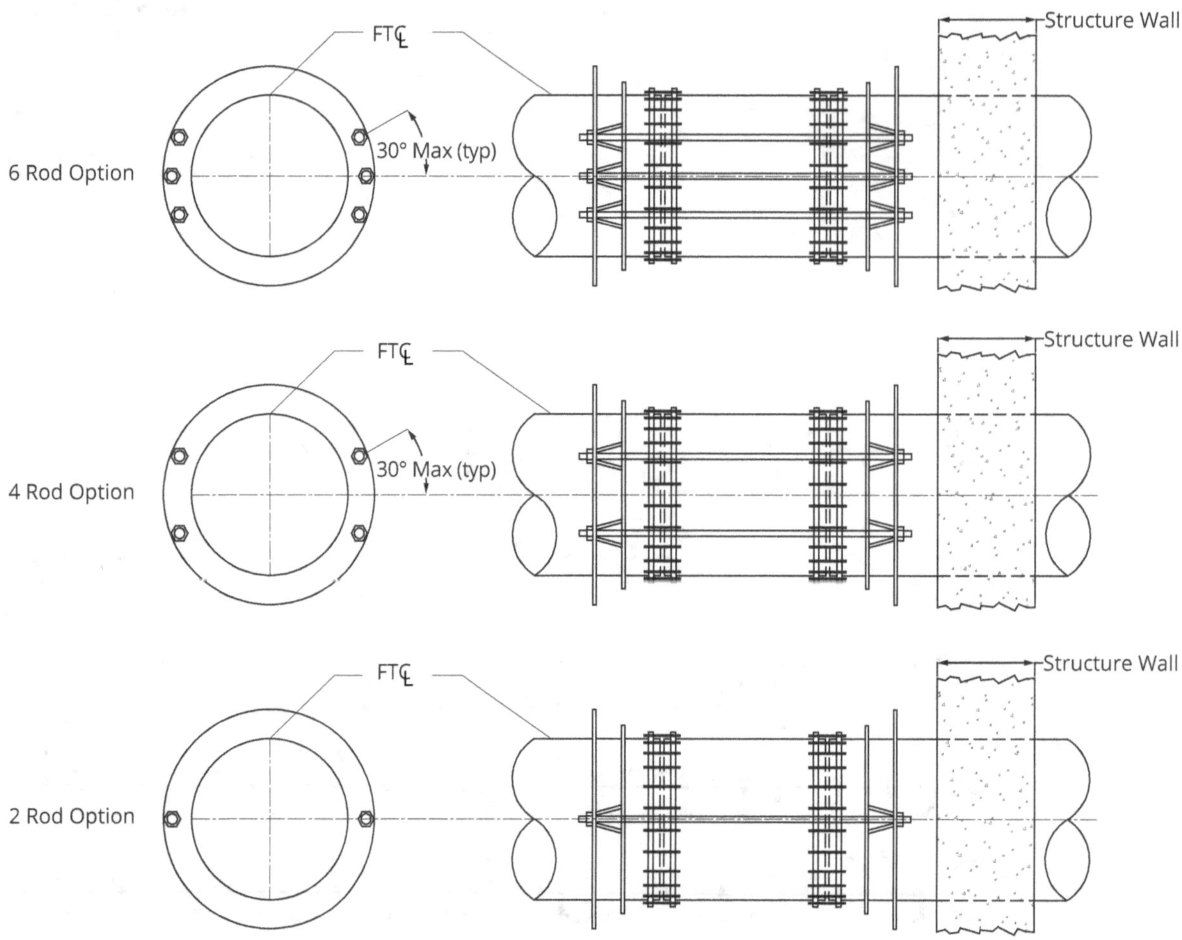

Figure C-1 Harness rod placement for differential settlement across multiple harnessed joints (section view)

rods' centerlines to accommodate without interference the intersections with the front ring of the gusset plates of a single lug and other adjacent lugs.) The total number of rods used in the design analysis is not representative of the number of rods that will actually be installed, but rather the number required to ensure symmetry in the design analysis. For this special analysis, each rod load must be equal to the total thrust at design pressure divided by the number of rods that are actually installed. For instance, if the total thrust at design pressure is 500 kip and the maximum number of 6 installed rods is desired, the rod size chosen must be capable of accommodating at least 83.3 kip. In this case, the designer would use 12 rods in the analysis, each capable of accommodating at least 83.3 kip, but only require 6 installed rods, each capable of accommodating at least 83.3 kip. Again, this larger number of rod loads in the design procedure is only used to maintain symmetry and proper representation of actual rod loads on the harness rings in the design analysis, and is not representative of the number of installed rods.

Appendix D

Design of Steel Water Pipelines in Geohazard Areas

For pipelines constructed in geohazard areas, seismic action due to shaking waves (transient action) and the potential of permanent land movement, mainly from seismic activity (permanent ground-induced action), can be an important factor in the pipeline design (EERI 1999; Liang and Sun 2000). Both transient and permanent actions can be important for water pipelines (O'Rourke 2003).

Seismic actions, such as tectonic fault movement, liquefaction-induced settlement, or lateral spreading and landslide movement due to slope instability, have shown to be most critical for buried pipeline design because they are often associated with large deformations that introduce high strains in the pipeline and may lead to pipeline failure (Gresnigt 1986). Furthermore, one should notice that the above geohazards may not be necessarily associated with severe seismic events; e.g., a rather insignificant seismic event may trigger slope instability and the corresponding movement can completely destroy a pipeline segment located within the moving earth mass. Moreover, in few cases, permanent ground-induced actions, such as slope instabilities, might not be related to seismic action. Under certain circumstances, seismic wave action can be important for water pipeline design as well (O'Rourke 2003; Karamanos et al. 2014; Karamanos et al. 2015).

Safeguarding structural integrity after a seismic event in terms of "maintaining water content" is the major requirement of pipeline seismic design, especially for main transmission pipelines. This requirement is also referred to as "no loss of water content" (Vazouras et al. 2012). The use of steel pipe has proven to be the preferred material in areas with seismic activity because of steel material strength and ductility, i.e., the ability of steel material to sustain large deformations without rupture. In the course of a seismic design procedure, the designer has to quantify

1. Seismic actions in terms of ground-induced strains in the pipeline wall (strain demand calculation)

2. Pipe resistance in terms of all possible failure modes,

to ensure that the steel pipeline is capable of sustaining the imposed strains without failure; in any case, strain demand should not exceed pipe resistance (Gresnigt 1986).

An important issue needs to be clarified at this point: Seismic events are extreme loading situations, which differ from regular operating or construction loads. Those extreme actions are associated with local strains and deformations that may reach or exceed the yield strength limit of pipeline material. Therefore, the traditional design approach that limits the stress level in the steel by utilizing allowable specified stresses, which are usually fractions of yield strength, may not be adequate for seismic design. Instead, a strain-based design approach based on "no loss of containment" should be followed.

Design of buried steel pipelines against severe ground-induced actions is not addressed directly in this manual. The determination of loads generated by geohazards and, in particular, generated by seismic activity, as well as of the corresponding pipeline resistance, is beyond the scope of this manual. In particular, the design parameters analyzed in this manual may not be applicable in the above actions; as an example the main requirement in seismic design should be "loss of containment" rather than some factor of yield strength.

Designers should keep in mind that seismic engineering is an evolving scientific area, which has reached a good level of know-how in the seismic response of aboveground structures, such as buildings, bridges, or liquid storage tanks. However, the particularities of buried pipeline behavior in geohazard areas may not allow direct application of well-established seismic design principles developed for buildings or bridges. In any case, the seismic structural performance of buried pipelines is an area that has not received much attention over the past decades.

This issue of seismic performance of buried steel pipelines has motivated significant amount of research work during the last years, reflected in several relevant publications. A summary of the work, mainly in the United States and Canada has been presented in the ASCE workshop on "Seismic Design of Water Pipelines" (ASCE Workshop, 2013). Furthermore, the ALA guidelines (2005) are an extensive document that accounts for research conducted mainly in the United States up to 2005. More recently, the current technological know-how has allowed for (1) the performance of dedicated small-scale and large-scale experiments (e.g., Ha et al. 2010; Demofonti et al. 2013) and (2) the rigorous modelling and prediction of pipeline deformation within the moving ground and soil-pipe interaction (Vazouras et al. 2012; 2015). The 2011 GIPIPE project (GIPIPE 2014), funded by the European Commission (2011–2014), includes state-of-the-art experimental and numerical work on this subject. However, despite the recent publication of several research papers and recommendations on the seismic design of steel pipelines, there is a lack of a complete set of seismic design provisions in pipeline design codes and standards to be used by practicing engineers. The main reason is that there are still several open issues that require further investigation.

It is important to note that most of the available design guidelines (e.g., ASCE 1984; PRCI 2004; PRCI et al. 2009; JGA 2010) have been developed for buried steel pipelines for hydrocarbon transmission and distribution (oil and gas pipelines). However, hydrocarbon pipelines are quite different from water pipelines in terms of geometry (size and diameter-to-thickness ratio), manufacturing process, material type, joining, and internal pressure level. Therefore, despite some useful clauses, direct application of those guidelines to water pipelines is not recommended.

Although not exhaustive, a useful list of references on this topic is included at the end of this appendix. Because of on-going research in seismic analysis and design of buried steel water pipelines, these references should be considered a starting point for designers interested in gaining a better understanding of seismic behavior and design of buried steel pipelines. The recent papers by Karamanos et al. (2014 and 2015), in conjunction with the ALA guidelines (2005), may constitute a good starting point for this topic.

Finally, because of the above issues, in order to safeguard pipeline structural integrity, it is strongly recommended that in the case of pipelines crossing seismically active areas, an engineer experienced in seismic analysis and design of steel water pipelines be consulted.

REFERENCES

American Lifelines Alliance (ALA). 2005. *Seismic Guidelines for Water Pipelines*. Washington, DC: FEMA.

ASCE. 1984. *Guidelines for the Seismic Design of Oil and Gas Pipeline Systems*. Prepared by the ASCE Technical Council on Lifeline Earthquake Engineering, D. J. Nyman Principal Investigator. Reston, VA: ASCE.

ASCE Workshop. 2013. Seismic Design of Buried Pipelines, Pre-Conference Workshop. In *Proc. ASCE 2013 Pipelines Conference*, Fort Worth, Texas, June 2013. Reston, VA: ASCE.

Demofonti, G., J. Ferino, S.A. Karamanos, P. Vazouras, and P. Dakoulas. 2013. An Integrated Experimental-Numerical Approach to Predict Strain Demand for Buried Steel Pipelines in Geo-hazardous Areas. In *Proc. Rio Pipeline Conference & Exposition*, Rio, Brazil. Rio de Janeiro: Brazilian Petroleum, Gas and Biofuels Institute.

Earthquake Engineering Research Institute (EERI). 1999. Kocaeli, Turkey Earthquake of August 17. *EERI Special Earthquake Report*. Oakland, CA: EERI.

GIPIPE Project. 2014. *Safety of Buried Steel Pipelines Under Ground-Induced Deformations*. RFSR-CT-2011-00027. Sponsored by the European Commission 2011-2014, RFCS programme, Brussels, Belgium. www.mie.uth.gr/gipipe.

GIPIPE Workshop. 2014. *Geohazards and Pipelines*. Delft, The Netherlands, June 2014. http://www.gipipe.eu/

Gresnigt, A.M. 1986. Plastic Design of Buried Steel Pipes in Settlement Areas. *HERON*, 31(4): 1–113.

Ha, D., T.H. Abdoun, M.J. O'Rourke, M.D. Symans, T.D. O'Rourke, M.C. Palmer, and H.E. Stewart. 2010. Earthquake Faulting Effects on Buried Pipelines—Case History and Centrifuge Study. *Journal of Earthquake Engineering*, 14(5):646–669.

Japan Gas Association (JGA). 2010. *Seismic Design for Gas Pipelines, Recommended Practice for Earthquake-Resistant Design of Gas Pipelines*. Tokyo: JGA.

Karamanos, S.A., B. Keil, and R.J. Card. 2014. Seismic Design of Buried Steel Water Pipelines. In *Proc. ASCE 2014 Pipelines Conference*, Portland, Oregon. Reston, VA: ASCE.

Karamanos, S.A., E. Koritsa, B. Keil, and R.J. Card. 2015. Analysis and Behavior of Steel Pipe Welded Lap Joints in Geohazard Areas. In Proc. *ASCE 2015 Pipelines Conference*, Baltimore, MD. Reston, VA: ASCE.

Liang, J., and S. Sun. 2000. Site Effects on Seismic Behavior of Pipelines: A Review. *ASME Journal of Pressure Vessel Technology,* 122(4):469–475.

O'Rourke, M.J. 2003. Buried Pipelines. In *Earthquake Engineering Handbook,* C Scawthorn and W-F Chen, eds. Boca Raton, FL: CRC Press.

Pipeline Research Council International (PRCI), D.J. Nyman, and D.G. Honegger. 2004. *Guidelines for the Seismic Design and Assessment of Natural Gas and Liquid Hydrocarbon Pipelines.* Houston, TX: PRCI.

Pipeline Research Council International. 2009. *Guidelines for Constructing Natural Gas and Liquid Hydrocarbon Pipelines Through Areas Prone to Landslide and Subsidence Hazards.* Houston, TX: PRCI.

Vazouras, P., S.A. Karamanos, and P. Dakoulas. 2012. Mechanical Behaviour of Buried Steel Pipelines Under Strike-Slip and Normal/Reverse Fault Displacements. *Soil Dynamics and Earthquake Engineering,* Vol. 41, pp. 164–180.

Vazouras, P., P. Dakoulas, and S.A. Karamanos. 2015. Pipe-Soil Interaction and Pipeline Performance Under Strike-Slip Fault Movements. *Soil Dynamics and Earthquake Engineering,* Vol. 72, pp. 48–65.

Appendix E

Useful Equations and Conversions

Note: Each equation in this appendix uses its own distinct nomenclature.

EQUATIONS

Equation 1: Cross-sectional Area of a Partially Full Cylinder

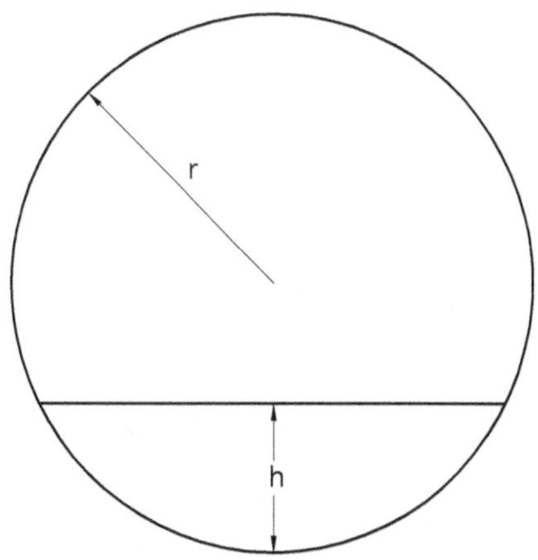

Figure E-1 Cross-section area of partially full cylinder

$$\text{Area} = r^2\, acos\left(\frac{r-h}{r}\right) - (r-h)\sqrt{2rh - h^2}$$

Where:

r = radius of pipe, in.
h = height of fluid from the invert of the pipe, in.
Area = cross-section area of fluid, in.^{2}

Note: acos is in radians.

Equation 2: Measuring Radius of Curvature

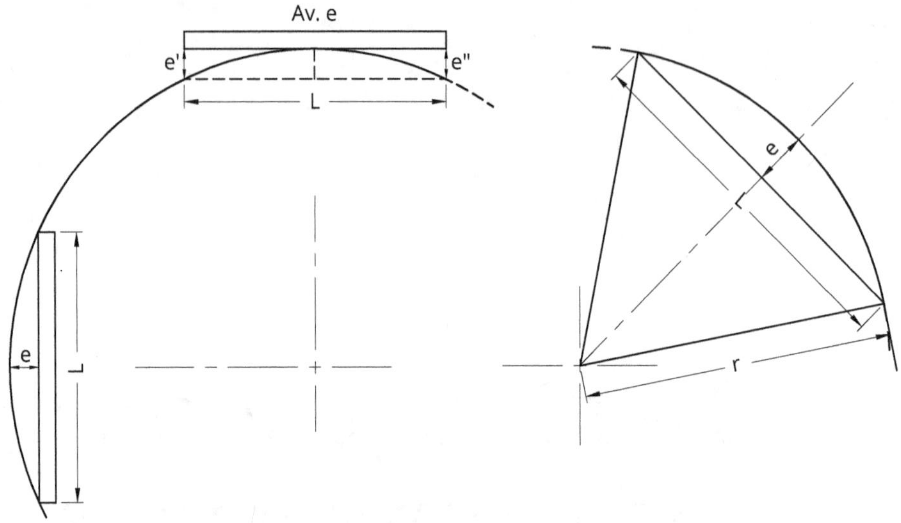

Figure E-2 Measuring radius of curvature

$$r = \frac{(4e^2 + L^2)}{8e}$$

Where:

r = radius of measured area, in.

e = offset distance, in. (for OD measurement, use the average e, i.e., $e = (e' + e'')/2$

L = chord length, in.

To simplify field measurements and calculations, set the measuring device length (chord length) to 12⅝ in., then the equation can be approximated to

$$r = \frac{20}{e}$$

Equation 3: Long Radius Curves

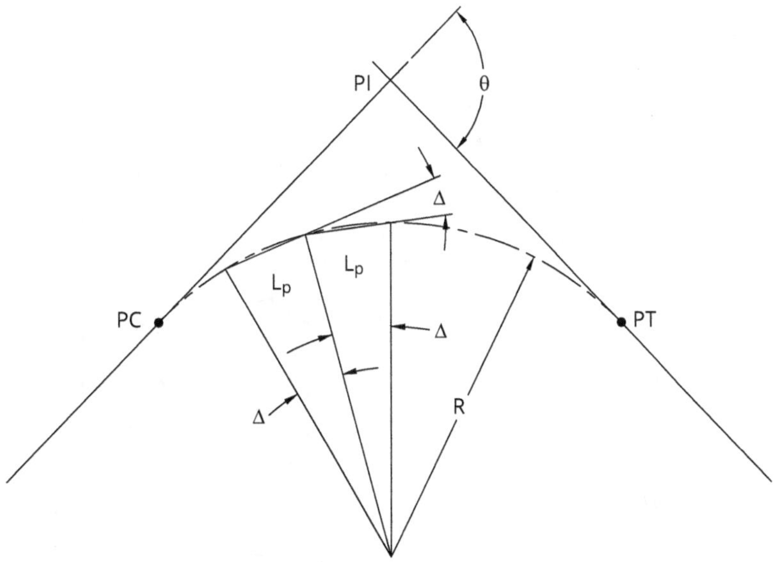

Figure E-3 Pipe with deflected joints or mitered ends in a long-radius curve

Where:

For deflected joints $R = \dfrac{L_p}{2\sin\dfrac{\Delta}{2}}$

For mitered joints $R = \dfrac{L_p}{2\sin\dfrac{\beta}{2}}$

R = pipeline centerline radius, ft
L_p = calculated centerline pipe laying length, ft
Δ = angular deflection per pipe section, deg
θ = total deflection angle of curve, deg
β = pipe end miter angle, deg
PT = point of tangent
PI = Point of intersection
PC = point of curve

Equation 4a: Spiral Pipe Helix Angle

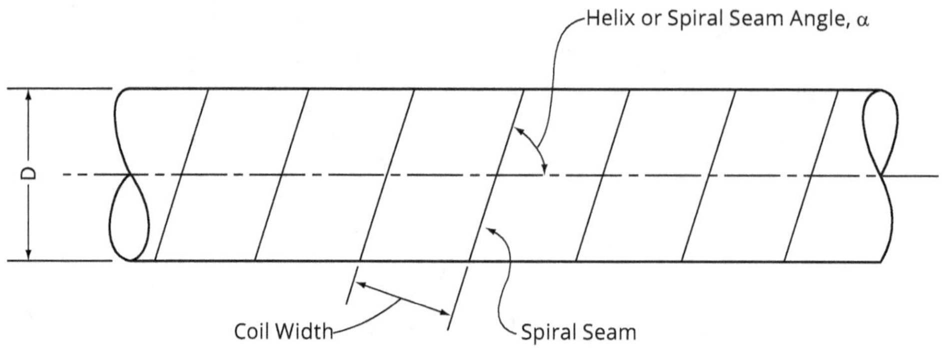

Figure E-4 Spiral pipe helix angle

$$\alpha = acos\left(\frac{w}{\pi D}\right)$$

Where:

α = helix or spiral seam angle, deg
w = coil width, in.
D = pipe outside diameter, in.

Equation 4b: Evaluation of Stress Across a Spiral Weld

$$\sigma_n = \sigma_1(\sin^2 \alpha) + \sigma_2(\cos^2 \alpha)$$

Where:

σ_n = stress normal to the axis of the weld, psi
α = helix or spiral seam angle, deg
σ_1 = longitudinal stress parallel to the longitudinal axis of the pipe, psi
σ_2 = circumferential or hoop stress, psi

For Equations 5 through 7, the following abbreviations apply:

CMC = cement-mortar coating
CML = cement-mortar lining
E = modulus of elasticity
I = moment of inertia

Equation 5: Pipe Deflection Under Own Weight

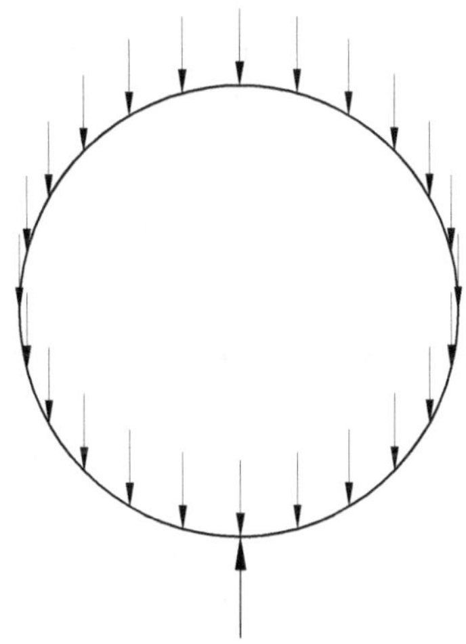

Figure E-5 Vertical deflection of a pipe under its own weight on a flat surface

$$D_v = \frac{-0.4253wr^4}{EI}$$

Where

D_v = vertical pipe deflection, in.
r = pipe radius, in.
w = weight of pipe per square inch, psi (for CML and CMC, w of steel and mortar are additive)
EI = pipe stiffness (for CML and CMC, EI of steel and EI of mortar are additive)

Equation 6: Deflection of a Pipe Full of Water on a Flat Surface

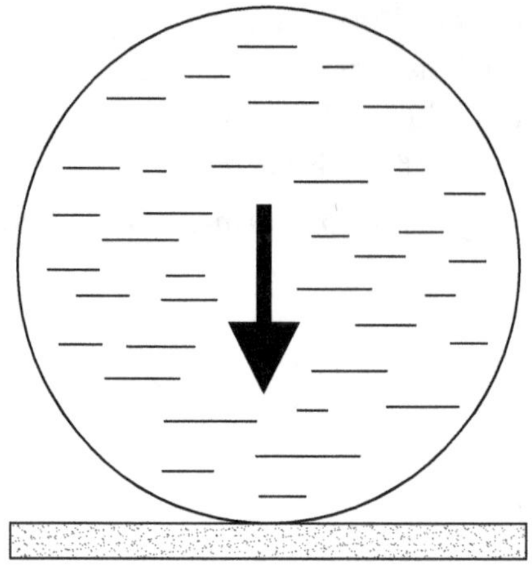

Figure E-6 Vertical deflection of pipe full of water without support

$$D_v = \frac{-0.007677r^5}{EI}$$

Where:

D_v = Vertical pipe deflection, in.
r = Pipe radius, in.
EI = Pipe stiffness (for CML and CMC, EI of steel and EI of mortar are additive)

Equation 7: Ring Deflection of a Circular Ring Under Load With Rigid Bottom Arc and No Side Support

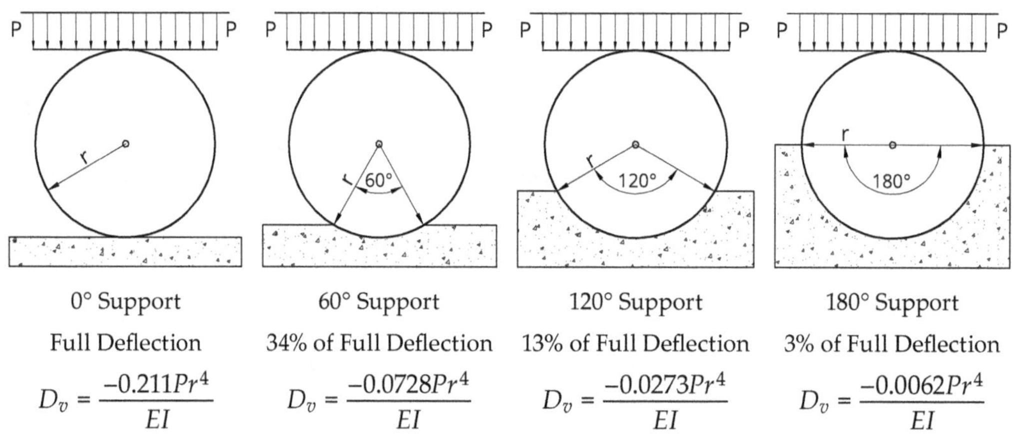

Figure E-7 Various ring deflections of a circular ring under load with rigid bottom arc and no side support

Where:

D_v = vertical pipe deflection, in.
r = pipe radius, in.
P = external load above the pipe, psi
EI = pipe stiffness (for CML and CMC, EI of steel and EI of mortar are additive)

Equation 8: Combined Elbows

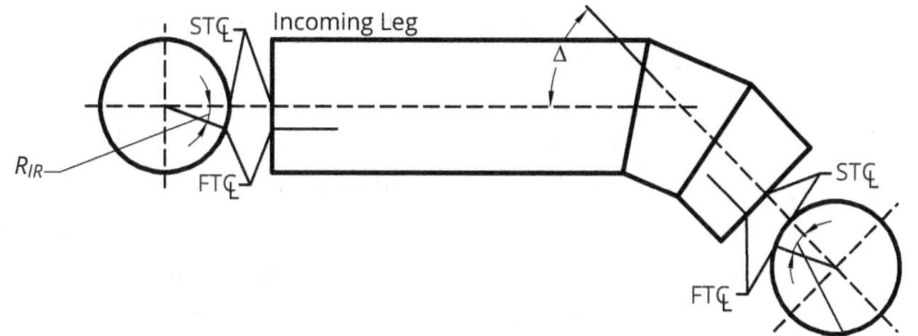

Combined Elbow FT℄ Location — Elbow Right

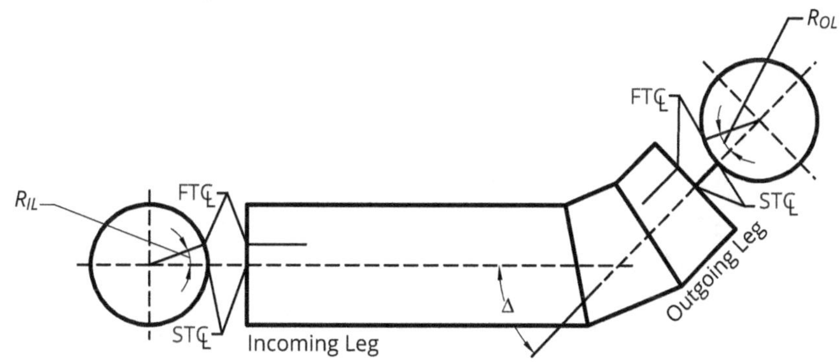

Combined Elbow FT℄ Location — Elbow Left

Figure E-8 Combined Elbows FT℄ Location

For combined elbows where both horizontal deflection and vertical deflection occur at the same location, the following equations allow for calculation of the resultant combined deflection angle and location of field top centerline (FTcL) on each end of the elbow.

$$\Delta = \cos^{-1}\left[\cos I \cos O \cos \Delta_p + \sin I \sin O\right]$$

$$R_{IR} = 0.01745 \sin^{-1}\left(\frac{\cos \Delta \sin I - \sin O}{\cos I \sin \Delta}\right) \text{ for plan angle to the right}$$

$$R_{IL} = 0.01745 \sin^{-1}\left(\frac{\cos \Delta \sin I - \sin O}{\cos I \sin \Delta}\right) \text{ for plan angle to the left}$$

$$R_{OR} = 0.01745 \sin^{-1}\left(\frac{\cos \Delta \sin O - \sin I}{\cos O \sin \Delta}\right) \text{ for plan angle to the right}$$

$$R_{OL} = 0.01745 \sin^{-1}\left(\frac{\cos \Delta \sin O - \sin I}{\cos O \sin \Delta}\right) \text{ for plan angle to the left}$$

$$C_I = \frac{D_O}{2} R_{IR} \text{ (right plan angle)} \quad \text{or} \quad C_I = \frac{D_O}{2} R_{IL} \text{ (left plan angle)}$$

$$C_O = \frac{D_O}{2} R_{OR} \text{ (right plan angle)} \quad \text{or} \quad C_O = \frac{D_O}{2} R_{OL} \text{ (left plan angle)}$$

Where:

Δ = resultant combined deflection angle of elbow, deg

I = incoming leg slope angle, deg

O = outgoing leg slope angle, deg

Δ_p = plan (horizontal) deflection angle of elbow, deg

R_{IR} = rotation angle for incoming leg with plan angle to the right, rad*

R_{IL} = rotation angle for incoming leg with plan angle to the left, rad*

R_{OR} = rotation angle for outgoing leg with plan angle to the right, rad*

R_{OL} = rotation angle for outgoing leg with plan angle to the left, rad*

C_I = incoming leg circumferential rotation distance from field top to shop top, in.

D_O = elbow outside diameter, in.

C_O = outgoing leg circumferential rotation distance from field top to shop top, in.

* Orientation of FT℄ relative to shop top centerline (ST℄) is based on the numerical sign of R_{IR}, R_{IL}, R_{OR}, and R_{OL}. Positive values represent clockwise rotation from FT℄ location to ST℄ location, looking into each respective pipe end. Generic representative diagrams above reflect positive values for R_{OR} and R_{IL} and negative values for R_{IR} and R_{OL}. An opposite numerical sign for any of these values results in the FT℄ location associated with that value moving to the opposite side of the ST℄ location.

CONVERSIONS

Conversions for Water

1 ft^3	= 7.48 gal	= 62.4 lb
1 gal	= 231 $in.^3$	= 8.333 lb
1 lb	= 27.72 $in.^3$	
1 acre-ft	= 43,560 ft^3	= 325,851 gal
1 ft^3/sec	= 448.8 gpm	= 1.98 acre-ft / day
1 gpm	= 1,440 gpd	= 1.61 acre-ft / year
1 mil gal	= 3.07 acre-ft	
1 mgd	= 1,122 acre-ft / year	
1,000 gpm	= 2.23 ft^3/sec	= 4.42 acre-ft / day
1 ppm	= 2.7 lb / acre-ft	= 0.0000623 lb / ft^3

acre-ft	= acre foot
gpm	= gallons per minute
gpd	= gallons per day
ft^3/sec	= cubic feet per second
ppm	= part per million

Standard Steel Sheet Gage Thickness

Gage Number	Decimal Equivalent (in.)
8	0.1644
10	0.1345
12	0.1046
14	0.0747

Steel Unit Weight

490 lb / ft^3

Conversions—English to Metric

1 in.	= 25.4 mm
1 $in.^2$	= 645 mm^2
1 $in.^3$	= 16,387 mm^3
1 $in.^4$	= 416,231 mm^4
1 micro in.	= 0.0254 micro mm
1 ft	= 0.305 m
1 ft^2	= 0.093 m^2
1 lb_f	= 4.45 N
1 lb	= 0.454 kg
1 lb/ft	= 1.488 kg/m

1 lb_f/ft^2	= 47.88 N/m^2
1 lb/ft^2	= 4.882 kg_f/m^2
1 lb_f/ft^3	= 157.09 N/m^3
1 lb_f/in.	= 0.175 N/mm = 0.0179 kg/mm
1 lb_f/linear in.	= 0.0179 kg/linear mm
1 lb/linear ft.	= 1.488 kg/linear m
1 in.-lb_f	= 0.113 N-m
1 in.-kip_f	= 112.98 N-m
1 psi	= 6.895 kPa
1 psi	= 0.006895 MPa
1 ksi	= 1000 psi
1 ksi	= 6894.757 kPa
1 ksi	= 6.895 MPa
1 kip	= 1000 lbs
1 Kip/in.	= 175.127 N/mm
°C	= (°F – 32) × 5/9
1 ft-lb	= 1.356 joules (J)
1 fps	= 0.3048 m/s
1 ft/s^2	= 0.3048 m/s^2
1 ft^2/s	= 0.0929 m^2/s
1 ft^3/s	= 0.0283 m^3/s
1 gpm	= 3.785 L/min
1 MGD	= 833.993 gpm

Index

Note: *f.* indicates figure; *t.* indicates table

D

E

J

K

L

M

T

U

V

W

Y

This page intentionally blank.

AWWA Manuals

M1, *Water Rates, Fees, and Charges,* #30001
M2, *Instrumentation and Control,* #30002
M3, *Safety Management for Water Utilities,* #30003
M4, *Water Fluoridation Principles and Practices,* #30004
M5, *Water Utility Management,* #30005
M6, *Water Meters—Selection, Installation, Testing, and Maintenance,* #30006
M7, *Problem Organisms in Water: Identification and Treatment,* #30007
M9, *Concrete Pressure Pipe,* #30009
M11, *Steel Pipe—A Guide for Design and Installation,* #30011
M12, *Simplified Procedures for Water Examination,* #30012
M14, *Backflow Prevention and Cross-Connection Control: Recommended Practices,* #30014
M17, *Installation, Field Testing, and Maintenance of Fire Hydrants,* #30017
M19, *Emergency Planning for Water Utilities,* #30019
M20, *Water Chlorination/Chloramination Practices and Principles,* #30020
M21, *Groundwater,* #30021
M22, *Sizing Water Service Lines and Meters,* #30022
M23, *PVC Pipe—Design and Installation,* #30023
M24, *Planning for the Distribution of Reclaimed Water,* #30024
M25, *Flexible-Membrane Covers and Linings for Potable-Water Reservoirs,* #30025
M27, *External Corrosion Control for Infrastructure Sustainability,* #30027
M28, *Rehabilitation of Water Mains,* #30028
M29, *Water Utility Capital Financing,* #30029
M30, *Precoat Filtration,* #30030
M31, *Distribution System Requirements for Fire Protection,* #30031
M32, *Computer Modeling of Water Distribution Systems,* #30032
M33, *Flowmeters in Water Supply,* #30033
M36, *Water Audits and Loss Control Programs,* #30036
M37, *Operational Control of Coagulation and Filtration Processes,* #30037
M38, *Electrodialysis and Electrodialysis Reversal,* #30038
M41, *Ductile-Iron Pipe and Fittings,* #30041
M42, *Steel Water-Storage Tanks,* #30042
M44, *Distribution Valves: Selection, Installation, Field Testing, and Maintenance,* #30044
M45, *Fiberglass Pipe Design,* #30045
M46, *Reverse Osmosis and Nanofiltration,* #30046
M47, *Capital Project Delivery,* #30047
M48, *Waterborne Pathogens,* #30048
M49, *Quarter-Turn Valves: Torque, Head Loss, and Cavitation Analysis,* #30049
M50, *Water Resources Planning,* #30050
M51, *Air-Release, Air/Vacuum, and Combination Air Valves,* #30051
M52, *Water Conservation Programs—A Planning Manual,* #30052
M53, *Microfiltration and Ultrafiltration Membranes for Drinking Water,* #30053
M54, *Developing Rates for Small Systems,* #30054
M55, *PE Pipe—Design and Installation,* #30055
M56, *Nitrification Prevention and Control in Drinking Water,* #30056
M57, *Algae: Source to Treatment,* #30057
M58, *Internal Corrosion Control in Water Distribution Systems,* #30058
M60, *Drought Preparedness and Response,* #30060
M61, *Desalination of Seawater,* #30061
M63, *Aquifer Storage and Recovery,* #30063
M65, *On-Site Generation of Hypochlorite,* #30065
M66, *Cylinder and Vane Actuators and Controls—Design and Installation,* #30066

This page intentionally blank.

www.ingramcontent.com/pod-product-compliance
Lightning Source LLC
LaVergne TN
LVHW080030210926
848481LV00011B/3146